Fundamentals of *In Vivo* Magnetic Resonance

Fundamentals of *In Vivo* Magnetic Resonance

Spin Physics, Relaxation Theory, and Contrast Mechanisms

Daniel M. Spielman and Keshav Datta
Stanford University, Stanford, CA, USA

Library of Congress Cataloging-in-Publication Data

Names: Spielman, Daniel (Daniel Mark), author. | Datta, Keshav, author.
Title: Fundamentals of *in vivo* magnetic resonance : spin physics, relaxation theory, and contrast mechanisms / Daniel M. Spielman and Keshav Datta.
Description: Hoboken, New Jersey : Wiley, [2024] | Includes index.
Identifiers: LCCN 2023051874 (print) | LCCN 2023051875 (ebook) | ISBN 9781394233090 (paperback) | ISBN 9781394233106 (adobe pdf) | ISBN 9781394233113 (epub)
Subjects: MESH: Magnetic Resonance Spectroscopy | Magnetic Resonance Imaging | Magnetic Phenomena
Classification: LCC RC386.6.M34 (print) | LCC RC386.6.M34 (ebook) | NLM QC 762.6.M34 | DDC 616.07/548–dc23/eng/20231220
LC record available at https://lccn.loc.gov/2023051874
LC ebook record available at https://lccn.loc.gov/2023051875

Cover Design: Wiley
Cover Image: Courtesy of Daniel Spielman, Professor of Radiology, Stanford University

Set in 9.5/12.5pt STIXTwoText by Straive, Chennai, India

Contents

Preface

Magnetic Resonance Imaging (MRI) plays a fundamental role in medicine, particularly for the evaluation of brain, spine, heart, muscle, and other soft tissues. Indeed, modern MRI techniques and associated hardware have become increasingly sophisticated, resulting in the acquisition of large volumes of clinically invaluable information within ever-decreasing scan times. Unlike X-ray imaging, computed tomography (CT), and ultrasound (US), wherein an energy source and detector pair are used to form images, the signal for MRI originates intrinsically within the body through the manipulation of a fundamental property of matter – the nuclear spin. In addition, MR signals strongly depend not only on the physical properties of the targeted nuclei themselves but also on processes linking nuclei, the molecules containing them, and interactions with the surrounding environment, providing a rich variety of image contrasts not seen in other modalities. Via these interactions, *in vivo* MRI can obtain unique biological information vital to our understanding of health and disease. This textbook focuses on developing a physical and mathematical understanding of these *in vivo* processes and how they can be utilized by Magnetic Resonance Spectroscopy (MRS) to measure individual biochemicals in the body or by MRI to generate unique image contrast.

Although there are multiple excellent MRI textbooks, the material presented here addresses what we think is a largely unmet need. The material bridges the gap between the physics of magnetic resonance (MR) image formation and the *in vivo* processes that influence the detected signals, equipping the reader with the mathematical tools essential to study the spin interactions leading to various contrast mechanisms. Specifically, the material arises from the lecture notes for a Stanford University Department of Radiology course taught for over 15 years, typically taken by engineering and biomedical physics PhD students as their second graduate class in MRI. The first course, based on the classical description of MRI starts with the Bloch equations followed by discussions of radiofrequency (Rf) excitation in combination with the linear gradient fields used for image formation. Although immensely powerful, this traditional approach based on following the evolution of a bulk magnetization vector arising from the sum of a very large number of independent nuclear spins, does not fully explain many important *in vivo* processes. For example, T_1 and T_2 relaxation times are typically included as phenomenological constants, whereas the physics driving these processes provides important insights connecting *in vivo* data and underlying anatomy and physiology. Indeed, individual spins have no T_1 or T_2, as these relaxation parameters are only emergent properties of a large collection or ensemble of spins!

More generally, interactions between spins (*e.g.*, J coupling, dipolar coupling, and chemical exchange) provide fundamental contributions to image contrast. This textbook aims to introduce the reader to the tools needed to analyze these spin interactions with a goal to answer many intriguing questions including the following:

- Do tissues behave more like liquids or solids (liquid-state versus solid-state Nuclear Magnetic Resonance [NMR] requires distinctly different analysis tools)?
- If gray matter, white matter, and cerebrospinal fluid (CSF) in brain have similar water content, why do these tissue types show such marked contrast differences depending on the chosen MRI sequence?
- Why is fat bright in Fast Spin Echo (FSE) imaging?
- How does the presence of oxy- versus deoxy-hemoglobin affect the MRI of blood?
- Why is the T_2 (and not the T_1) of tendons strongly dependent on the angular orientation of the tendon with respect to the main magnetic field B_0?
- Why is the chemical shift of water, but not fat, temperature-dependent?
- Why is the MRI contrast agent gadolinium-diethylenetriamine penta-acetic acid (Gd-DTPA) used to shorten T_1, whereas dysprosium-DTPA (Dy-DTPA) has almost no T_1 shortening effects but is useful as a T_2 agent?

To analyze spin-spin interactions, we have chosen to start from a quantum mechanical (QM) formulation. It is not the case that the QM derivation is any more rigorous than that from classical physics. Rather, both static and dynamic spin-spin interactions are most easily incorporated using a QM approach. An emphasis is placed on understanding the associated math and physics while maintaining the critical intuition needed to make such knowledge useful in practice, and topical questions for the reader have been added throughout the text to encourage students to think critically and develop a deeper understanding of the material. We have also included exercises (with solutions provided in a companion volume) and biographical sketches to best capture the individual contributions to the rich history of the development of MR theory and practice. The chapters are organized as described below.

- Chapter 1 introduces the subject of MR with some of the important historical developments.
- Chapter 2 contains a brief description of the source of nuclear magnetism followed by a classical description of MR starting from the Bloch equations and then covering the important topics of Rf excitation, signal reception, transverse and longitudinal magnetization, and spatial localization via gradients. This leads to the fundamental imaging equation of MRI.
- Chapter 3 outlines the physics and mathematical concepts underlying the aspects of quantum mechanics most relevant to analyzing MR. Descriptions of MR are then provided in Hilbert and Liouville spaces with a focus on maintaining the vector formulation of MR that provides much of the physical intuition underlying classical MRI. As part of this development, the density operator, the fundamental quantity underlying the QM description of a collection of spins, is introduced.
- Chapter 4 covers the primary terms in the nuclear spin Hamiltonian, *i.e.*, the QM operator corresponding to the energy of a spin system. Internal and external interactions are separately described followed by a discussion of the key concepts of spin populations and phase coherences.
- Chapter 5 describes the Product Operator Formalism (POF) of MR, the key analysis tool allowing a robust, yet intuitive, description of MR pulse sequences via a series of rotations in a vector space. POF is the basic tool for analyzing modern MRS sequences that exploit both J coupling and multiple quantum coherences. In particular, a solid understanding of J coupling is fundamental to understanding *in vivo* MRS.
- Chapter 6 gives an overview of *in vivo* magnetic resonance spectroscopy (MRS) techniques and clinical applications. MRS is a key technique for measuring *in vivo* metabolism and, hence, very

important in its own right. Furthermore, a robust understanding of J coupling is a key first step before moving on to the more difficult-to-analyze dipolar coupling phenomenon that is the primary driver of *in vivo* relaxation.

- Chapter 7 introduces the fundamental mechanisms dominating *in vivo* MR relaxation, including high-level descriptions of dipolar coupling and chemical exchange phenomena.
- Chapter 8 develops Redfield theory, a rigorous mathematic framework for deriving relaxation rates in liquids. This formulation is particularly relevant in that most MR-detectable *in vivo* tissues are well modeled as liquids, with some notable exceptions. Redfield theory also provides a convenient approach for deriving a third MR relaxation parameter beyond T_1 and T_2, namely, $T_{1\rho}$ (also known as spin-lattice relaxation in the rotating frame).
- Chapter 9 discusses the various types of MRI contrast agents and their associated physics, focusing on the clinical agents used to shorten T_1, T_2, and T_2^*.
- Chapter 10 is the final chapter and focuses on a representative set of *in vivo* tissues, highlighting concepts introduced in the earlier chapters.

As we embark on this journey into the fascinating realm of *in vivo* magnetic resonance, it is with great pleasure and gratitude that we acknowledge the collective efforts and support that have shaped this textbook. We extend our sincere thanks to the myriad trainees and colleagues, especially those at the Stanford University Radiological Sciences Laboratory, whose contributions have been invaluable. While it's impractical to individually list everyone, we express our heartfelt gratitude to all. Thank you.

We also extend our appreciation to Dr. Martin Preuss for his expert guidance through the intricate publication process, Neena Ganjoo for skillfully managing publishing timelines, and a special commendation to Sindhuraj Kuttappan and Sakthivel Kandaswamy for their exceptional assistance in editing and manuscript preparation.

Finally, a project of this magnitude demands not only intellectual collaboration but also the unwavering support of our spouses and families. We are profoundly grateful for their infinite patience and enduring encouragement, without which this volume would not have come to fruition.

Note by DMS: I am especially grateful to my wife, Frances Sun, for providing me with never ending patience and encouragement, and my children, Eleanor and Max, for their curiosity and inspiring sense of wonder.

Note by KD: I am immensely grateful to Prof. Daniel Spielman for his inspiring mentorship, guidance, and expert teaching that sparked my passion for *in vivo* magnetic resonance. I sincerely thank him for this incredible opportunity to be part of an enriching learning experience presented by the creation of this book. I would also like to express my gratitude to my parents, Dr. A. Gajanana and Gayathri Devi, for instilling the value of knowledge, and my brother Shashi Bhushan for his unwavering encouragement. I am profoundly thankful to my wife, Jayshree Narasimhan, for her steadfast support and for nurturing both my physical well-being and intellectual pursuits. Finally, I am appreciative of my sons, Shreyas, Tejas, and Varchas, whose curiosity and candid questions have inspired me to learn and teach better.

In memoriam by DMS: My father recently passed away before this textbook could be published, so I wanted to include a special dedication. He would often ask me, in all seriousness, whether "any of my research was destined to win a Nobel Prize?". I can only answer most likely not, but the least I could do is help train the next generation of scientists and researchers.

February 10, 2024

Daniel M. Spielman and Keshav Datta
Stanford University, Stanford, USA

About the Companion Website

This book is accompanied by a companion website:

www.wiley.com/go/Spielman

This website include:
- Exercises and Solutions

1

Introduction

Certain nuclei, as well as electrons, behave much like small spinning magnets and will accordingly interact with an applied magnet field. Oscillating magnetic fields applied at the appropriate frequency can then alter this state, a phenomenon known as nuclear magnetic resonance (NMR), giving rise to detectable signals at frequencies characteristic of each NMR-active nucleus. By examining these signals, information regarding the presence of specific compounds and their chemical environments can be measured using a technique known as magnetic resonance spectroscopy (MRS), and spatially mapped using magnetic resonance imaging (MRI).

In general, spectroscopy is the study of materials via their interactions with electromagnetic (EM) fields, and the energy of these fields determines the associated physics (see Figure 1.1). At typical magnetic field strengths (0.1–10 T), NMR spectroscopy operates in the radiofrequency band. In contrast, electron spin resonance (ESR), which involves electrons rather than nuclei, utilizes frequencies in the microwave region. Dubbed NMR by the physicists and chemists who first discovered and applied the phenomenon, the "nuclear" label was later dropped in the medical field due to unfounded fears of ionizing radiation. For our purposes, we will use the general term magnetic resonance (MR), reserving the labels MRI, MRS, or ESR for when a distinction is being drawn with respect to a specific application or experiment.

A critically important factor to keep in mind is that the MR energies for *in vivo* studies are several orders of magnitude below 37 °C thermal vibrations, leading to fundamentally insensitive measurements. For example, in contrast to other techniques, such as X-ray imaging or computed tomography (CT), where energies are sufficient for the detection of individual photons, MR only effectively observes net signals summed over a very large number of nuclei.

Despite this sensitivity problem and our inability to distinguish individual nuclei, MR-detectable nuclei provide ideal molecular probes. Nuclei are spatially localized, and their magnetic properties are highly sensitive to local magnetic fields. Yet they interact extremely weakly with their physical environment permitting measurements with minimal to no macroscopic effects. Consequently, MR has a rich history with multiple important scientific and medical applications.

1.1 A Brief History of MR

NMR was first detected in 1937. By sending a molecular beam of lithium chloride through an apparatus employing a combination of static and oscillating magnetic fields, I.I. Rabi and colleagues were able to observe resonance peaks of both the lithium and chloride nuclei (Rabi et al. 1938). Often missing from histories of MR but deserving mention is Y.K. Zavoisky, a Soviet scientist whose

Fundamentals of In Vivo Magnetic Resonance: Spin Physics, Relaxation Theory, and Contrast Mechanisms, First Edition. Daniel M. Spielman and Keshav Datta.
© 2024 John Wiley & Sons, Inc. Published 2024 by John Wiley & Sons, Inc.
Companion website: www.wiley.com/go/Spielman

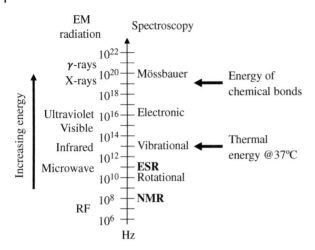

Figure 1.1 Spectroscopy is the study of materials via interactions with electromagnetic (EM) fields. An important feature of NMR is that the associated energy is several orders of magnitude below thermal energy under typical laboratory or *in vivo* conditions.

work was largely overlooked in the West during the Cold War. Since his discovery in 1944 (Zavoisky 1945), Electron spin resonance (ESR), also known as electron paramagnetic resonance (EPR), has been exploited as a highly sensitive method for the investigation of different kinds of paramagnetic species in both solid and liquid states.

In 1945, Felix Bloch and Edward Purcell, working independently at Stanford and Harvard University, respectively, first demonstrated the detection of NMR in condensed matter. Bloch's analysis followed the time evolution of the net magnetic moment via nuclear induction processes (Bloch 1946; Bloch et al. 1946), whereas Purcell emphasized the energy absorption lines in corresponding spectra (Purcell et al. 1946).

Starting in the 1960s, Richard Ernst discovered that the sensitivity of NMR techniques could be dramatically increased by replacing the slowing sweeping radiofrequency magnetic fields traditionally used in NMR spectroscopy (known as continuous-wave NMR) with brief, intense pulses (Ernst and Anderson 1966). This discovery, known as pulsed NMR (later, the "pulsed" descriptor was routinely dropped), greatly expanded the utility of this scientific tool. Ernst later extended this method to enable high-resolution studies of larger molecules, resulting in many of the "two-dimensional" NMR methods used today. Building on the two-dimensional pulsed NMR methods introduced by Ernst, Kurt Wüthrich, and coworkers developed novel nuclear MRS method for mapping the three-dimensional structure of large biological molecules (Wuthrich 1995).

From the perspective of medical imaging, the field of MR split in the 1970s and 1980s, from being primarily a chemical analysis tool to also include imaging, with medical applications soon following. The key concept of augmenting the large uniform primary magnetic field with small deliberate spatially varying components (known as gradients), as proposed independently by Paul Lauterbur (1973) and Peter Mansfield and Grannell (1973), allowed the spatial localization of NMR signals, leading directly to MRI. Lacking the harmful side effects of ionizing radiation present in X-ray imaging, computed tomography (CT), and positron emission tomography (PET), and having exquisite soft tissue contrast, MRI rapidly filled an important unmet need in medicine (Figure 1.2).

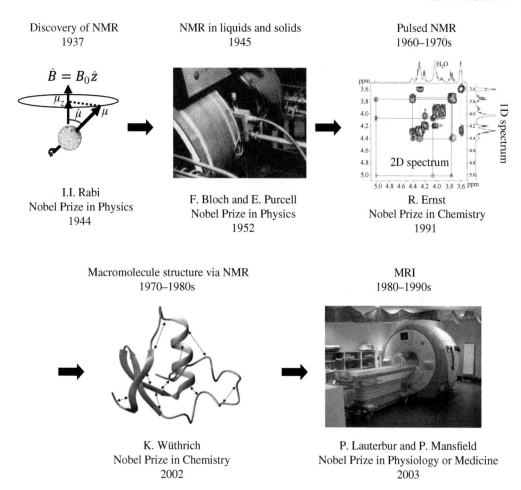

Figure 1.2 A history of Nobel Prizes awarded for development in magnetic resonance. Source: Adapted with permission in part from Bloch (2007) and Foss and Krane (2004).

1.2 NMR versus MRI

The practice of MR in the chemistry laboratory is generally quite different from that employed for *in vivo* imaging applications, and a brief comparison is worthwhile. As a rather broad generalization, typical chemistry experiments involve highly purified samples and ask detailed questions regarding the chemical structure of the materials under study. Further, many laboratories focus on either liquid-state versus solid-state NMR analyses as the performance and associated analysis methods for these two general classes of experiments are typically quite different; rapid molecular tumbling in liquids results in the averaging out of important spin–spin interactions that cannot be ignored in solid or crystalline samples (Figure 1.3).

In contrast, *in vivo* tissues generally contain a complex mixture of materials undergoing multiple physiological processes. Although tissues exhibit both liquid-like and solid-like properties, liquid-state NMR analysis is generally the most appropriate (with some notable exceptions). However, for *in vivo* applications, very different questions are typically asked than those presented in the chemistry laboratory.

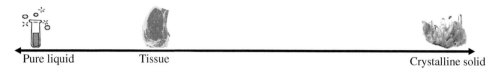

Pure liquid   Tissue   Crystalline solid

Figure 1.3 *In vivo* tissues exhibit both liquid-like and solid-like properties; however, with some notable exceptions, liquid-state NMR analysis is the most appropriate.

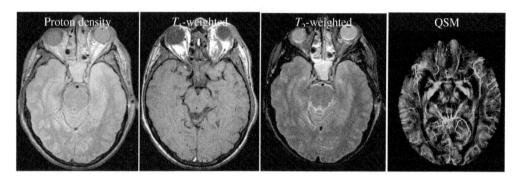

Figure 1.4 Brain MRI. Although all ^{1}H MRI acquisitions target water (and fat), different imaging techniques yield widely varying contrasts. Images show representative proton density, T_1-weighted, T_2-weighted, and quantitative susceptibility mapping (QSM) axial brain images.

Due to a combination of ubiquity and relatively large magnetic moment, hydrogen nuclei (here also referred to as protons) in water molecules are the primary target with inquiries pertaining to location, concentration, and dynamics (e.g., flow or chemical exchange processes). Conventional MRI analysis (to be presented in Chapter 2) is based on assuming the water signal is made up of many independent (*i.e.*, noninteracting) hydrogen nuclei with the addition of phenomenological T_1 and T_2 relaxation times. The combination of a large primary magnetic field, B_0, with smaller linearly varying gradient fields results in a very powerful spatial frequency (k-space) signal analysis framework. The result has been the proliferation of many MRI acquisition schemes capable of assessing a wide spectrum of tissue-specific information, as illustrated in Figure 1.4.

However, this figure raises several fundamental questions. Given that clinical MRI primarily images water, why do different tissues exhibit different MRI signal intensities? Are differences solely due to varying water content (typically referred to as "proton density")? If not, why do different tissues have different T_1s and T_2s? What is the effect of magnetic susceptibility or dynamic processes such as molecular tumbling, chemical exchange, and diffusion?

QUESTION: Does it matter that water has two ^{1}H nuclei? How would MRI be different if water had only one? What about three?

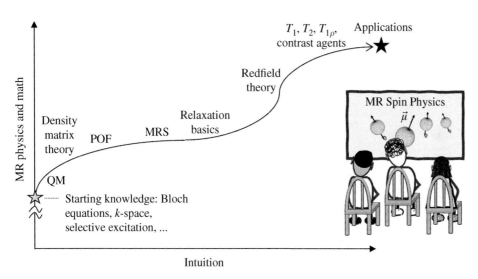

Figure 1.5 An MR roadmap highlighting the balance between physics and mathematical theory, and the development of critical intuitive understanding.

1.3 The Roadmap

This textbook is meant to be a journey. We assume the reader has a basic understanding of classical MRI and the Bloch equations (this material is reviewed briefly in Chapter 2). We next present the quantum mechanical description of MR, with the goal of developing both the analytical tools needed to analyze interactions among nuclear spins and the intuition necessary for putting these principles into practice. J coupling is introduced first, being both fundamental to spectroscopy applications and equally important as the most mathematically tractable spin coupling mechanism. The basic principles of *in vivo* relaxation are then introduced, followed by a deeper mathematical analysis. With these tools in hand, endogenous contrast mechanisms (*e.g.*, T_1, T_2, $T_{1\rho}$, chemical exchange saturation transfer, and magnetization transfer contrast) are discussed in detail, followed by an introduction to the physical principles underlying the most common MRI contrast agents. The last chapter presents a series of representative tissues, dominant contrast mechanisms, and common imaging methods. Figure 1.5 shows the overall roadmap for this textbook. The overall goal is to provide the reader with a deeper understanding of *in vivo* MR than typically provided by a first course in MRI, which will hopefully be useful in the development and evaluation of novel contrast mechanisms with associated imaging methods, basic science studies of physiology, and clinical applications.

Historical Notes

Many chapters end with brief biographies of some of the key scientific contributors. While readily acknowledging that marking scientific progress based on Nobel Prizes inevitably leaves out many significant investigators, the ubiquity of Nobel Prizes for MR (in physics, chemistry, and medicine) at least provides a place to start. Other notable contributors are added as appropriate.

Isidor Isaac Rabi, a prominent American physicist born on July 29, 1898, in Rymanów, which was part of Austria–Hungary and is now in Poland, has left a lasting impact on the field of NMR through his groundbreaking contributions. These significant achievements culminated in his receipt of the Nobel Prize in Physics in 1944, recognizing his development of the atomic and molecular beam MR technique.

Rabi's most noteworthy accomplishment involved his work on devising a meticulous and innovative approach to gauge the magnetic properties of atoms, nuclei, and molecules. His research in this field commenced during the 1930s, with a primary focus on measuring nuclear magnetic moments, particularly the spin of protons within an atom's nucleus. This technique enabled him to derive various mechanical and magnetic characteristics, offering insights even into the structural configuration of atomic nuclei. This work paved the way for groundbreaking applications like the atomic clock, maser, and laser, thereby ushering in a new era of scientific and technological advancement.

After relocating to New York City in 1899, Rabi pursued education at Cornell University, achieving a bachelor's degree in chemistry in 1919. However, driven by his passion for physics, he redirected his focus and went on to secure a Ph.D. from Columbia University in 1927. Following this, he became a faculty member at Columbia University in 1929 after completing postgraduate studies in Europe.

In the tumultuous years marked by World War II, Rabi played a pivotal role in leading a distinguished group of scientists at MIT, making substantial contributions to the development of radar technology, which held immense strategic importance during that critical period. Furthermore, he assumed a significant role on the General Advisory Committee of the Atomic Energy Commission from 1946 to 1956, eventually ascending to the position of its chairman from 1952 to 1956. This not only underscores his commitment to advancing scientific progress but also his dedication to responsible governance in the realm of atomic energy.

Beyond his research endeavors, Rabi's visionary outlook extended to the conceptualization of CERN – an international high-energy physics laboratory situated in Geneva, Switzerland – an idea that later materialized into one of the globe's foremost scientific institutions. Furthermore, he played a pivotal role as a founding figure in the Brookhaven National Laboratory located in Upton, New York, further solidifying his indelible mark on the scientific landscape.

(Source: Photo from https://commons.wikimedia.org/wiki/File:II_Rabi.jpg, public domain.)

Yevgeny Konstantinovich Zavoisky, a well-known Soviet physicist, left a notable impact in the field of ESR. Born on September 28, 1907, in Mohyliv-on-Dniester, Ukraine, which was part of the Russian Empire at the time, he made lasting contributions through his groundbreaking discoveries. Despite not being honored with a Nobel Prize, his work in ESR had profound implications across various scientific domains.

Zavoisky embarked on his academic journey at Kazan State University, completing his studies in 1930. Driven by a deep enthusiasm for physics, he engaged in teaching at the university from 1933 to 1947. During this period, he immersed himself in research related to radio and microwave spectroscopy. In 1944, a pivotal moment came with his breakthrough discovery of EPR, also recognized as ESR. This technique opened avenues for investigating the magnetic properties of paramagnetic materials, revolutionizing research possibilities for scientists in various fields.

Between 1947 and 1971, Zavoisky became a part of Laboratory #2, later renamed the Kurchatov Institute of Atomic Energy, situated in Moscow. During his tenure, he contributed to the development of the Soviet atomic bomb and conducted crucial experimental studies in nuclear physics.

By 1958, Zavoisky had redirected his focus toward nuclear fusion research, particularly concentrating on plasma physics. His notable discovery centered on the phenomenon of turbulent heating. The application of a strong electric field to plasma led to increased resistivity, inducing substantial turbulence. As a result, thermal energy transitioned from macroscale to microscale, consequently raising plasma temperatures to extraordinary levels. This finding held immense importance for the pursuit of controlled nuclear fusion – an area of research with profound implications for both energy production and fundamental physics.

Throughout his illustrious career, Zavoisky's accomplishments garnered recognition from fellow scientists and was the U.S.S.R. Academy of Sciences in 1953. Subsequently, in 1964, he achieved the status of full membership, underscoring the significance of his contributions in the eyes of his peers.

(Source: Photo of Zavoisky's university ID from https://en.wikipedia.org/wiki/Yevgeny_Zavoisky#/media/File:Zavoisky.JPG, public domain)

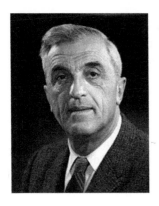

Felix Bloch, an American physicist originally from Switzerland, was born on October 23, 1905, in Zürich. His impact in the fields of NMR and the quantum theory of solids has led to significant advancements in both physics and medical diagnostics. Bloch's journey in the realm of physics traversed continents, culminating in his recognition as a Nobel laureate and a respected figure within the scientific community.

Bloch's academic pursuits commenced with his doctoral research at the University of Leipzig in 1928. During this time, he formulated a quantum theory of solids that offered a foundational understanding of how electrical conduction occurs in materials. This early work laid the groundwork for subsequent explorations in the field of solid-state physics.

In 1933, due to the escalating political situation in Germany, Bloch sought refuge in the United States. He found his academic home at Stanford University in Palo Alto, California, where he joined the faculty in 1934. One of his notable contributions during this period was the proposal of a method to split a beam of neutrons into two components based on their orientations within a magnetic field. Collaborating with Luis Alvarez, Bloch utilized this method in 1939 to measure the magnetic moment of neutrons.

During World War II Bloch conducted atomic energy research at Los Alamos, New Mexico, as well as radar countermeasure work at Harvard University. His role in advancing military technology and nuclear physics was significant. Postwar, he returned to Stanford and collaborated with physicists W.W. Hansen and M.E. Packard to conceptualize the principle of NMR in condensed matter. This groundbreaking idea introduced a precise method for measuring the magnetic field of atomic nuclei, establishing a connection between the magnetic and crystalline properties of materials.

In 1952, Felix Bloch received the Nobel Prize in Physics, an honor shared with Edward M. Purcell. This recognition was based on their independent yet complementary contributions to the development of NMR. This accolade further solidified Bloch's reputation as a pioneering figure in the field of physics.

(Source: Photo from https://commons.wikimedia.org/wiki/File:Felix_Bloch,_Stanford_University.jpg, Licensed under the Creative Commons Attribution 3.0 Unported).

Edward Mills Purcell, was born on August 30, 1912, in Taylorville, Illinois. His groundbreaking advancements in the realm of NMR earned him the Nobel Prize in Physics in 1952, an honor he shared with Felix Bloch.

During World War II, Purcell played a vital role in leading a research group focused on radar-related matters at the radiation laboratory situated within the Massachusetts Institute of Technology (MIT). In a notable turn of events in 1946, Purcell independently formulated the NMR technique, applicable to both liquids and solids. This phenomenon involved certain atomic nuclei placed within a magnetic field absorbing and subsequently emitting EM radiation, enabling the examination of molecular structures in both pure substances and mixtures. It swiftly emerged as an indispensable tool in diverse scientific domains, ranging from chemistry and physics to biology and medicine.

Purcell's NMR detection approach demonstrated remarkable precision and represented a significant advancement over the previously introduced atomic-beam method by Isidor I. Rabi. This pivotal contribution marked a turning point in NMR exploration, effectively laying the groundwork for its widespread adoption across a multitude of scientific fields.

By 1949, Purcell had assumed the role of a physics professor at Harvard University. It was during this same year that he achieved another notable feat. By detecting the 21-centimeter-wavelength radiation emitted by neutral atomic hydrogen in interstellar space – an occurrence initially predicted by Dutch astronomer H.C. van de Hulst in 1944 – Purcell made a substantial impact. His study of these radio waves not only aided astronomers in understanding the distribution and placement of hydrogen clouds within galaxies but also facilitated the measurement of the Milky Way's rotation.

Throughout his career, Purcell continued to excel. His work served as a major source of inspiration for students and fellow researchers alike, leaving an enduring impression on the scientific community as a whole. In recognition of his exceptional achievements, Purcell was honored with the National Medal of Science in 1980.

(Source: Photo from https://commons.wikimedia.org/wiki/File:Edward_Mills_Purcell.jpg, public domain)

Richard Robert Ernst, a distinguished Swiss chemist with a passion for education, left an indelible mark on NMR spectroscopy. Born on August 14, 1933, in the picturesque town of Winterthur, Switzerland, Ernst spearheaded transformative breakthroughs in NMR methodologies, culminating in his well-deserved receipt of the 1991 Nobel Prize in Chemistry. This accolade solidified NMR spectroscopy as an indispensable cornerstone across an array of scientific domains.

Ernst earned a Bachelor of Arts degree in chemistry in 1957 and his Ph.D. degree in physical chemistry from the illustrious Federal

Institute of Technology in Zürich in 1962. After graduation, Ernst embarked on an enriching tenure at Varian, nestled in the heart of Palo Alto, California. Serving as a dedicated research chemist until 1968, he contributed significantly to Varian's pioneering scientific endeavors.

A pivotal chapter in Ernst's career unfolded in 1966. Traditionally, NMR techniques grappled with constraints imposed by the slow and incremental sweeping of radio waves, limiting analyses to a select few nuclei. Ernst's ingenious innovation entailed substituting these conventional radio waves with succinct pulses. The key contribution of this idea was a remarkable enhancement in the sensitivity of NMR techniques. Consequently, the analytical scope expanded, encompassing a broader array of nuclei even with minuscule sample sizes. By 1968, Ernst had returned to Switzerland, at his cherished alma mater. Ascending to the rank of full professor in 1976, he became an inspiring beacon for countless students.

Ernst's second significant contribution to NMR spectroscopy was the conceptualization of a groundbreaking technique facilitating high-resolution "two-dimensional" exploration of hitherto elusive larger molecules. This innovation opened doors for scientists to delve into the intricate three-dimensional world of both organic and inorganic compounds. The technique proved invaluable in the study of biological macromolecules, affording insights into their interactions and reaction kinetics.

(Source: Photo from https://commons.wikimedia.org/wiki/File:Richard_R._Ernst_15.10.2020 .jpg, licensed under the Creative Commons Attribution-Share Alike 4.0 International license.)

2

Classical Description of MR

This chapter contains an overview of the classical description of magnetic resonance imaging (MRI) based on following a net (or bulk) magnetization vector, $\vec{M}$, as it rotates, precesses, shrinks, and grows over time. We first describe the behavior of nuclear spins in a magnetic field, culminating in the formulation of the well-known Bloch equations. The chapter concludes with a description of the basic image formation process central to MRI.

2.1 Nuclear Magnetism

Subatomic particles, such as electrons, protons, and neutrons, possess intrinsic angular momentum called "spin" (see Figure 2.1). Unlike mass and charge, whose macroscopic values are the sum of masses and charges of individual particles, respectively, the spin angular momentum of a system is a more complicated aggregate of the individual spin angular momenta. Using non-trivial combination rules, one can derive the spins of the neutron and proton (each consisting of 3 quarks) to be $\frac{1}{2}$. In a similar fashion, the spin of various nuclei can be derived from the contributions of the constituent protons and neutrons by applying these rules. Some common examples of nuclei with spin-$\frac{1}{2}$ are hydrogen (^{1}H), carbon-13 (^{13}C), and phosphorous (^{31}P), in contrast, deuterium (^{2}H) and nitrogen (^{14}N) have spin of 1, and sodium (^{23}Na) is a spin-3/2 nucleus.

To start, however, it is useful, though not entirely accurate, to think of such particles as a spinning or rotating charge generating a current, which, in turn, produces a magnetic dipole moment, $\vec{\mu}$. From electromagnetic (EM) theory – a current loop in the far field looks just like a magnetic dipole with magnetic moment μ given by the current times the area of the loop. Hence,

$$\mu = \frac{qv}{2\pi r} \cdot \pi r^2 = \frac{q}{2m} mvr, \tag{2.1}$$

where q is the charge, m is the mass, $L = mvr$ is the angular momentum, and the gyromagnetic ratio is defined as $\gamma = q/2m$. As vector quantities, this equation can be written as:

$$\vec{\mu} = \gamma \vec{L}. \tag{2.2}$$

The gyromagnetic ratio γ is often expressed as $\gamma = g\mu_b/\hbar$ where g is the spin factor (≈ 2 for electrons) and μ_b is known as the Bohr magneton, which for an electron with charge e and mass m_e, is:

$$\mu_b = \frac{e\hbar}{2m_e}. \tag{2.3}$$

Fundamentals of In Vivo Magnetic Resonance: Spin Physics, Relaxation Theory, and Contrast Mechanisms, First Edition. Daniel M. Spielman and Keshav Datta.
© 2024 John Wiley & Sons, Inc. Published 2024 by John Wiley & Sons, Inc.
Companion website: www.wiley.com/go/Spielman

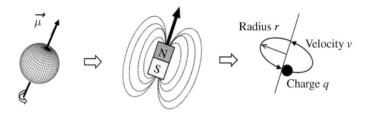

Figure 2.1 The spin of a subatomic particle gives rise to a magnetic dipole moment, $\vec{\mu}$, which can be modeled classically as a rotating charge generating a loop of current.

Note the presence of $\hbar$, Planck's constant divided by 2π, indicates that the full derivation requires quantum mechanics (to be discussed in the next chapter). For protons, $g \approx 5.6$.

From Eq. (2.2), it is evident that the magnetic moment is directly proportional to the spin angular momentum with proportionality constant γ, a measure of the strength of the magnetic dipole and, therefore, its interactions with any applied magnetic fields. For example, the gyromagnetic ratio of the hydrogen nucleus (^{1}H) is $\gamma_H = 2.67522 \times 10^8$ rad/s/T as compared to $\gamma_C = 6.72828 \times 10^7$ rad/s/T for ^{13}C, signifying that a hydrogen nucleus (which we will often simply refer to as a "proton" due to the presence of a single proton in the ^{1}H nucleus) is about a 4× stronger magnetic dipole than a ^{13}C nucleus. Note, nuclei such as ^{12}C have no spin or net magnetic moment and are, therefore, not detectable using NMR. On the other hand, the gyromagnetic ratio of an electron is $\gamma_e = -1.76085 \times 10^{11}$ rad/s/T, highlighting that the smaller mass of electron makes its magnetic moment ~650 times larger than that of a proton. This large difference is important for ESR experiments as well as analyzing MRI contrast agents containing unpaired electrons.

When placed in a uniform magnetic field, a magnetic dipole will experience a torque given by $\vec{\tau} = \vec{\mu} \times \vec{B}$. The potential energy is given by $\mathcal{E} = -\vec{\mu} \cdot \vec{B} = -\mu B \cos\theta$, where θ is the angle between $\vec{\mu}$ and $\vec{B}$ (Note, we are using a script $\mathcal{E}$ for energy instead of the usual E to save the symbol E for an alternate use as an identity operator in later chapters). Classically, the energy can take on any value between $\pm\mu B$, and the resulting torque is why magnetized compass needles align with the Earth's magnetic field (the lowest energy state).

Let us now derive the equations of motion for a nuclear particle with a magnetic moment $\vec{\mu}$ when placed in a magnetic field. The rotational formulation of Newton's second law of motion is $d\vec{L}/dt = \vec{\tau}$. Combining this with our previous equations ($\vec{\mu} = \gamma \vec{L}$ and $\vec{\tau} = \vec{\mu} \times \vec{B}$) yields:

$$\frac{d\vec{\mu}}{dt} = \gamma \vec{\mu} \times \vec{B} \tag{2.4}$$

For a single spin in a uniform z-directed magnetic field $\vec{B} = B_0 \vec{z}$ ($\vec{z}$ = unit vector in z direction), $d\vec{\mu}/dt = \gamma \vec{\mu} \times B_0 \vec{z}$. First note that $|\vec{\mu}|$ is constant, indicating that the change in $\vec{\mu}$ is always perpendicular to the plane containing the vectors $\vec{\mu}$ and $\vec{B}$. Therefore, the resulting motion is a precession of the spin about the axis defined by the applied magnetic field at angular frequency γB_0 (see Figure 2.2). The precession frequency is known as the Larmor frequency, which we will define to be $\omega_0 \equiv \gamma B_0$ and is a left-handed rotation. Note that many NMR references define the Larmor frequency to be $\omega_0 \equiv -\gamma B_0$, but our definition is the more common one used in the MRI literature.

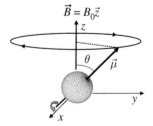

Figure 2.2 A nuclear spin in a magnetic field.

QUESTION: A compass needle has a magnetic moment and sits in the Earth's magnetic field. Why does it not precess?

2.2 Net Magnetization and the Bloch Equations

In tissue, or other physical samples, we are always dealing with many spins, and the net, or "bulk," magnetization, $\vec{M}$, is the sum of the contributions from all of them.

$$\vec{M} = \sum_{volume} \vec{\mu}. \tag{2.5}$$

Substituting Equation 2.5 into 2.4 results in the net magnetization being proportional to the applied field. Hence,

$$\frac{d\vec{M}}{dt} = \gamma \vec{M} \times \vec{B}. \tag{2.6}$$

This equation is valid for non-interacting spins.

It is important to remember that thermal energies, which are several orders of magnitude greater than the energies associated with nuclear spins in typical MRI scanners, result in the individual spin alignments constantly fluctuating, and the average net magnetization at standard *in vivo* or laboratory temperatures (even with large and expensive magnets) is actually minute, typically on the order of several parts-per-million (ppm). The small magnetic dipole moments of nuclear spins, in combination with their tiny net alignment at *in vivo* temperatures, is what accounts for the fundamental insensitivity of MRI. MRI can only robustly detect signals summed over a huge number of individual spins, as compared to an imaging modality such as positron emission tomography (PET) where the energies are sufficiently high, so that individual photons can be recorded (Figure 2.3).

If the bulk magnetization is ever perturbed, for example by using Rf excitation as discussed below, multiple thermal processes (collectively known as MR relaxation) act to restore the magnetization to its thermal equilibrium value, M_0. The simplest approach to account for these thermal processes is to introduce exponential transverse (T_2) and longitudinal (T_1) relaxation time constants

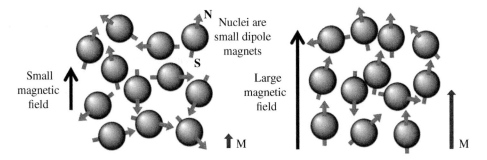

Figure 2.3 Net magnetization arises from the combined contribution of large number of spins and is proportional to the strength of the applied magnetic field.

for the $x-y$ and z components of the magnetization, respectively. Namely,

$$\frac{d\vec{M}}{dt} = \gamma\vec{M} \times \vec{B} - \frac{M_x\vec{x} + M_y\vec{y}}{T_2} - \frac{(M_z - M_0)\vec{z}}{T_1}, \tag{2.7}$$

where M_0 is the thermal equilibrium value for the net magnetization. Equation (2.7) is the famous Bloch equation first introduced in 1946, and it plays a fundamental role in NMR.

QUESTION: What is the fractional alignment for ^{1}H nuclei in a 3T clinical MRI scanner?

2.3 Rf Excitation and Reception

If we now apply a small magnetic field rotating in the $x-y$ plane, namely $\vec{B}_1 = B_1 \cos(\omega_{RF}t)\vec{x} - B_1 \sin(\omega_{RF}t)\vec{y}$, the net magnetization reacts by nutating away from $B_0\vec{z}$ via rotation around the net applied field. As shown in Figure 2.4, this process can be viewed both in the stationary laboratory $(\vec{x}, \vec{y}, \vec{z})$ frame of reference, or in a more convenient frame of reference, $(\vec{x}', \vec{y}', \vec{z})$, rotating about the z-axis at the Larmor frequency. In general, the magnitude of $\vec{B}_1$ is several orders of magnitude smaller than B_0, and would have minimal effect on the net magnetic field vector, except in the case where when $\omega_{RF} = \omega_0$. Under this condition, known as on-resonance, $\vec{M}$, as depicted in the rotating frame of reference, rotates about the axis defined by the Rf field.

QUESTION: What happens if the B_1 field is off-resonance?

The oscillating B_1 field required for resonance is in the radiofrequency (Rf) band, and the process of transitioning the net magnetic moment away from its equilibrium condition is known as Rf excitation. If the oscillating magnetic field is applied for a duration of τ seconds, the angle θ formed

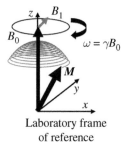

Laboratory frame
of reference

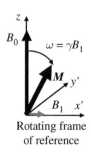

Rotating frame
of reference

Figure 2.4 Net magnetization vector during Rf excitation as depicted in the laboratory and rotating frames of reference.

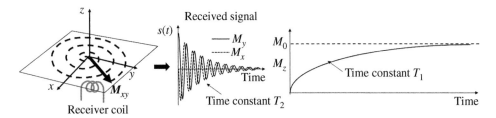

Figure 2.5 A free induction decay (FID). The magnetization vector's time evolution following a 90° Rf excitation pulse, the received signal, $S(t)$, from the transverse magnetization M_{xy} oscillates at ω_0 and decays at a rate of $1/T_2$. In contrast, the M_z component regrows at a slower rate of $1/T_1$.

between the z-axis and the net magnetic moment $\vec{M}$, called the flip or tip angle, is given by $\theta = \gamma B_1 \tau$. Transverse magnetization is maximized for a 90° flip angle.

If an Rf-tunned resonant coil is now brought near the spins, the time-varying magnetic flux will induce a detectable voltage proportional to the time derivative of M_x. In practice, the receiving coil is designed such that the two orthogonal components, M_x and M_y, where $M_{xy} \equiv M_x + iM_y$, are detected independently (quadrature sensing) to improve the signal-to-noise ratio (SNR). The observed signal following a 90° excitation is known as a free induction decay (FID), and, as shown in Figure 2.5, the M_x and M_y components will decay with an exponential time-constant T_2, while the M_z component regrows with time constant T_1.

The observed decay of M_{xy} arises from the individual spin orientations losing phase coherence over time due to additional perturbing magnetic fields experienced by the nuclei. Simultaneously, a regaining of phase coherence along the z axis ultimately returns M_z to its thermal equilibrium value of M_0, if given sufficient time.

While a 90° Rf pulse is commonly used to generate transverse magnetization, a 180° Rf pulse is often used to eliminate dephasing due to static field inhomogeneities (deviations from B_0) in what is known as a ***spin echo***. As shown in Figure 2.6a, a 180° excitation, for example along the $y-$axis, flips the spins in the transverse plane, causing them to rephase at a time known as the echo time (TE). A second use for 180° Rf pulses is to invert M_z magnetization. Figure 2.6b follows the evolution of the longitudinal magnetization M_z after the repeated 180° Rf pulses. This technique, known as ***inversion recovery***, is often employed to either measure T_1 values or generate T_1-based image contrast.

2.4 Spatial Localization

So how do we go from the Bloch equation given in Eq. (2.7) to forming images of the spatial distribution of the nuclear spins, $M(x, y, z)$? Let us begin with the definition of the linear magnetic field gradients, G_x, G_y, and G_z, wherein the amplitudes of the z-component of these fields vary linearly with x, y, and z coordinates, respectively. Adding such gradients to a large static field B_0 field oriented in the z-direction results in the net magnetic field at any point (x, y, z) of $B = B_0 + G_r r$ for $r = x, y$, or z. The effect is to generate a mapping between spatial position and the frequency of the received signal (see Figure 2.7).

Similarly, linear gradients can be used to selectively excite spins within a chosen plane. This process is known as ***selective excitation*** and exploits a linear mapping that exists in the presence of a gradient field between the frequency content of an Rf pulse and the resonant frequency of the spins in the sample (see Figure 2.8).

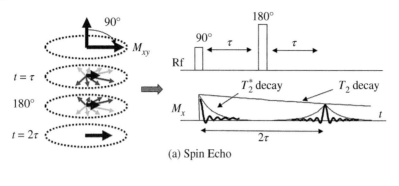

(a) Spin Echo

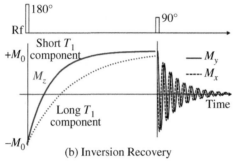

(b) Inversion Recovery

Figure 2.6 Commonly used MRI techniques: (a) spin echo and (b) inversion recovery. T_2^* is the observed transverse magnetization decay rate including both thermal processes and static B_0 field inhomogeneities, while T_2 is the M_{xy} decay rate in the absence of these inhomogeneities.

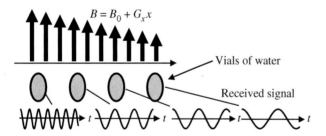

Figure 2.7 MR signal from vials of water in a linear gradient magnetic field.

2.5 The MRI Signal Equation

Using the Bloch equations (ignoring relaxation), the transverse magnetization following exci-tation from each small tissue volume looks like a tiny oscillating magnetic dipole $M_{xy}(x, y, z)e^{-i\varphi(x, y, z, t)}$ (see Figure 2.9). Assuming a uniformly sensitive RF coil, the received signal is given by:

$$s(t) = \iiint\limits_{x\,y\,z} M_{xy}(x, y, z)e^{-i\varphi(x,y,z,t)}\,dxdydz. \tag{2.8}$$

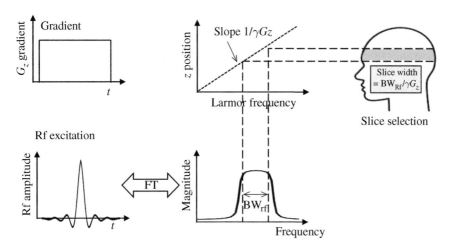

Figure 2.8 Selective excitation is used to excite a single spatial slice of spins.

Figure 2.9 The signal from each volume of tissue is represented by an oscillating magnetic dipole.

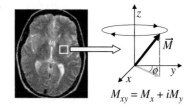

The instantaneous frequency is $\omega = d\varphi/dt$; hence,

$$\varphi(x, y, z, t) = \int_0^t \omega(x, y, z, t')dt' = \gamma \int_0^t B(x, y, z, t')dt'. \tag{2.9}$$

Then, in the presence of linear gradients, $B = B_0 + G_x(t)x + G_y(t)y + G_z(t)z$, the received signal is:

$$s(t) = \iiint_{xyz} M_{xy}(x, y, z)e^{-i\gamma B_0 t}e^{-i\gamma \int_0^t (G_x(t')x + G_y(t')y + G_z(t')z)dt'} \, dxdydz. \tag{2.10}$$

Demodulating at frequency $\omega = \gamma B_0$ yields:

$$s_d(t) = \iiint_{xyz} M_{xy}(x, y, z)e^{-i\gamma \int_0^t (G_x(t')x + G_y(t')y + G_z(t')z)dt'} \, dxdydz. \tag{2.11}$$

Comparing Eq. (2.11) to the Fourier transform of M_{xy}, denoted by $\mathcal{M}$,

$$\mathcal{M}(k_x, k_y, k_z) = \iiint_{xyz} M_{xy}(x, y, z)e^{-i2\pi(k_x x + k_y y + k_z z)} \, dxdydz, \tag{2.12}$$

yields a direct spatial frequency, typically called k-space, interpretation of MRI. Namely,

$$s_d(t) = \mathcal{M}\left(\frac{\gamma}{2\pi}\int_0^t G_x(t')dt', \frac{\gamma}{2\pi}\int_0^t G_y(t')dt', \frac{\gamma}{2\pi}\int_0^t G_z(t')dt'\right), \tag{2.13}$$

where the spatial frequencies are:

$$k_x(t) = \frac{\gamma}{2\pi}\int_0^t G_x(t')dt', k_y(t) = \frac{\gamma}{2\pi}\int_0^t G_y(t')dt', \text{ and } k_z(t) = \frac{\gamma}{2\pi}\int_0^t G_z(t')dt'. \tag{2.14}$$

Hence, one way to view MRI is that, following Rf excitation, the gradients simply trace a trajectory through k-space, and the detected signal can be converted into an image via an inverse Fourier transform. In most cases, owing to relaxation, multiple excitations are needed to fully sample k-space (see Figure 2.10). There are innumerable MRI sequences characterized by the chosen Rf pulses and gradient waveforms, with sequences typically optimized for acquisition speed and/or image contrast.

The various MRI pulse sequences are typically depicted by what are known as **pulse sequence diagrams**, showing the amplitudes and timings of waveforms for all the applied magnetic fields as well as the data acquisition window. Figure 2.11 shows an example pulse sequence diagram for a simple gradient-echo imaging acquisition.

We end this chapter with a brief discussion of the sensitivity of an MR experiment. The thermal equilibrium value for the longitudinal magnetization, to be derived in the next chapter, is:

$$M_0 = \rho \frac{\gamma^2 \hbar^2 B_0}{4kT}, \tag{2.15}$$

where ρ = spins/unit volume, $\hbar$ = Planck's constant/$2\pi = 1.05 \times 10^{-34}$ Joule s, k is the Boltzmann's constant $= 1.38 \times 10^{-23}$ J/K, and T the absolute temperature. The received signal, *i.e.*, the voltage induced in Rf receiver coil, is:

$$s(t) \propto \frac{dM_{xy}}{dt} \propto \omega M_{xy} \propto \rho \frac{\gamma^3 \hbar^2 B_0^2}{4kT}. \tag{2.16}$$

But what about the noise? Consider the simple model of a tuned Rf receiver coil shown in Figure 2.12, where L and C are the coil inductance and capacitance, respectively. The resistance seen by the coil arises from two components. First, there is the inherent electrical resistance of the coil itself, R_{coil}, that is proportional to $\sqrt{\omega}$ due to Rf skin depth effects. There is also noise (which can be modeled as a resistance) coming from the subject being imaged. This component, known

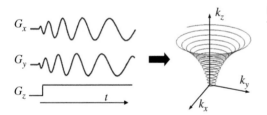

Figure 2.10 The gradients trace out a trajectory through k-space.

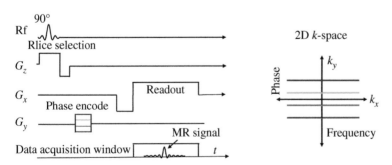

Figure 2.11 MRI pulse sequence diagram for an acquisition consisting of slice-selective 90° pulse along z, phase encoding in y, and a readout in x. One line of k-space is acquired for each value of the phase-encode gradient.

Figure 2.12   Model for an Rf receiver coil.

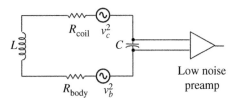

as R_{body}, is proportional to ω^2 due to inductive losses (ignoring dielectric losses that are usually much smaller).

Any resistor can be viewed as a noise source, which, using the Johnson noise spectral density function of $v_n^2 = 4kTR$ (where T is the temperature and R is the resistance), results in a body-plus-coil MR noise factor of:

$$\text{noise} \propto \sqrt{\alpha\gamma^2 B_0^2 + \beta\sqrt{\gamma B_0}}, \tag{2.17}$$

where α and β are constants depending on the coil construction and object being imaged. Combining with Eq. (2.16),

$$SNR \propto \rho \frac{\gamma^3 \hbar^2 B_0^2}{4kT\sqrt{\alpha\gamma^2 B_0^2 + \beta\sqrt{\gamma B_0}}}. \tag{2.18}$$

For most human imaging applications conducted at relatively high field strengths (e.g., $\gamma B_0 \gg 10$ Mz), body noise typically dominates coil noise, resulting in:

$$SNR \propto \rho \frac{\gamma^2 \hbar^2 B_0}{4kT}. \tag{2.19}$$

This contrasts with some small-animal imaging applications or when performing chemical assays in a high-field spectrometer, for which noise due to the coil resistance can dominate the noise from the sample.

2.6   Summary

To summarize this chapter, we have described the Bloch equation governing the behavior of spins in a magnetic field, how Rf excitation can be used to convert between longitudinal magnetization and transverse magnetization, and the use of linear gradients to provide the spatial information needed for imaging applications. So, what is missing?

T_1 and T_2 can be included as k-space weightings. Namely,

$$s_d(t) = \iiint_{xyz} M_{xy}(x, y, z, T_1, T_2) e^{-t/T_2} e^{-i\gamma \int_0^t (\vec{G}\cdot\vec{r}) dt'} dx dy dz, \tag{2.20}$$

and, as shown in Chapter 6, off-resonant effects, due to chemical shift or spatial B_0 inhomogeneities, can be included by adding time evolution as an additional k-space axis, $t = k_f$. However, what are the underlying mechanisms driving T_1 and T_2 relaxation? Why does water in different tissues exhibit different relaxation times? How do contrast agents work, and what about couplings between spins, or chemical exchange effects?

We have now finished our brief overview of classical MRI, and we will address these interesting questions in the rest of this text (see Figure 2.13).

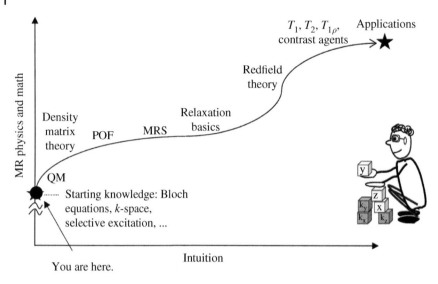

Figure 2.13 The roadmap at the end of Chapter 2.

Exercises

E.2.1 A Simple MRI Sequence

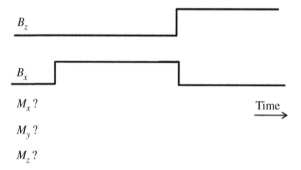

A sample of water is left to reach thermal equilibrium in zero magnetic field. A magnetic field is then turned on in the x-direction. After some time, the magnetic field is switched from the x-direction to the z-direction. Can you describe the behavior of the net ^{1}H magnetic moment from the sample over time? What about the behavior of an individual nuclear spin? Can you explain T_1 and T_2?

E.2.2 Bottle Flipping and Inversion Recovery

a) A student is first asked to collect a free induction decay (pulse sequence: 90° – acquire) from a bottle of water, and he/she generates the data shown below. The student is then asked to acquire data using an inversion recovery sequence: 180° – 90° – acquire. However, instead of reprogramming the scanner to add a 180° inversion pulse, he/she decides to just physically flip the bottle immediately prior to collecting an FID (90° – acquire). Sketch the expected signal response and compare the two experiments.

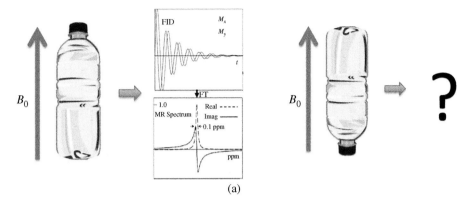

(a)

b) An alternative solution is to rapidly rotate B_0 from +z to −z. Does this work or is this the same as part (a)? If it does work, what would be the constraint on the flipping time, T, to achieve a 180° inversion?

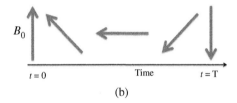

(b)

c) A third alternative is to rapidly ramp B_0 down going through zero and stopping at $-B_0$. Does this work? If so, what would be the constraint on T to achieve a 180° inversion?

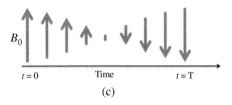

(c)

E.2.3 Deriving the Bloch Equations

a) Starting with basic equations of motion,

$$\frac{d\vec{M}(t)}{dt} = \vec{M}(t) \times \gamma\vec{B}(t),$$

and $\vec{B}(t) = B_x\vec{x} + B_y\vec{y} + B_0\vec{z}$, derive the Bloch equations in the laboratory frame without relaxation. Namely,

$$\frac{dM_x(t)}{dt} = \gamma[M_y(t)B_0 - M_z(t)B_y]$$

$$\frac{dM_y(t)}{dt} = \gamma[M_z(t)B_x - M_x(t)B_0]$$

$$\frac{dM_z(t)}{dt} = \gamma[M_x(t)B_y - M_y(t)B_x]$$

b) Starting from the full Bloch equations in the laboratory frame,

$$\frac{dM_x(t)}{dt} = \gamma[M_y(t)B_0 - M_z(t)B_y] - \frac{M_x(t)}{T_2}$$

$$\frac{dM_y(t)}{dt} = \gamma[M_z(t)B_x - M_x(t)B_0] - \frac{M_y(t)}{T_2}$$

$$\frac{dM_z(t)}{dt} = \gamma[M_x(t)B_y - M_y(t)B_x] - \frac{M_z(t) - M_0}{T_1}$$

Derive the Bloch equations for a frame of reference rotating at a frequency ω clockwise about the z axis (left-handed rotation).

c) Show that the following are solutions to the Bloch equations in the rotating frame, assuming initial magnetization in the transverse plane and a free induction decay, i.e., $\vec{B}(t) = B_0\vec{z}$ with an off-resonance term of ω are:

$$M_x(t) = M_0 \cos((\gamma B_0 - \omega)t + \varphi)e^{-1/T_2}$$

$$M_y(t) = -M_0 \sin((\gamma B_0 - \omega)t + \varphi)e^{-1/T_2}$$

$$M_z(t) = M_0 \left(1 - e^{-1/T_1}\right)$$

E.2.4 Nuclear Induction

This question concerns the paper by Bloch (1946).

a) Verify Eq. (2.16),

$$\tan \theta = \frac{\pm \gamma H_1}{\omega \pm \gamma H_0} = \frac{\gamma H_1}{\gamma H_0 \mp \omega}.$$

What is the physical interpretation for this value of θ?

b) Which of the following concepts are discussed in Bloch's (1946) paper?
1. Continuous wave (CW) NMR, 2. pulsed NMR, 3. prepolarized NMR, 4. quadrature excitation, 5. adiabatic excitation, 6. T_1, 7. T_2, 8. doping samples with small amounts of paramagnetic material to reduce T_1, and 9. imaging.

E.2.5 Steady-State NMR and Saturation

a) Find the steady-state solution to the Bloch equation for the case of a continuous Rf excitation, i.e., (in the rotating frame), $\vec{B} = [B_1, 0, B_0 - \omega/\gamma]$.

b) How large does B_1 have to be in order for the steady-state magnetization to be driven to approximately zero (i.e., $M_x \approx M_y \approx M_z \approx 0$), a condition known as "saturation" (assume RF excitation is on resonance)?

E.2.6 MRI Sensitivity and SNR

PhD graduate student **A**, who works in a medical imaging lab, claims that the SNR for MRI is linearly proportional to B_0. Whereas student **B**, who works in a chemistry lab studying protein structure, claims that the SNR for NMR goes as $B_0{}^{7/4}$. Derive an expression for SNR and determine who is correct.

E.2.7 SNR revisited

The quality factor (Q) of an Rf coil is the ratio of the stored and dissipated energy.

$$Q = \frac{\text{Maximum energy stored}}{\text{Average energy dissipated per cycle}} = \frac{\omega L}{R},$$

where L is the coil inductance and R the resistance, and the performance of a coil can be measured by the ratio of the loaded to unloaded Q. Namely:

$$\frac{Q_L}{Q_U} = \frac{R_{\text{coil}}}{R_{\text{coil}} + R_{\text{body}}}.$$

Show that the SNR for a 90°-acquire MRI measurement of a collection of N spin-½ particles with gyromagnetic ratio γ can be expressed as:

$$SNR \propto N\frac{\gamma \hbar P}{2}\sqrt{1 - \frac{Q_L}{Q_U}},$$

where P = polarization, Q_U = unloaded quality factor (Q) of the Rf coil, and Q_L = loaded Q of the Rf coil. Note, in general there is also a term to account for the noise factors of the preamp. However, this term is typically less that 0.5 dB, and for our purposes can be neglected.

E.2.8 Sensitivity

This question concerns the paper by Hoult and Lauterbur (1979).

a) What is the purpose of a Faraday shield in an RF coil? Why is the shield absent in many commonly used MRI coils (*e.g.*, standard bird-cage head coil)?

b) Conclusion 1 (p. 432) implies clinical MRI scanners need to operate at fields ≤ 0.24 T (10 MHz). Where is the flaw in this reasoning?

Historical Notes

Sir Joseph Larmor, a distinguished mathematical physicist, was born on 11 July 1857, in Magheragall, Country Antrim, Ireland. He received his education at the Royal Academical Institution and later pursued his BA and MA at Queen's College, Belfast. Larmor's brilliance earned him a fellowship at Cambridge University in 1876, where he eventually became Senior Wrangler in 1880. He later served as a professor of natural philosophy at Queen's College, Galway, and returned to Cambridge as a mathematics lecturer. In 1903, he achieved the prestigious position of Lucasian Professor of Mathematics.

Larmor's remarkable contributions to physics were recognized through his election as a fellow of the Royal Society in 1892. In 1909, he was knighted, and he also served as the unionist member of parliament for the University of Cambridge from 1911 to 1922. After his death in 1942, a lunar crater on the far side of the moon was named in his honor to commemorate his contributions to science.

Larmor's work in physics was deeply influenced by James Clerk Maxwell's groundbreaking research, particularly in electrodynamics. Despite living during the emergence of relativity and quantum theory, Larmor remained at the forefront of classical physics. He explored the concept of the ether, which played a central role in his thinking. In his book "Aether and Matter," published in 1900, Larmor focused on multiple ethers and their interactions with matter mediated through electric charges, particularly electrons embedded in the ether. One of Larmor's significant contributions to physics came in his 1897 paper, where he derived equations describing how observers in relative motion perceive physical events. These equations, later restated by Lorentz, became a crucial component of Albert Einstein's theory of relativity in 1905. Unfortunately, they are now more commonly associated with Lorentz, despite Larmor's earlier work in this area.

Larmor's enduring legacy lies in the field of nuclear magnetic resonance (NMR). In 1897, while studying the splitting of optical spectral lines caused by an external magnetic field (as observed by Pieter Zeeman), Larmor derived a fundamental equation governing the resonance frequency of an NMR system, realizing that the precession frequency of nuclear spins in a magnetic field is directly proportional to the strength of the applied field multiplied by a constant related to the particle's charge/mass ratio.

(Source: photo from https://commons.wikimedia.org/wiki/File:Joseph_Larmor.jpeg, public domain)

Sir Peter Mansfield, born on October 9, 1933, in London, England, was a pioneering English physicist who made groundbreaking contributions to NMR and played a pivotal role in the development of MRI. Alongside American chemist Paul Lauterbur, he received the Nobel Prize for physiology or medicine in 2003, a testament to his revolutionary work in medical imaging.

After obtaining a Ph.D. in physics from the University of London in 1962, Mansfield spent two transformative years as a research associate in the United States. He then returned to the United Kingdom and joined the faculty of the University of Nottingham, where he made significant contributions to the scientific community.

Mansfield's award-winning research is built upon the foundation of NMR, a powerful technique that involves the selective absorption of high-frequency radio waves by specific atomic nuclei in a strong stationary magnetic field. NMR had already proven to be an essential tool in chemical analysis, providing valuable insights into the molecular structure of various substances.

In the early 1970s, Paul Lauterbur's innovative idea of introducing intentional nonuniformities in the magnetic field opened the door to medical applications of NMR and laid the groundwork for MRI. Building on Lauterbur's work, Peter Mansfield extended the concept to rapidly generate detailed and clear images of internal organs and soft tissues, revolutionizing medical diagnostics.

Mansfield's contributions extended beyond technical implementation of MRI acquisition methods; he developed new mathematical techniques for rapid data analysis, enabling efficient processing of vast amounts of information obtained from NMR signals. Thanks to his efforts, MRI imaging became considerably faster and more accessible, greatly enhancing its practicality as a diagnostic tool. In recognition of his exceptional contributions to science and medicine, Sir Peter Mansfield was knighted in 1993.

(Source: photo from https://www.nobelprize.org/prizes/medicine/2003/mansfield/facts/, public domain.)

Paul Christian Lauterbur was born on May 6, 1929, in Sidney, Ohio, and developed a deep fascination with scientific exploration from a young age, pursuing his passion by earning a Ph.D. in chemistry from the University of Pittsburgh in 1962. Lauterbur's groundbreaking work in NMR would go on to revolutionize the field of medicine and earn him the prestigious Nobel Prize.

As a professor at the University of New York at Stony Brook in 1969, Lauterbur conducted innovative research in NMR. This technique involved the selective absorption of high-frequency radio waves by specific atomic nuclei when subjected to a strong stationary magnetic field. It proved to be a powerful tool in chemical analysis, offering valuable insights into the molecular structures of solids and liquids.

During the early 1970s, Lauterbur had a visionary idea – intentionally creating a nonuniform magnetic field in NMR experiments to exploit signal distortions and generate two-dimensional images of a sample's internal structure. This groundbreaking concept laid the foundation for the development of MRI, a technology that would revolutionize the medical world.

In 1985, Lauterbur's remarkable work led to his appointment as a professor at the University of Illinois at Urbana-Champaign and director of its Biomedical Magnetic Resonance Laboratory. Here, he continued to push the boundaries of MRI research, refining the technique, and exploring new possibilities for noninvasive medical imaging.

The true transformation of MRI into a practical medical tool was achieved through a collaborative effort. Lauterbur's innovative ideas caught the attention of English physicist Sir Peter Mansfield, who was conducting his own research in NMR imaging. Recognizing the immense potential of Lauterbur's work, Mansfield played a crucial role in translating it into a viable medical application.

Together, Lauterbur and Mansfield made groundbreaking advancements, perfecting the MRI technology. Their joint efforts resulted in the development of computerized scanning technology capable of producing detailed images of internal body structures, with a particular focus on soft tissues. Unlike X-ray and computed tomography (CT) examinations, MRI was free from harmful side effects, making it an attractive option for medical diagnosis and research.

(Source: photo from https://www.nobelprize.org/prizes/medicine/2003/lauterbur/facts/, Nobel Foundation archive, public domain.)

3

Quantum Mechanical Description of MR

3.1 Introduction

3.1.1 Why Quantum Mechanics for Magnetic Resonance?

Quantum mechanics (QM) is the only theory that correctly predicts the behavior of matter on the atomic scale, and QM effects are seen *in vivo*. In general, systems of isolated nuclei can be described with the intuitive picture of a classical magnetization vector rotating in three-dimensional (3D) space (as discussed in Chapter 2). However, systems of interacting nuclei generally require a more complete QM description. Fortunately, we will only need a small piece of the complete theory of QM found in most physics textbooks. We first develop an understanding of MR based on a branch of QM known as density matrix theory. Then, to retain the intuitive concepts of classical vector models while retaining the advantages of the more complete QM analysis, we will extend this framework to what is known as the product operator formalism (POF) of MR (Sorensen et al. 1983).

3.1.2 Historical Developments

At the end of the 19th century, physicists divided the world into two entities – matter, as described by Newtonian mechanics, and electromagnetic fields and waves, as described by Maxwell's equations (see Figure 3.1). By the early 20th century, the inability of classical physics to explain an increasing number of experimental results led to a revolution in physics with the introduction of both QM and relativity. Paul Dirac, in 1928, was able to merge QM and special relativity (characterized by maximum velocities capped by the speed of light); however, the ultimate merger with general relativity (Einstein's theory including gravity via mass/energy warping the geometry of space and time) remains an open problem.

The beginning of QM is typically marked by Max Planck's theory of black-body radiation as published in 1901 (Planck 1901). In physics, a "black-body" is an idealized object that absorbs all incident electromagnetic radiation; no radiation is reflected, but rather the only emitted radiation is produced via thermal processes. Although nature has created materials approaching such a theorized object, the most practical experimental object is a cavity with a blackened interior and only a small hole permitting incident radiation to enter (see Figure 3.2). The basic concept is that light entering the hole will be absorbed by the cavity with minimal light reflected. Light exiting the hole is assumed to be solely emitted radiation, dependent only on the temperature of the object.

The problem, in what became known as the "ultraviolet catastrophe," involved the study of the color of objects as they were heated. According to the theoretical laws of classical physics, such an

Fundamentals of In Vivo Magnetic Resonance: Spin Physics, Relaxation Theory, and Contrast Mechanisms, First Edition.
Daniel M. Spielman and Keshav Datta.
© 2024 John Wiley & Sons, Inc. Published 2024 by John Wiley & Sons, Inc.
Companion website: www.wiley.com/go/Spielman

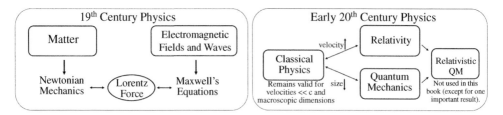

Figure 3.1 The transition from classical physics (left) to include relativity and quantum mechanics (right).

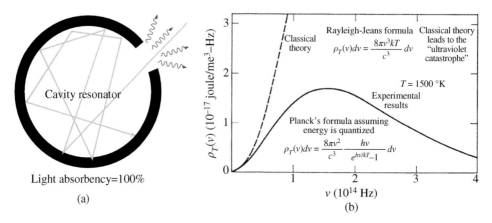

Figure 3.2 Black-body radiation. (a) An idealized cavity resonator modeling a black-body and (b) Planck's solution to the black-body radiation problem via energy quantization. Source: (b) Adapted with permission from Eisberg and Resnik (1974), Figure 1–8, p. 15.

object should emit an increasing amount of energy as a function of frequency, in direct contrast to intuitive and observed results. Planck was able to solve this problem by heuristically assuming energy is only emitted to finite-sized packets, given by the formula:

$$\mathcal{E} = nh\nu, \tag{3.1}$$

where $\mathcal{E}$ is energy, n is an integer, ν the frequency in Hz, and $h = 6.63 \times 10^{-34}$ J s (a constant known as Planck's constant).

Albert Einstein, in explaining the photoelectric effect, generalized Planck's results and proposed a return to the particle theory of light (Einstein 1905a). That is, light is composed of photons each with energy $h\nu$. Hence EM waves, *i.e.*, radiation, exhibit both wave and particle features with the characteristic parameters of a photon linked by the Planck–Einstein relations:

$$\mathcal{E} = h\nu = \hbar\omega \text{ and } \vec{p} = \hbar\vec{k}, \tag{3.2}$$

where $\vec{p}$ is the momentum and $\vec{k}$ is related to the wavelength λ via the relation $|\vec{k}| = 2\pi/\lambda$. Note, we denote vector quantities by the symbol $\rightarrow$. Interestingly, photons were not experimentally shown to exist until almost 10 years later via the Compton effect (Compton 1923). In 1923, Louis de Broglie proposed that material particles (electrons, protons, and neutrons), just like photons, can have wavelike aspects in what is now known as wave–particle duality (de Broglie 1923). The wave properties of matter were later demonstrated via interference patterns obtained in diffraction experiments (see Figure 3.3).

Hence, every particle can be associated with an energy $\mathcal{E}$, momentum $\vec{p}$, angular frequency $\omega = 2\pi\nu$, wave number $\vec{k}$, and associated de Broglie wavelength $\lambda = 2\pi/|\vec{k}| = h/|\vec{p}|$ (later we will

Figure 3.3 Diffraction patterns from a NaCl crystal bombarded with (a) X-rays (electromagnetic wave) and (b) neutrons (matter). Source: Adapted with permission from Eisberg and Resnik (1974), Figures 3–5, p. 68.

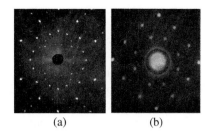

(a)   (b)

add spin). However, it is important to remember that the wavelengths of macroscopic particles, even tiny ones, are exceedingly small. For example, consider a speck of dust with mass $m = 10^{-15}$ kg and velocity $v = 10^{-3}$ m/s. The de Broglie wavelength is only:

$$\lambda = \frac{h}{p} = \frac{h}{mv} = 6.6 \times 10^{-6} \, \text{Å}. \tag{3.3}$$

From which we can reasonably conclude that living in a macroscopic world provides little intuition regarding the behavior of matter on the atomic scale.

3.1.3 Wave Functions

QM, unlike classical physics, does not deal directly with observable physical quantities such as position or momentum (or, in the case of an MR experiment, M_x, M_y, and M_z). QM deals with the state of the system as described by a function known as a wavefunction, independent of the observable to be detected. In this formulation, probability is fundamental. Whereas classical physics is described by the sentence, "If we know the present exactly, we can predict the future," QM is based on the proposition, "We *cannot* know the present exactly, as a matter of principle."

In place of the classical concept of a trajectory (succession in time of the state of a classical particle), we substitute the concept of the quantum state of a particle characterized by a wave function $\psi(\vec{r}, t)$. The function $\psi(\vec{r}, t)$ contains all the information possible to obtain about the particle, with the magnitude $|\psi(\vec{r}, t)|^2$ interpreted as the probability of detecting the particle's presence at position $\vec{r}$ and time t. Note, wave functions are generally complex; hence, $|\psi(\vec{r}, t)|^2 = \psi(\vec{r}, t)\psi^*(\vec{r}, t)$, where * denotes the complex conjugate. Given the sum over all probabilities must add to one,

$$dP(\vec{r}, t) = C|\psi(\vec{r}, t)|^2 d^3\vec{r}, \tag{3.4}$$

where C is a constant such that $\int dP(\vec{r}, t) = 1$. This, in turn, implies,

$$\frac{1}{C} = \int |\psi(\vec{r}, t)|^2 d^3\vec{r} < \infty, \tag{3.5}$$

naturally leading to the study of square-integrable functions. Wave functions are typically normalized, so, unless explicitly stated otherwise, we shall assume $\int |\psi(\vec{r}, t)|^2 d^3\vec{r} = 1$.

A key question is: How does $\psi(\vec{r}, t)$ change with time? The answer, as postulated by Erwin Schrödinger, is given by the now famous Schrödinger's equation:

$$i\hbar\frac{\partial}{\partial t}\psi(\vec{r}, t) = -\frac{\hbar^2}{2m}\nabla^2\psi(\vec{r}, t) + V(\vec{r}, t)\psi(\vec{r}, t), \tag{3.6}$$

where ∂ denotes the partial derivative, $\nabla^2 = \frac{\partial^2}{\partial x^2} + \frac{\partial^2}{\partial y^2} + \frac{\partial^2}{\partial z^2}$ the Laplacian, and V corresponds to potential energy (Schrodinger 1926). This equation is often written in the form:

$$\frac{\partial}{\partial t}\psi(\vec{r}, t) = -\frac{i}{\hbar}H\psi(\vec{r}, t), \tag{3.7}$$

where,

$$H = -\frac{\hbar^2}{2m}\nabla^2 + V(\vec{r}, t). \tag{3.8}$$

In QM, physical quantities are expressed as operators; H, the Hamiltonian, being the operator for total energy. Note, classically, total energy is given by the sum of kinetic and potential terms: $H = -p^2/2m + V$ with the momentum p defined as $-\hbar\nabla$. Hereafter, all operators will be denoted by $\hat{}$, for example, $\hat{H}$, to avoid confusion with their classical equivalents.

As an aside, it is of some interest to provide a plausibility argument for Schrödinger's equation (presented here for the one-dimensional case). Given Newton's equations for force and energy,

$$F = \frac{d}{2dt}p, \tag{3.9}$$

and,

$$\mathcal{E} = \frac{p^2}{2m} + V. \tag{3.10}$$

In combination with the de Broglie–Einstein relations $k = 2\pi/\lambda = p/\hbar$ and $\mathcal{E} = \omega\hbar$, it is reasonable to look for a QM wave of the form:

$$\psi(x, t) = e^{i(kx-\omega t)}. \tag{3.11}$$

This expression represents a traveling wave with constant wave number and frequency (equivalently momentum and energy) that satisfies both the Newton and de Broglie–Einstein relations for constant V.

The spatial and temporal derivatives of ψ are then given by:

$$\frac{\partial}{\partial x}\psi(x, t) = i\frac{p}{\hbar}e^{i(kx-\omega t)}, \tag{3.12}$$

and

$$\frac{\partial}{\partial t}\psi(x, t) = -i\frac{\mathcal{E}}{\hbar}e^{i(kx-\omega t)}. \tag{3.13}$$

These two equations imply:

$$-i\hbar\frac{\partial}{\partial x}\psi(x, t) = p\psi(x, t), \tag{3.14}$$

and

$$i\hbar\frac{\partial}{\partial t}\psi(x, t) = \mathcal{E}\psi(x, t), \tag{3.15}$$

respectively. Defining the operators for momentum and energy, respectively, as $\hat{p} = -i\hbar\frac{\partial}{\partial x}$ and $\hat{\mathcal{E}} = i\hbar\frac{\partial}{\partial t}$ and substituting into the energy Eq. (3.10), we obtain:

$$\hat{\mathcal{E}}\psi(x, t) = \frac{\hat{p}^2}{2m}\psi(x, t) + \hat{V}\psi(x, t). \tag{3.16}$$

Equivalently,

$$i\hbar\frac{\partial}{\partial t}\psi(x, t) = \frac{-\hbar^2}{2m}\frac{\partial^2}{\partial x^2}\psi(x, t) + \hat{V}\psi(x, t). \tag{3.17}$$

Schrödinger's equation is obtained by postulating that Eq. (3.17) holds for V being a function of space and time, and not just a constant.

Using wave functions, the quantum description of a free particle is readily found. For a particle subject to no external forces, namely setting $V(\vec{r}, t) = 0$,

$$i\hbar \frac{\partial}{\partial t} \psi(\vec{r}, t) = -\frac{\hbar^2}{2m} \nabla^2 \psi(\vec{r}, t), \tag{3.18}$$

which can be shown to be satisfied by:

$$\psi(\vec{r}, t) = A e^{i(\vec{k} \cdot \vec{r} - \omega t)}. \tag{3.19}$$

This equation represents a plane wave with wave number k and frequency ω. Because $|\psi(\vec{r}, t)|^2 = |A|^2$, a constant, the probability of finding the particle is uniform throughout space. Note, strictly speaking, $\psi(\vec{r}, t)$ in this case is not square integrable, but this presents no more a problem than the use of delta functions in Fourier theory.

The linearity of Schrödinger's equation implies that superposition holds. That is, a general linear combination of plane waves is also a solution.

$$\psi(\vec{r}, t) = \frac{1}{(2\pi)^{3/2}} \int g(\vec{k}) e^{i(\vec{k} \cdot \vec{r} - \omega t)} d^3\vec{k}. \tag{3.20}$$

Consider the one-dimension case evaluated at a fixed time, say $t = 0$,

$$\psi(x, 0) = \frac{1}{\sqrt{2\pi}} \int g(k) e^{ikx} dk. \tag{3.21}$$

Subject to some potential V, the probability of finding the particle at a given point in space might look something like that shown in Figure 3.4.

In quantum terms, the prior example of a dust speck would be described by a wave packet with group velocity v, average momentum p, with the maximum of the wave function typically called the "position." But how accurately can we measure the dust speck's position at a given moment in time, say $t = 0$? The general wavefunction $\psi(x, 0)$ can be written as a Fourier transform,

$$\psi(x) = \frac{1}{\sqrt{2\pi}} \int \overline{\psi}(k) e^{ikx} dk, \tag{3.22}$$

where the momentum, from Eq. (3.2), is given by $p = \hbar k$.

The Fourier theory of equivalent widths immediately yields the most common form of the uncertainty principle, namely $\Delta x \cdot \Delta p \geq \hbar/2$. In the dust speck example, if the position is measured to an accuracy of, for example, $\Delta x = 0.01$ μm, then $\Delta p = \hbar/\Delta x \cong 10^{-26}$ kg $\cdot$ m/s. Given no current momentum measuring device can achieve this level of accuracy, both Δx and Δp are negligible, and we can treat the dust speck as a classical particle. At the atomic level, however, Δx and Δp are not negligible. Note, later when discussing MR, we will not be dealing with position and momentum, but rather another intrinsic property of matter known as spin angular momentum.

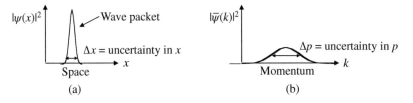

Figure 3.4 The probabilities of finding a particle at (a) a given point in space x with (b) a given momentum p is described by sums of wave functions known as "wave packets," where $\psi(x)$ and $\overline{\psi}(k)$ satisfy the relationship given by Eq. (3.22).

To summarize, the classical concept of a trajectory is replaced by that of a wavefunction $\psi(\vec{r}, t)$ containing all possibly obtainable information about a particle. The probability of finding the particle in a differential volume $d^3\vec{r}$ at time t is

$$dP(\vec{r}, t) = C|\psi(\vec{r}, t)|^2 d^3\vec{r}, \tag{3.23}$$

where C is a normalization constant, and the time evolution is given by Schrödinger's equation

$$\frac{\partial}{\partial t}\psi(\vec{r}, t) = -\frac{i}{\hbar}\hat{H}\psi(\vec{r}, t), \tag{3.24}$$

where $\hat{H}$ is the Hamiltonian representing the total energy of interaction.

3.2 Mathematics of QM

3.2.1 Linear Vector Spaces

The mathematics of QM, as needed to understand MR, is essentially linear algebra and associated familiar (and some maybe not too familiar) topics. Linear algebra is often taught to engineers as a method for solving linear equations (see Figure 3.5). Analyses then involve terms such as inverse, transpose, determinant, row and column spaces, eigenvectors, and eigenvalues. Note, we shall denote all matrices with an underbar _; the reason for this notation will become clearer later in this chapter.

Alternatively, think of an $n \times n$ matrix $\underline{A}$ as mapping an n-dimensional vector x into a new vector $\underline{A}\vec{x}$. This is graphically depicted in Figure 3.6, where the vector $\vec{x} = [a, b, c]^T$ is mapped into the vector $\vec{x}' = [0, b, 0]^T$ via the transformation $\vec{x}' = \underline{A}\vec{x}$. Linear transformations satisfy the equation $\underline{A}(a\vec{x} + b\vec{y}) = a(\underline{A}\vec{x}) + b(\underline{A}\vec{y})$, where a and b are scalars.

It is important to become familiar with the concept of treating functions, e.g., $\psi_1(x, y, z)$ and $\psi_2(x, y, z)$, as elements in a vector space. For example, consider $f(n)$ for $1 \le n \le N$, n integer. This function can be thought of as being equivalent to a N-dimension column vector as depicted in Figure 3.7.

We can also write $\vec{f}$ as a linear combination of other vectors, as shown in Figure 3.8.

$$\begin{matrix} x_1 + x_2 + x_3 = 5 \\ 3x_1 + x_2 + 7x_3 = 3 \\ x_1 + 2x_2 + 2x_3 = 3 \end{matrix} \implies \begin{bmatrix} 1 & 1 & 1 \\ 3 & 1 & 7 \\ 1 & 2 & 2 \end{bmatrix}\begin{bmatrix} x_1 \\ x_2 \\ x_3 \end{bmatrix} = \begin{bmatrix} 5 \\ 3 \\ 2 \end{bmatrix} \implies \underline{A}\vec{x} = \vec{y}$$

Figure 3.5 Linear algebra as a method for solving sets of linear equations.

Matrices $\Longleftrightarrow$ Linear transformations

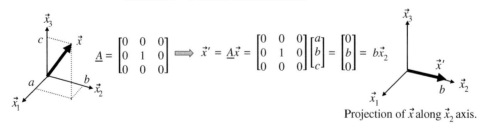

$$\underline{A} = \begin{bmatrix} 0 & 0 & 0 \\ 0 & 1 & 0 \\ 0 & 0 & 0 \end{bmatrix} \implies \vec{x}' = \underline{A}\vec{x} = \begin{bmatrix} 0 & 0 & 0 \\ 0 & 1 & 0 \\ 0 & 0 & 0 \end{bmatrix}\begin{bmatrix} a \\ b \\ c \end{bmatrix} = \begin{bmatrix} 0 \\ b \\ 0 \end{bmatrix} = b\vec{x}_2$$

Projection of $\vec{x}$ along $\vec{x}_2$ axis.

Figure 3.6 Mapping the vector $\vec{x}$ into the vector $\vec{x}'$ via the matrix $\underline{A}$, where $\vec{x}_1$, $\vec{x}_2$, and $\vec{x}_3$, are orthogonal unit vectors along the three spatial axes.

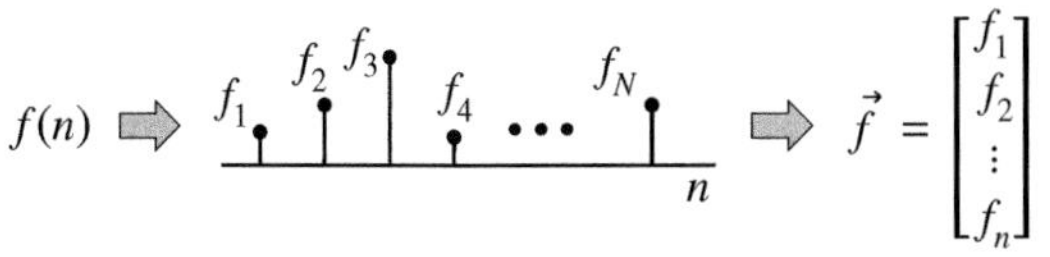

$$f(n) \implies \text{(figure)} \implies \vec{f} = \begin{bmatrix} f_1 \\ f_2 \\ \vdots \\ f_n \end{bmatrix}$$

Figure 3.7 The function f can be associated with an N-dimensional vector $\vec{f}$.

$$\vec{f} = f_1 \begin{bmatrix} 1 \\ 0 \\ 0 \\ 0 \\ \vdots \\ 0 \end{bmatrix} + f_2 \begin{bmatrix} 0 \\ 1 \\ 0 \\ 0 \\ \vdots \\ 0 \end{bmatrix} + \cdots + f_N \begin{bmatrix} 0 \\ 0 \\ \vdots \\ 0 \\ 0 \\ 1 \end{bmatrix} \quad \text{or} \quad \vec{f} = f_1' \begin{bmatrix} 1 \\ \vdots \\ 1 \\ 1 \\ \vdots \\ 1 \end{bmatrix} + f_2' \begin{bmatrix} 1 \\ \vdots \\ 1 \\ 0 \\ \vdots \\ 0 \end{bmatrix} + \cdots + f_n' \begin{bmatrix} 1 \\ 0 \\ 0 \\ 0 \\ 1 \\ 0 \end{bmatrix}$$

Figure 3.8 Two possible choices for writing N-dimensional vector $\vec{f}$ written as a linear combination of N other vectors, where $f_1, f_2, \ldots f_n$ and $f_1', f_2', \ldots f_n'$ are scalars (possibly complex).

QUESTION: How would you find a vector for the two-dimensional function $f(n, m)$, or if $f(x)$ was continuous instead of discrete?

At this point, we need to say a little more about linear vector spaces. First, the set of all square-integrable functions belongs to the vector space known as L^2. Every wavefunction $\psi(\vec{r})$ is associated with an element of a linear vector space F, and the vector space F has some important properties. First, F is a subspace of L^2, in that not all square-integrable functions qualify as wavefunctions. The temporal and spatial derivatives of $\psi(\vec{r})$ must exist to satisfy Schrödinger's equation. Further, F is a linear vector space, meaning if ψ_1 and ψ_2 are both elements of F then $\psi_3 = \lambda_1 \psi_1 + \lambda_2 \psi_2 \in F$, where λ_1 and λ_2 are arbitrary complex numbers. In addition, there are other mathematical niceties such as the existence of a zero element, defined by $\psi(\vec{r})$ plus the zero element corresponds to the same element in the vector space F.

3.2.2 Dirac Notation and Hilbert Space

A useful QM notation, known as Dirac notation, was introduced appropriately by Paul Dirac and is defined as follows: An element of F, denoted by $|\psi\rangle$, is called a **ket**. A **linear functional**, χ, is a linear operation that associates a complex number with every ket $|\psi\rangle$, that is $\chi(|\psi\rangle) =$ number. Then, the set of all such linear functionals constitute a vector space F^*, called the **dual space** of F, and an element of F^* is denoted as $\langle\chi|$. We can then write:

$$\chi(|\psi\rangle) = \langle\chi|\psi\rangle, \tag{3.25}$$

where $\langle\chi|$ is called a "bra" and $|\psi\rangle$ is called a **ket**, which together forms a "bracket." For those more familiar with vectors and matrices, bras, and kets play analogous roles to that of column and row vectors.

The expression $\langle\chi|\psi\rangle$ is called the **scalar** or **inner product**, a quantity that plays a fundamental role in QM. The scalar product satisfies several important properties, namely,

$$\langle\chi|\psi\rangle = \langle\psi|\chi\rangle^*, \tag{3.26}$$

$$\langle\chi|(|\psi\rangle + |\zeta\rangle) = \langle\chi|\psi\rangle + \langle\chi|\zeta\rangle, \tag{3.27}$$

and

$$\langle \psi | \psi \rangle \geq 0. \tag{3.28}$$

From Eq. (3.28), $\langle \psi | \psi \rangle = 0$ if and only if $|\psi\rangle = |0\rangle$. In addition, the inner product can be used to define a ***norm*** of the ket $|\psi\rangle$ as defined as $\|\psi\| = \sqrt{\langle \psi | \psi \rangle}$, as well as the distance between any two kets. A linear vector space with an inner product and associated norm is called a ***Hilbert space***.

Using the property of linearity, a linear combination of kets is also a ket. If it follows that

$$\sum_{i=1}^{N} c_i |f_i\rangle = 0 \tag{3.29}$$

implies $c_i = 0$ for all i, then the kets $|f_i\rangle$ are called ***linearly independent***. If N is the maximum number of linearly independent kets in a Hilbert space F, then F is ***N-dimensional***.

We can now define a very important quantity known as a ***basis set*** or simply a ***basis***. In an N-dimensional Hilbert space, N linearly independent kets constitute a basis, and every element $|\psi\rangle$ in F can be written as:

$$|\psi\rangle = \sum_{i=1}^{N} \psi_i |f_i\rangle, \tag{3.30}$$

where ψ_i is the scalar coefficient of the ket $|f_i\rangle$. A special type of particularly useful basis set is the ***orthonormal basis set***. Namely, if the basis kets $|f_i\rangle$ are normalized (*i.e.*, $\langle f_i | f_i \rangle = 1$), and

$$\langle f_i | f_j \rangle = \delta_{ij} = \begin{cases} 1 & \text{if } i = j \\ 0 & \text{if } i \neq j \end{cases} \tag{3.31}$$

then they constitute an orthonormal basis, and the following identify (known as ***closure***) also holds:

$$|\psi\rangle = \sum_{i=1}^{N} \langle f_i | \psi \rangle |f_i\rangle = \sum_{i=1}^{n} |f_i\rangle \langle f_i | \psi \rangle, \tag{3.32}$$

where we note the convenient use of Dirac notation. The closure property simply states that a vector is equal to the sum of its projections.

Now that we have generalized vectors to kets, let us generalize the concept of a matrix to that of an operator. We previously mentioned operators when discussing Schrödinger's equation, but let us now provide a more formal definition. An operator, $\hat{O}$, generates one ket from another. Namely, $\hat{O}|\psi\rangle = |\zeta\rangle$, and linear operators obey the following equation,

$$\hat{O}(\lambda |\psi\rangle + \mu |\zeta\rangle) = \lambda \hat{O}|\psi\rangle + \mu \hat{O}|\zeta\rangle, \tag{3.33}$$

where λ and μ are scalars. A Hilbert space, F, is ***closed*** under $\hat{O}$ if $|\psi\rangle \in F$ and $\hat{O}|\psi\rangle = |\zeta\rangle$ implies $|\zeta\rangle \in F$. The adjoint of $\hat{O}$ is denoted $\hat{O}^\dagger$ and defined as:

$$(\hat{O}|\psi\rangle)^\dagger = \langle \psi | \hat{O}^\dagger, \tag{3.34}$$

where we note that $\langle \psi | \hat{O} | \zeta \rangle = \langle \zeta | \hat{O}^\dagger | \psi \rangle^*$. Finally, $\hat{O}$ is ***Hermitian*** if $\hat{O} = \hat{O}^\dagger$.

We will encounter many different linear operators in our study of MR, so a few more details and examples will prove helpful. While the inner product $\langle \phi | \psi \rangle$ of two kets $|\phi\rangle$ and $|\psi\rangle$ produces a scalar, the outer product $|\phi\rangle\langle\psi|$ results in an operator. The projection of any ket $|\zeta\rangle$ onto ket $|\psi\rangle$ is achieved via the linear operator $\hat{P}_\psi = |\psi\rangle\langle\psi|$. Namely,

$$\hat{P}_\psi |\zeta\rangle = |\psi\rangle\langle\psi|\zeta\rangle = \lambda |\psi\rangle \tag{3.35}$$

where we have again made use of Dirac notation by noting that the quantity $\langle \psi | \zeta \rangle = \lambda$ is a scalar. For a set of orthonormal basis kets, $|f_i\rangle$, and their corresponding set of projection operators $\hat{P}_i = |f_i\rangle \langle f_i|$,

$$\sum_i \hat{P}_i = \sum_i |f_i\rangle \langle f_i| = \hat{E}, \tag{3.36}$$

where $\hat{E}$ is the identity operator defined by $\hat{E} |\psi\rangle = |\psi\rangle$. Note, we are saving the more common linear algebra notation for the identity operator $\hat{I}$ for the spin operators to come later. Equation (3.36) makes use of the closure relationship given in Eq. (3.32). Another common operator used in MR is the **transition** operator given by the expression:

$$\hat{T}_{kl} = |f_k\rangle \langle f_l| . \tag{3.37}$$

QUESTION: For what might transition operators be useful?

Operators have additional interesting properties. In contrast to kets or bras, the product of two operators is well defined. The expression $\hat{O}_1 \hat{O}_2$ simply means that first $\hat{O}_2$ is applied to a ket, and then $\hat{O}_1$ is applied to the result,

$$\hat{O}_1 \hat{O}_2 |\psi\rangle = \hat{O}_1 (\hat{O}_2 |\psi\rangle). \tag{3.38}$$

A function of an operator can thus be defined via, for example, a polynomial expansion. For example,

$$e^{\hat{O}} = \hat{E} + \hat{O} + \frac{1}{2!} \hat{O}\hat{O} + \frac{1}{3!} \hat{O}\hat{O}\hat{O} + \cdots . \tag{3.39}$$

In general, $\hat{O}_1 \hat{O}_2 \neq \hat{O}_2 \hat{O}_1$, and a quantity used frequently in QM is the **commutator** defined as:

$$[\hat{O}_1, \hat{O}_2] = \hat{O}_1 \hat{O}_2 - \hat{O}_2 \hat{O}_1. \tag{3.40}$$

We now come to an important distinction between vectors and matrices versus kets and operators. Let a ket, $|\psi\rangle$, be expressed as a linear combination of basis kets $|f_i\rangle$, then a corresponding column vector, $\vec{\psi}$, can be constructed as follows: Starting with

$$|\psi\rangle = \sum_{i=1}^{N} \psi_i |f_i\rangle , \tag{3.41}$$

we can associate it with the following column vector,

$$\vec{\psi} = \begin{bmatrix} \psi_1 \\ \psi_2 \\ \vdots \\ \psi_N \end{bmatrix} . \tag{3.42}$$

Note, the corresponding row vector would be given by $\vec{\psi}^{\dagger} = [\psi_1^*, \psi_2^*, \dots \psi_n^*]$.

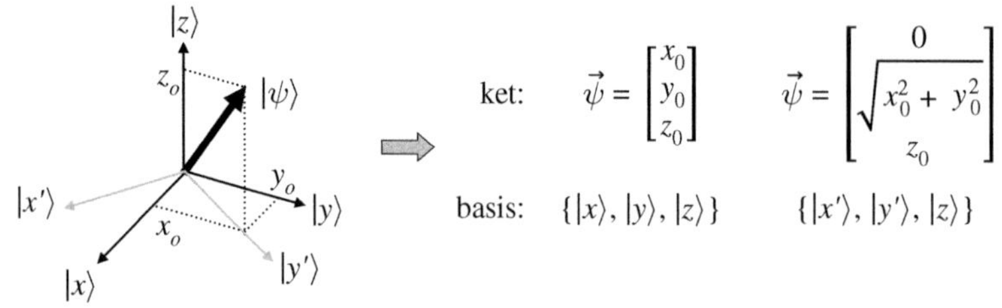

Figure 3.9 The vector and matrix representations of a given ket and operator depend on the chosen basis set. Changing the basis results in different vectors, both corresponding to the same ket.

Similarly, operators have matrix representations. Let $|x\rangle = \hat{O}|y\rangle$, then we can write $\vec{x} = \underline{O}\vec{y}$ where $\underline{O}$ is a $n \times n$ matrix with elements:

$$O_{ij} = \langle f_i | \hat{O} | f_j \rangle. \tag{3.43}$$

As shown in Figure 3.9, the key concept to keep in mind is that a given vector or matrix representation depends on the chosen basis set! Note also that the ket $|\psi\rangle$ in the left-hand side of Figure 3.9 is depicted as an element in Hilbert space where the axes are kets rather than the more conventional vectors used in Figure 3.6.

If the kets $|f_k\rangle$ constitute one orthonormal basis, and the kets $|g_k\rangle$ constitute another, then a **change of basis** (also called a **basis transformation**) is accomplished by:

$$O_{kl} = \langle f_k | \hat{O} | f_l \rangle = \sum_i \sum_j \langle f_k | g_i \rangle \langle g_i | \hat{O} | g_j \rangle \langle g_j | f_l \rangle. \tag{3.44}$$

Defining the unitary transformation matrix $\underline{U}$ as $U_{ij} = \langle f_i | g_i \rangle$, then

$$\underline{O}_f = \underline{U}\,\underline{O}_g\underline{U}^\dagger, \tag{3.45}$$

where $\underline{O}_f$ and $\underline{O}_g$ represent the matrices of operator $\hat{O}$ with respect to the basis sets $|f_k\rangle$ and $|g_k\rangle$.

A final important Hilbert space concept is that of eigenkets and eigenvalues. If $\hat{O}|\psi\rangle = \lambda|\psi\rangle$ where λ is a scalar, then $|\psi\rangle$ is called an **eigenket** of $\hat{O}$ with **eigenvalue** λ. Further, the **trace** of a matrix is defined as the sum of the diagonal elements,

$$\text{Tr}(\underline{O}) = \sum_i O_{ii}. \tag{3.46}$$

We note that although the trace operation is only defined for matrices, the trace of a matrix is invariant under a change of basis (see Exercise E.3.1). Hence, we can, and will, regularly refer to the trace of an operator. Technically, starting with a given operator, one would need to first choose a basis, express the operator as a matrix in this basis, and then compute the sum of the diagonal elements. A convenient identity is that the trace of a matrix is equal to the sum of its eigenvalues. Hence, we can write,

$$\text{Tr}(\underline{O}) = \sum_i \lambda_i, \tag{3.47}$$

where λ_i are the eigenvalues of $\hat{O}$.

3.2.3 Liouville Space

Let us now continue with our exploration of linear algebra into some, perhaps, less familiar (and perhaps more abstract) topics. Operators defined on an n-dimensional Hilbert space are themselves elements of an n^2-dimensional vector space known as **Liouville space** (sometimes called operator space or, more generally, an **algebra**). We need a metric, and the metric in Liouville space is defined as:

$$(\hat{A}|\hat{B}) = \mathrm{Tr}(\hat{A}^\dagger \hat{B}). \tag{3.48}$$

In contrast to Hilbert space, the product of two Liouville space elements is well defined. Namely, let $\mathcal{L}$ be a Liouville space. If $\hat{A}, \hat{B} \in \mathcal{L}$, then $\hat{A}\hat{B} = \hat{C} \in \mathcal{L}$. Product is associative, but not necessarily commutative, *i.e.*, in general $\hat{A}\hat{B} \neq \hat{B}\hat{A}$.

But what about operators? Operators that work on elements of a Liouville space are called **superoperators**, which we shall denote with a double hat, $\hat{\hat{\ }}$. For example, the superoperator $\hat{\hat{A}}$ could be used in the following expression: $\hat{\hat{A}}\hat{O}_1 = \hat{O}_2$. There are actually many different types of superoperators, but we will restrict our discussion to just one, namely the commutator:

$$\hat{\hat{A}}\hat{B} = [\hat{A}, \hat{B}] = \hat{A}\hat{B} - \hat{B}\hat{A}, \tag{3.49}$$

where $\hat{\hat{A}}$ is the superoperator notation and $[\hat{A}, \hat{B}]$ is the nonsuperoperator notation used in many other texts. Superoperators have an important role in describing **rotations** in multidimensional vector spaces. Let $[\hat{A}, \hat{B}] = \hat{C}$. $\hat{A}$ and $\hat{B}$ are said to **cyclically commute** if $[\hat{A}, \hat{C}] = \hat{B}$. In superoperator notation $[\hat{A}, \hat{C}] = [\hat{A}, [\hat{A}, \hat{B}]] = \hat{\hat{A}}\hat{\hat{A}}\hat{B} = \hat{B}$. Given $\hat{A}$ and $\hat{B}$ cyclically commute, solve: $e^{i\theta\hat{\hat{A}}}\hat{B} = ?$

$$e^{i\theta\hat{\hat{A}}}\hat{B} = \left(1 + i\theta\hat{\hat{A}} + \frac{1}{2!}(i\theta\hat{\hat{A}})^2 + \frac{1}{3!}(i\theta\hat{\hat{A}})^3 + \ldots\right)\hat{B}$$

$$= \hat{B} + i\theta[\hat{A}, \hat{B}] + \frac{(i\theta)^2}{2!}\hat{B} + \frac{(i\theta)^3}{3!}[\hat{A}, \hat{B}] + \cdots$$

$$= \hat{B}\cos\theta + i\hat{C}\sin\theta. \tag{3.50}$$

The geometric interpretation of Eq. (3.50) as a rotation is shown in Figure 3.10. It is important to note that this figure is drawn in Liouville space where the axes are operators rather than the kets used in the Hilbert space diagram of Figure 3.9.

The fact that good references for superoperators and Liouville space can be found in F. van de Ven (1996) and the second chapter of R. Ernst (1987) is a solid hint that these concepts might actually be useful for understanding MR. Figure 3.11 presents a summary of the relevant vector spaces.

After covering the basic postulates of QM, we will first describe MR in Hilbert space, followed by a discussion in Liouville space. The key concept to keep in mind is, given a background in classical physics, MR is easiest to understand and further extend this understanding to include interactions between spins, when described in Liouville space!

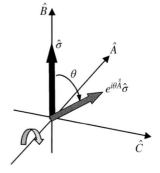

Figure 3.10 Superoperator exponential $e^{i\theta\hat{\hat{A}}}$ applied to $\hat{B}$ produces a rotation in the $\hat{B} \times [\hat{A}, \hat{B}]$ plane of operator space! The rotation is about $\hat{A}$ axis. Note, some books define a **rotation superoperator** as $e^{i\theta\hat{\hat{A}}} \equiv \hat{\hat{R}}_{\theta\hat{A}}$, but we will just stick with $e^{i\theta\hat{\hat{A}}}$, and all superoperators in this text will simply refer to the commutator.

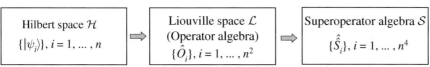

Figure 3.11 A summary of vector spaces, where n is the dimension of the initial Hilbert space.

QUESTION: For experts only: What are the remaining digits in this famous number 0.1100010000000000000000000100...?

3.3 The Six Postulates of QM

We will now present the six postulates of QM, followed by a description of the Stern–Gerlach experiment to illustrate some of the primary implications. Note, proof of many of the statements elucidating these postulates and their implications are relegated to exercises at the end of the chapter. For simplicity, we will just consider what is known as the discrete nondegenerate case.

3.3.1 Postulate 1

First Postulate: At any point in time, t_0, the state of a physical system is defined by a ket $|\psi\rangle$ that belongs to the state space F.

F is a Hilbert space, *i.e.*, a linear vector space with the inner product as the metric. An important additional point to keep in mind is that superposition holds. Namely, linear combinations of kets are also kets in F.

3.3.2 Postulate 2

Second Postulate: Every measurable physical quantity $\mathcal{A}$ is described by an operator $\widehat{A}$ acting in F. This operator is observable.

This postulate highlights the major difference between classical versus QM. In classical mechanics, the state of a system is described by a set of physical parameters (*e.g.*, $\vec{r}, \vec{p}$). Whereas in QM, the state of the system is described by a ket and physical quantities by operators. The question at hand is what is special about **observable** operators?

Observables, being operators corresponding to physical quantities, have several important properties. Let $\widehat{A}$ be a Hermitian operator with eigenvalues a_n and corresponding normalized eigenkets $|u_n\rangle$ for $n = 1, 2...N$. Then, $\widehat{A}$ is an observable if its eigenkets form a basis in state space. Given that the eigenkets for an orthonormal basis, $\langle u_i | u_j \rangle = \delta_{ij}$, the closure relationship holds:

$$\sum_{n=1}^{N} |u_n\rangle \langle u_n| = \widehat{E}. \tag{3.51}$$

Finally, the set of eigenvalues, $\{a_n : n = 1, 2, ...N\}$, is called the **spectrum** of $\widehat{A}$.

3.3.3 Postulate 3

> **Third Postulate**: The only possible result of the measurement of a physical quantity $\mathcal{A}$ is one of the eigenvalues of the corresponding observable $\hat{A}$.

It follows from $\hat{A}$ being Hermitian, that any measurement of $\mathcal{A}$ always yields a real number. Furthermore, if the spectrum of $\hat{A}$ is discrete, the results that can be obtained are quantized.

3.3.4 Postulate 4

> **Fourth Postulate**: When a physical quantity $\mathcal{A}$ is measured, the probability $\mathcal{P}(a_n)$ of obtaining the eigenvalue a_n of the corresponding observable $\hat{A}$ is $\mathcal{P}(a_n) = |\langle u_n|\psi\rangle|^2$, where $|u_n\rangle$ is the eigenket of $\hat{A}$ associated with a_n.

Consider the following two normalized kets, $|\psi\rangle$ and $|\psi'\rangle = e^{i\theta}|\psi\rangle$. As stated in the fourth postulate, if a system is described by the state $|\psi\rangle$, then the probability of obtaining a_n is $\mathcal{P}(a_n) = |\langle u_n|\psi\rangle|^2$. Similarly, for $|\psi'\rangle$, $\mathcal{P}(a_n) = |\langle u_n|e^{i\theta}\psi\rangle|^2 = |e^{i\theta}|^2|\langle u_n|\psi\rangle|^2 = |\langle u_n|\psi\rangle|^2$. Thus, $|\psi\rangle$ and $|\psi'\rangle$ represent equivalent states, and the phase factor $e^{i\theta}$ makes no difference. However, let $|\psi\rangle = |\psi_1\rangle + |\psi_2\rangle$ and $|\zeta\rangle = e^{i\theta_1}|\psi_1\rangle + e^{i\theta_2}|\psi_2\rangle$. It does NOT follow that $|\psi\rangle$ and $|\zeta\rangle$ represent the same state. Namely,

$$|\psi\rangle: \mathcal{P}(a_n) = |\langle u_n|\psi_1\rangle|^2 + |\langle u_n|\psi_2\rangle|^2 + 2\text{Re}\left\{\langle u_n|\psi_1\rangle\langle u_n|\psi_2\rangle^*\right\}, \tag{3.52}$$

but

$$|\zeta\rangle: \mathcal{P}(a_n) = |\langle u_n|\zeta\rangle|^2 = \left|e^{i\theta_1}\langle u_n|\psi_1\rangle + e^{i\theta_2}\langle u_n|\psi_2\rangle\right|^2$$
$$= |\langle u_n|\psi_1\rangle|^2 + |\langle u_n|\psi_2\rangle|^2 + 2\text{Re}\left\{e^{i(\theta_1-\theta_2)}\langle u_n|\psi_1\rangle\langle u_n|\psi_2^*\rangle\right\}. \tag{3.53}$$

In general, interference terms **cannot** be ignored. Hence, global phase factors do not affect physical predictions, but the relative phases of the kets are important.

3.3.5 Postulate 5

> **Fifth Postulate**: If the measurement of the physical quantity $\mathcal{A}$ on a system in state $|\psi\rangle$ gives the result a_n, the state of the system immediately after the measurement is $|\psi\rangle = |u_n\rangle$, the eigenket of $\hat{A}$ corresponding to eigenvalue a_n.

This postulate embodies the nonclassical concept of a fundamental interaction between a system and the measurement of that system. What is presented here is known as the Copenhagen interpretation of QM, but the "measurement problem" leaves many unanswered questions. What constitutes a measurement? Does a human need to be involved? Is the measurement device itself treated as a QM or classical object? Does the wavefunction really collapse? Though beyond the scope of this text, interested readers can pursue many articles on this topic that now go under the general label of "The Foundations of QM."

From Postulates 4 and 5, quantum mechanical calculations involve computing probabilities. Hence, probability is fundamental to the world in which we live. For the case of *in vivo* MR (and indeed many other areas of physics), a given measurement of a physical quantity does not involve just a single particle, but rather, most often, a large number of particles. Hence, the more meaningful metric is the **expected** (or **mean**) value of an observable. Denoted as $\langle \hat{A} \rangle_\psi$ or simply $\langle \hat{A} \rangle$, the expected value of an observable A is defined as the average obtained from a large number of measurements performed on systems all in the same quantum state.

$\langle \hat{A} \rangle$ provides the critical connection between classical and quantum mechanical physics and is straightforward to compute. We make the following claim,

$$\langle \hat{A} \rangle_\psi = \langle \psi | \hat{A} | \psi \rangle, \tag{3.54}$$

and the proof is as follows. Using Postulate 4,

$$\langle \hat{A}_\psi \rangle = \sum_i a_i P(a_i) = \sum_i a_i |\langle u_i | \psi \rangle|^2,$$

$$= \sum_i a_i \langle u_i | \psi^* \rangle \langle u_i | \psi \rangle = \sum_i a_i \langle \psi | u_i \rangle \langle u_i | \psi \rangle. \tag{3.55}$$

Noting that $\hat{A} |u_i\rangle = a_i |u_i\rangle$ and using closure,

$$\langle \hat{A} \rangle_\psi = \langle \psi | \left[\sum_i a_i |u_i\rangle\langle u_i| \right] |\psi\rangle,$$

$$= \langle \psi | \left[\sum_i \hat{A} |u_i\rangle\langle u_i| \right] |\psi\rangle$$

$$= \langle \psi | \hat{A} \left[\sum_i |u_i\rangle\langle u_i| \right] |\psi\rangle$$

$$= \langle \psi | \hat{A} | \psi \rangle. \tag{3.56}$$

Two observables that can be simultaneously determined are called **compatible**. From which it follows that if $\hat{\hat{A}}\hat{B} = 0$, then A and B are compatible (remember $\hat{\hat{A}}\hat{B}$ is superoperator notation for the commutator $[\hat{A}, \hat{B}] = \hat{A}\hat{B} - \hat{B}\hat{A}$). If $\hat{\hat{A}}\hat{B} \neq 0$, then A and B are incompatible and cannot be measured simultaneously.

3.3.6 Postulate 6

Sixth Postulate: The time evolution of the state ket $|\psi(t)\rangle$ is governed by Schrödinger's equation:

$$\frac{\partial}{\partial t} |\psi(t)\rangle = -i\hat{H}(t) |\psi(t)\rangle$$

where $\hat{H}(t)$ is the operator for the observable $\mathcal{H}(t)$ associated with the total energy of the system.

We have previously introduced the operator for the total energy of a system (see Eq. (3.24)), and here we are making a slight modification. It is often most convenient to work with energy in units of $\hbar$. Hence, let $\hat{H}(t)$, the Hamiltonian operator of the system, be redefined as the operator corresponding to $\mathcal{H}(t) = \text{total energy}/\hbar$.

Note from Postulate 6, given an initial state $|\psi(t_0)\rangle$, the state at any subsequent time is deterministic (*i.e.*, not probabilistic). Probability only enters when a physical quantity is measured, upon which the state vector undergoes a probabilistic change (Postulate 5).

Solving Schrödinger's equation is a critical part of analyzing MR. In this text, we will primarily be interested in addressing three special cases: (1) $\hat{H}(t)$ is independent of time, in which case it is easy to verify the solution is $|\psi(t)\rangle = e^{-it\hat{H}}|\psi(0)\rangle$. (2) $\hat{H}(t)$ is periodic. Here, the solution will be found by changing to a rotating frame of reference (*i.e.*, change of basis) such that Schrödinger's equation becomes time independent. An important example will be for analyzing Rf excitation. (3) $\hat{H}(t)$ is random in time. The solution in this case will be found using perturbation theory, for example, to analyze the behavior of an ensemble of spins; the quintessential MR example being relaxation theory.

The Stern–Gerlach experiment serves an excellent vehicle to illustrate some of the consequences arising from the postulates of QM. In 1922, Otto Stern and Walther Gerlach measured the magnetic dipole moment of stream of silver atoms emitted from an oven (Gerlach and Stern 1922) (see Figure 3.12). It was later shown that this moment was due to the intrinsic spin angular momentum of an unpaired electron.

The magnet was designed such that there is a B_0 gradient resulting in a z-directed force on the atoms proportional to μ_z. Hence, this device measures the z component of angular momentum. Namely, the potential energy is given by $\mathcal{E} = -\vec{\mu} \cdot \vec{B}$, and the force on the silver atoms is proportional to the gradient of the magnetic field $F_z = \mu_z \frac{\partial B_z}{\partial z}$. Classically, since atoms can be oriented randomly in space, device should detect continuum of μ_z values. However, the actual result is depicted in Figure 3.13. From which, we are forced to conclude that μ_z is not continuous but quantized to two values $\pm \gamma \hbar/2$ (a result known as ***space quantization***).

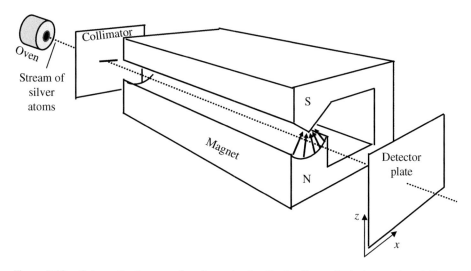

Figure 3.12 Schematic diagram of equipment setup for the Stern–Gerlach experiment. Source: Adapted with permission from Eisberg and Resnik (1974), Figures 8–5, p. 296.

Figure 3.13 Results from the Stern–Gerlach experiment. According to classical physics, silver atoms hitting the detector screen should form a continuous broad pattern, in contrast to the observed result of distinct upper and lower bars. Source: Adapted with permission from Eisberg and Resnik (1974) Figures 8–6, p. 297.

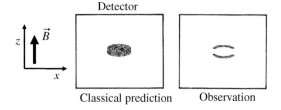

Although Stern and Gerlach were looking to demonstrate space quantization, this result was very surprising. It can be shown that the angular momentum for any physically spinning object must be quantized. However, for rotating objects, the spin quantum number must be an integer (van de Ven 1996). In a great triumph of theoretical physics, Paul Dirac, in 1928, using a relativistic version of Schrödinger's equation, derived the spin of an electron (and a positron) to be equal to ½ (Dirac 1928). Hence, atomic particles such as electrons have spin angular momentum, even though they are not actually "spinning."

Let us now provide a QM description of a silver atom, as an example of a spin-½ particle. Associated with the physical quantity $\mathcal{L}_z$ (z-component of angular momentum) is the observable angular momentum operator $\hat{L}_z$ with eigenvalues $\pm\hbar/2$. Hence,

$$\hat{L}_z|+\rangle = +\frac{\hbar}{2}|+\rangle \text{ and } \hat{L}_z|-\rangle = -\frac{\hbar}{2}|-\rangle , \tag{3.57}$$

where a shorthand notation for associated eigenkets is $|+\rangle$ for "spin up" and $|-\rangle$ for "spin down." Take as given that the state space is two dimensional. Hence, a general form for $|\psi\rangle$ is:

$$|\psi\rangle = c_+e^{i\theta_+}|+\rangle + c_-e^{i\theta_-}|-\rangle , \tag{3.58}$$

with $|c_+|^2 + |c_-|^2 = 1$. For simplicity, let's ignore θ_+ and θ_-, and write down the spin operators $(\hat{I}_{x,y,z} = \hat{L}_{x,y,z}/\hbar)$ and their corresponding normalized matrix representations in the $\{|+\rangle, |-\rangle\}$ basis. Namely,

$$\underline{I}_x = \frac{1}{\sqrt{2}}\begin{bmatrix} 0 & 1 \\ 1 & 0 \end{bmatrix}$$

$$\underline{I}_y = \frac{1}{\sqrt{2}}\begin{bmatrix} 0 & -i \\ i & 0 \end{bmatrix}$$

$$\underline{I}_z = \frac{1}{\sqrt{2}}\begin{bmatrix} 1 & 0 \\ 0 & -1 \end{bmatrix} \tag{3.59}$$

QUESTION: What are the vector representations for $|+\rangle$ and $|-\rangle$?

These are known as the Pauli matrices. One important thing to note is that $\hat{I}_x$, $\hat{I}_y$, and $\hat{I}_z$ do not commute! Further, the expected value $\langle \hat{I}_z \rangle = \langle\psi|\hat{I}_z|\psi\rangle = \frac{1}{2}|c_+|^2 - \frac{1}{2}|c_-|^2$, but for any given atom we only measure $+\frac{1}{2}$ or $-\frac{1}{2}$.

Assume all silver atoms, upon first leaving the oven, are described by the same wavefunction, $|\psi\rangle = c_+|+\rangle + c_-|-\rangle$ (note, we will deal with ensemble of spins having different wavefunctions later) and consider the following two devices called a spin polarizer and a spin detector. The "detector" is as described above, and the "polarizer" is just a modified detector with a small slit cut to allow the passage of the upper bar of spins through undetected (see Figure 3.14).

Consider the configurations shown in Figure 3.15. In experiment (a), we get the expected result that filtering the spin having an initial wavefunction $|\psi\rangle = c_+|+\rangle + c_-|-\rangle$ results in the detection of only spins with the up, $|+\rangle$, state. Experiment (b) is just the standard Stern–Gerlach apparatus

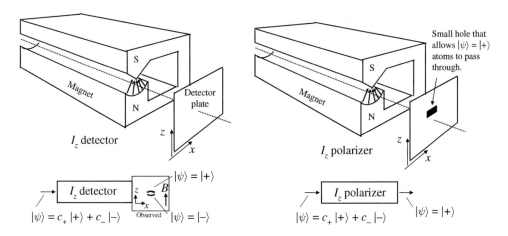

Figure 3.14 A Stern–Gerlach I_z detector (left) which measures the z-component of spin, and Stern–Gerlach-like I_z polarizer (right) that only allows spins with $+\frac{1}{2}$ z spin to pass through undisturbed. (Note, in practice, $|\psi\rangle$ is described as an ensemble of atoms with different c_+ and c_-, but we will deal with ensembles later.) Source: Adapted with permission from Eisberg and Resnik (1974), Figure 8–5, p. 296.

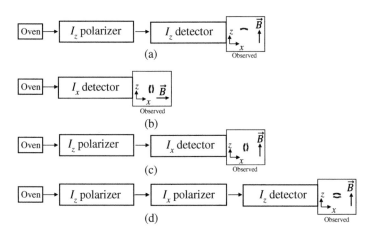

Figure 3.15 Schematic diagrams for the following Stern–Gerlach experiments: (a) detection of the z-spin component of atoms from an oven first sorted to have a z-spin equal to $+\frac{1}{2}$, (b) detection of x-spin component of atoms from an oven, (c) detection of the x-spin component of atoms from an oven first sorted to have a z-spin equal to $+\frac{1}{2}$, and (d) detection of the z-spin component of atoms from an oven first sorted to have a z-spin equal to $+\frac{1}{2}$, and then sorted to have an x-spin equal to $+\frac{1}{2}$.

rotated by 90°. In experiment (c), the spins are first sorted into the $|\psi\rangle = |+\rangle$ state by the I_z polarizer and then passed into an I_x detector which results in an equal number of spins having angular momentum in the $\pm x$-direction. The wavefunction of these spins is $|\psi\rangle = \frac{\sqrt{2}}{2}(|+\rangle + |-\rangle)$, a linear combination of spin up and spin down. Experiment (d) is the most interesting. After sorting the spins to only contain those with $+\frac{1}{2}$ z-spin ($|\psi\rangle = |+\rangle$), then, of those, only keeping spins which have $+\frac{1}{2}$ x-spin, we find the remaining spins are equally likely to be $|+\rangle$ or $|-\rangle$. In effect, the "measurement" of the I_x perturbs our prior measurement of I_z. Because I_x and I_z do not commute, they do not share a common set of eigenvectors, and they cannot be measured simultaneously.

3.4 MR in Hilbert Space

In this section, we will first review spin operators, and then derive the equations describing the behavior of a single spin in a magnetic field. This will then be generalized to an ensemble of many spins. The overall goal is to find the appropriate wavefunction $|\psi(t)\rangle$ that describes a system of spin-$\frac{1}{2}$ nuclei in a uniform magnetic field, and we will take the following approach. Starting with Schrödinger's equation, we will first find the appropriate Hamiltonian $\widehat{H}(t)$, then solve for $|\psi(t)\rangle$, and finally compute the primary quantities of interest, namely the three components of magnetic moment $\langle\widehat{\mu}_x\rangle, \langle\widehat{\mu}_y\rangle$, and $\langle\widehat{\mu}_z\rangle$. Later we will show $\langle\widehat{\mu}_x\rangle, \langle\widehat{\mu}_y\rangle$, and $\langle\widehat{\mu}_z\rangle$ correspond to the familiar classical quantities M_x, M_y, and M_z.

3.4.1 Review of Spin Operators

Using the general properties of angular momentum and the operator forms of the Pauli matrices given in Eq. (3.59), it is easily shown that $\widehat{I}_x, \widehat{I}_y$, and $\widehat{I}_z$ satisfy the following commutator relationships: $[\widehat{I}_x, \widehat{I}_y] = i\widehat{I}_z$, $[\widehat{I}_y, \widehat{I}_z] = i\widehat{I}_x$, and $[\widehat{I}_z, \widehat{I}_x] = i\widehat{I}_y$. From which it follows that $\widehat{I}_x, \widehat{I}_y$, and $\widehat{I}_z$ cyclically commute. In superoperator notation,

$$(\widehat{\widehat{I}}_p)^n\widehat{I}_q = \begin{cases} [\widehat{I}_p, \widehat{I}_q], & n \text{ odd} \\ \widehat{I}_q, & n \text{ even} \end{cases} \quad p, q = x, y, z; \ p \neq q. \tag{3.60}$$

Let us also define a new operator $\widehat{I^2} = \widehat{I}_x\widehat{I}_x + \widehat{I}_y\widehat{I}_y + \widehat{I}_z\widehat{I}_z$ corresponding to the total angular moment, which is readily shown to satisfy $[\widehat{I^2}, \widehat{I}_x] = [\widehat{I^2}, \widehat{I}_y] = [\widehat{I^2}, \widehat{I}_z] = 0$. From the commutator relations given in Eq. (3.60), one can then derive the eigenkets and eigenvalues of $\widehat{I^2}$ and $\widehat{I}_z$ (note, since these operators commute, they have a common set of eigenkets). The spectrum of $\widehat{I^2} = I(I+1)$ for I an integer multiple of $\frac{1}{2}$, and the spectrum of $\widehat{I}_z = m$ for $m = -I, -I+1, \ldots, I-1, I$, where I and m are known as the spin and magnetic quantum numbers respectively. Hence, spin/angular momentum/magnetic moment of elementary particles (*e.g.*, electrons, protons, or, more generally, nuclei) are quantized in magnitude and along a projection onto any one axis. Formally, eigenkets of $\widehat{I}_z$ are written as $|I, m\rangle$, but, referring to our shorthand notation, we will use $\left|\frac{1}{2}, +\frac{1}{2}\right\rangle \equiv |+\rangle$ and $\left|\frac{1}{2}, -\frac{1}{2}\right\rangle \equiv |-\rangle$.

As stated in Chapter 2, spin, angular momentum, and magnetic moment operators are linearly related. In operator form, $\widehat{\mu}_p = \gamma\widehat{L}_p = \gamma\hbar\widehat{I}_p$, where $\widehat{\mu}_p, \widehat{L}_p$, and $\widehat{I}_p$ for $p = \{x, y, z\}$ are the operators for magnetic moment, angular momentum, and spin, respectively. For a spin-$\frac{1}{2}$ particle, the eigenkets and eigenvalues of $\widehat{I}_z$ are given by $\widehat{I}_z|+\rangle = \frac{1}{2}|+\rangle$ and $\widehat{I}_z|-\rangle = -\frac{1}{2}|-\rangle$. The matrix representations of these operators in the $\{|+\rangle, |-\rangle\}$ basis, are the familiar Pauli matrices given in Eq. (3.59).

3.4.2 Single Spin in a Magnetic Field

When a spin-$\frac{1}{2}$ particle is placed in a magnetic field $\vec{B} = B_0\vec{z}$, the quantized z-component of its magnetic moment is readily observed (see Gerlach and Stern 1922). Two of the more commonly used pictorial drawings for such a spin, called here the "cones" versus "polarization" pictures, are shown in Figure 3.16. Neither is perfect (for example the cones picture fails to depict the quantization in x or y) but both provide some useful intuition.

"Cones"

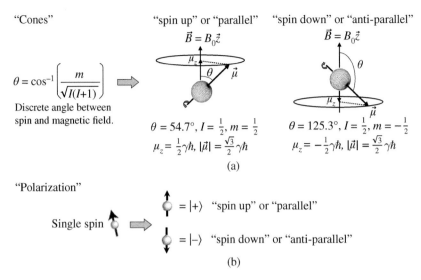

$$\theta = \cos^{-1}\left(\frac{m}{\sqrt{I(I+1)}}\right)$$

Discrete angle between spin and magnetic field.

"spin up" or "parallel"

$\vec{B} = B_0\vec{z}$

$\theta = 54.7°, I = \frac{1}{2}, m = \frac{1}{2}$

$\mu_z = \frac{1}{2}\gamma\hbar, |\vec{\mu}| = \frac{\sqrt{3}}{2}\gamma\hbar$

"spin down" or "anti-parallel"

$\vec{B} = B_0\vec{z}$

$\theta = 125.3°, I = \frac{1}{2}, m = -\frac{1}{2}$

$\mu_z = -\frac{1}{2}\gamma\hbar, |\vec{\mu}| = \frac{\sqrt{3}}{2}\gamma\hbar$

(a)

"Polarization"

Single spin

$= |+\rangle$ "spin up" or "parallel"

$= |-\rangle$ "spin down" or "anti-parallel"

(b)

Figure 3.16 Pictorial representations of a spin-½ particle in a magnetic field. (a) "Cones" picture of a spin in a magnetic field depicting quantization of μ_z via restricting the spin to lie in a cone making a 54.7° or 125.3° angle between the spin and applied field. (b) "Polarization" picture of a spin in a magnetic field in which the arrow depicts the direction in space for which measurement of the magnetic moment μ_z is $+\gamma\hbar/2$ with probability 1. Examples of the $\hat{I}_z$ eigenkets $|+\rangle$ and $|-\rangle$ are also shown. Note $|-\rangle \neq -|+\rangle$.

 QUESTION: If you were to measure μ_x, μ_y, and μ_z for an individual spin, what would you get?

Classically, assuming the magnetic field is in the z direction, the potential energy of a dipole in a magnetic field is $\mathcal{E} = -\vec{\mu} \cdot \vec{B} = -\mu_z B_0$. Substituting the operator corresponding to μ_z, yields the quantum mechanical Hamiltonian operator (remember we defined H as energy/$\hbar$)

$$\hat{H} = -\gamma B_0 \hat{I}_z = -\omega_0 \hat{I}_z. \tag{3.61}$$

Hence, the spectrum of $\hat{H}$ is discrete with eigenvalues $\mp\frac{1}{2}\omega_0$ corresponding to eigenkets $|+\rangle$ and $|-\rangle$ respectively. The energy difference $\Delta\mathcal{E} = \hbar\omega_0$ increases linearly with the magnetic field, an effect called ***Zeeman splitting***.

Noting that $\hat{H} = -\gamma B_0 \hat{I}_z$ is independent of time, Schrödinger's equation is easily solved by simple integration yielding $|\psi(t)\rangle = e^{-it\hat{H}}|\psi(0)\rangle$. Expanding this exponential in a Taylor's series,

$$|\psi(t)\rangle = \left(\hat{E} + (-it\hat{H}) + \frac{(-it\hat{H})^2}{2!} + \frac{(-it\hat{H})^3}{3!} + \ldots\right)|\psi(0)\rangle, \tag{3.62}$$

implies we are going to evaluate terms such as $|\psi(0)\rangle$, $\hat{H}|\psi(0)\rangle$, $\hat{H}^2|\psi(0)\rangle$, ... This suggests that expanding $|\psi(t)\rangle$ in terms of the eigenkets of $\hat{H}$ would be very helpful. A general solution is then

$$|\psi(t)\rangle = c_+ e^{i(\phi_+ + \gamma B_0 t/2)} |+\rangle + c_- e^{i(\phi_- - \gamma B_0 t/2)} |-\rangle, \tag{3.63}$$

where c_+, c_-, ϕ_+, and ϕ_- are real constants subject to the constraint $c_+^2 + c_-^2 = 1$.

The longitudinal magnetization is then found using:

$$\langle \hat{\mu}_z \rangle = \hbar\gamma \langle \psi|\hat{I}_z|\psi\rangle = \frac{\hbar\gamma}{2}\left(c_+^2 - c_-^2\right) = \frac{\hbar\gamma}{2}(P_+ - P_-), \tag{3.64}$$

where P_+ and P_- are the probability of measuring the system to be in the state $|+\rangle$ or $|-\rangle$ respectively. So how do we find P_+ and P_-?

From thermodynamics, the probability P_+ of finding a system in a specific state $|n\rangle$ is dependent on the energy $\mathcal{E}_n$ and is given by the Boltzmann distribution:

$$P_n = \frac{1}{Z}e^{-\mathcal{E}_n/kT}, \tag{3.65}$$

where T is temperature, k is the Boltzmann constant, and the normalization factor $Z = \sum_i e^{-\mathcal{E}_i/kT}$ is known as the **partition function** with the sum performed over all possible energies. Energies associated with *in vivo* MR ($\mathcal{E}_n = \pm\hbar\omega/2$) are typically much smaller than kT (see Figure 1.1). Thus, we can use what is known as the **high temperature approximation**

$$e^{-\mathcal{E}_n/kT} \approx 1 - \mathcal{E}_n/kT. \tag{3.66}$$

Hence,

$$\langle \hat{\mu}_z \rangle = \frac{\hbar\gamma}{2}(P_+ - P_-) = \frac{\hbar\gamma}{2}\left(\frac{\hbar\omega_0}{2kT}\right) = \frac{\hbar^2\gamma^2 B_0}{4kT}, \tag{3.67}$$

where the extra factor of 2 comes from the Z term.

To find the transverse magnetization, we make note of the following useful equations: $\hat{I}_x|+\rangle = \frac{1}{2}|-\rangle$, $\hat{I}_x|-\rangle = \frac{1}{2}|+\rangle$, $\hat{I}_y|+\rangle = \frac{i}{2}|-\rangle$, and $\hat{I}_y|-\rangle = -\frac{i}{2}|+\rangle$. Letting $\Delta\phi = \phi_- - \phi_+$, yields (after some algebra):

$$\langle \hat{\mu}_x \rangle = \hbar\gamma\langle\psi|\hat{I}_x|\psi\rangle = \frac{\hbar\gamma}{2}\left(c_+ c_- e^{-i(\omega_0 t + \Delta\phi)} + c_+ c_- e^{i(\omega_0 t + \Delta\phi)}\right),$$

$$= \hbar\gamma c_+ c_- \cos(\omega_0 t + \Delta\phi). \tag{3.68}$$

Similarly,

$$\langle \hat{\mu}_y \rangle = \hbar\gamma c_+ c_- \sin(\omega_0 t + \Delta\phi). \tag{3.69}$$

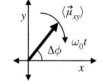

Figure 3.17 Precession of $\langle \hat{\mu}_x \rangle$ and $\langle \hat{\mu}_y \rangle$ in the transverse plane.

From Eqs. (3.68) and (3.69), the expected values of the x and y components of the magnetic moment, written as $\langle \vec{\mu}_{xy} \rangle$, simply undergo Larmor precession (see Figure 3.17)!

3.4.3 Ensemble of Spins in a Magnetic Field

In an *in vivo* MR experiment, we are never dealing with just a single spin. Consider an ensemble of N independent spins with ϕ_+ and ϕ_- (and by extension $\Delta\phi$) randomly distributed. Then, $\overline{\langle \hat{\mu}_x \rangle} = \overline{\langle \hat{\mu}_y \rangle} = 0$, where the overbar — is used to denote the average taken over the ensemble. Physical pictures for a collection of spins in states $|+\rangle$ or $|-\rangle$ are shown below in Figure 3.18.

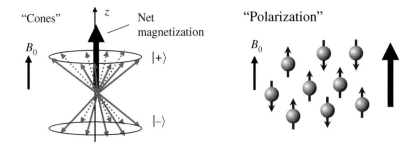

Figure 3.18 Net magnetization for an ensemble of spins in the spin up or spin down states as depicted by the cones and polarization pictures.

In this case, $\overline{\langle \hat{\mu}_z \rangle} = N \frac{\hbar^2 \gamma^2 B_0}{4kT}$, and the net z magnetization would typically be expressed as:

$$M_z = \frac{N}{V} \frac{\hbar^2 \gamma^2 B_0}{4kT} = \rho \frac{\hbar^2 \gamma^2 B_0}{4kT}, \tag{3.70}$$

where ρ is the number of spins per unit volume V.

To get transverse magnetization, we need to establish some nonzero phase relationship (namely a phase coherence, or more commonly referred to as a coherence) among the spins, and we do this via Rf excitation. In the presence of an additional magnetic field rotating in the $x - y$ plane, $\vec{B}_1 = B_1(\cos \omega t \, \vec{x} - \sin \omega t \, \vec{y})$, the Hamiltonian becomes:

$$\hat{H}(t) = -\omega_0 \hat{I}_z - \omega_1 (\hat{I}_x \cos \omega t - \hat{I}_y \sin \omega t), \tag{3.71}$$

where $\omega_1 = \gamma B_1$. $\hat{H}(t)$ is now periodic, and Schrödinger's equation can be solved by changing to a rotating frame of reference (equivalent to a change of basis), $| \psi' \rangle = e^{-i\omega t \hat{I}_z} | \psi \rangle$. The new Hamiltonian becomes:

$$\hat{H}' = e^{-i\omega t \hat{I}_z} \hat{H} e^{i\omega t \hat{I}_z}, \tag{3.72}$$

Substituting into Schrödinger's equation and using the chain rule for differentiation yields the time independent expression in the rotating frame of reference:

$$\frac{\partial}{\partial t} | \psi' \rangle = -i \hat{H}_{\text{eff}} | \psi' \rangle, \tag{3.73}$$

where $\hat{H}_{\text{eff}} = -(\omega_0 - \omega)\hat{I}_z - \omega_1 \hat{I}_x$. This is directly analogous to the classical case of the effective field in the rotating frame.

Assuming the Rf pulse is on resonance (i.e., $\omega = \omega_0$), $| \psi'(t) \rangle$ at the end of a constant pulse of length τ is given by the somewhat unwieldy expression:

$$| \psi'(\tau) \rangle = e^{-i\tau \hat{H}_{\text{eff}}} | \psi'(0) \rangle = c_+ e^{-i\phi_+} \left[\cos \left(\frac{1}{2} \omega_1 \tau \right) |+\rangle + i \sin \left(\frac{1}{2} \omega_1 \tau \right) |-\rangle \right]$$
$$+ c_- e^{-i\phi_-} \left[\cos \left(\frac{1}{2} \omega_1 \tau \right) |-\rangle + i \sin \left(\frac{1}{2} \omega_1 \tau \right) |+\rangle \right]. \tag{3.74}$$

Consider the special case of $\omega_1 t = 180°$, then $| \psi'(\tau_{180}) \rangle = i \left(c_+ e^{-i\phi_+} |-\rangle + c_- e^{-i\phi_-} |+\rangle \right)$, which has the simple solution of $\langle \hat{\mu}_z \rangle = \frac{\hbar \gamma}{2} \left(c_-^2 - c_+^2 \right)$. This is an inversion, and the pictorial diagrams are shown in Figure 3.19.

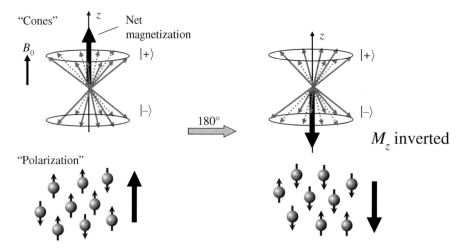

Figure 3.19 Pictorial diagrams shown the net magnetization at time $\omega_1 t = 180°$.

A second important case is for a $-90°$ excitation about the x axis, for which we have:

$$|\psi'(\tau_{90})\rangle = \frac{\sqrt{2}}{2}\left[\left(c_+ e^{-i\phi_+} + ic_- e^{-i\phi_-}\right)|+\rangle + \left(c_- e^{-i\phi_-} + ic_+ e^{-i\phi_+}\right)|-\rangle\right]. \qquad (3.75)$$

Here, for a single spin, we have $\langle\hat{\mu}_x\rangle = \hbar\gamma c_+ c_- \cos\Delta\phi$, $\langle\hat{\mu}_y\rangle = -\frac{1}{2}\hbar\gamma\left(c_+^2 - c_-^2\right)$, and $\langle\hat{\mu}_z\rangle = \hbar\gamma c_+ c_- \sin\Delta\phi$. Taking the ensemble average gives us the expected results for an 90° excitation. Namely, $\overline{\langle\hat{\mu}_x\rangle} = 0$, $\overline{\langle\hat{\mu}_y\rangle} = -\frac{1}{2}N\hbar\gamma\left(c_+^2 - c_-^2\right)$, and $\overline{\langle\hat{\mu}_z\rangle} = 0$.

In summary, a $90°_x$ Rf pulse causes: (1) equalization of probabilities of the $\{|+\rangle, |-\rangle\}$ states (yielding $M_z = 0$), and (2) a phase coherence between the $\{|+\rangle, |-\rangle\}$ states generating M_y (see Figure 3.20).

Note, here we have added the energy diagram on the right to honor Purcell, who unlike Bloch preferred not to follow the magnetization vector over time but rather view MR excitation as the emission or absorption of a photon at energy $\hbar\omega_0$. We will make more use of this alternative energy diagram picture when studying MR relaxation.

Now let us consider a slightly more sophisticated version of this problem, known as a ***linear superposition of states***. Let *System 1* consist of N_+ spins with $|\psi_+\rangle = |+\rangle$ and and N_- spins with $|\psi_-\rangle = |-\rangle$ respectively such that $N = N_+ + N_-$, $N_+/N = c_+^2$, and $N_-/N = c_-^2$. This implies that a given spin has probabilities c_+^2 and c_-^2 of being in state $|+\rangle$ or $|-\rangle$ respectively. This is precisely the system presented in Figure 3.18. The problem is that this system virtually **never** occurs in

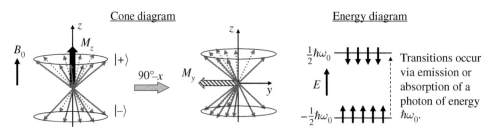

Figure 3.20 Cone and energy diagrams following a $90°_x$ excitation.

Figure 3.21 Polarization picture depicting a system of N spins polarized along a direction close to the z axis.

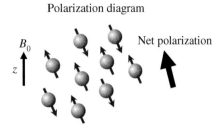

Polarization diagram

Net polarization

B_0

z

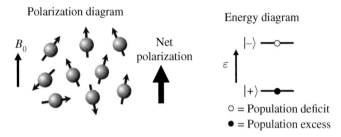

Polarization diagram

B_0

Net polarization

Energy diagram

ε

$|-\rangle$ ——○——

$|+\rangle$ ——●——

○ = Population deficit

● = Population excess

Figure 3.22 Pictorial representations of a realistic spin system.

practice! It is completely artificial to claim that all spins are either "spin up" or "spin down." For example, if we insist each spin is always either "spin up" or "spin down," then for all spins: $\{c_+, c_-\} = \{1, 0\}$ or $\{c_+, c_-\} = \{0, 1\}$, and the product $c_+ c_-$ is always zero. From Eqs. (3.68) and (3.69), $\langle \hat{\mu}_x \rangle = \langle \hat{\mu}_y \rangle = 0$ independent of any phase coherences! This system could *never* generate any transverse magnetization.

Consider instead *System 2*, where we have N spins each with wavefunction $|\psi\rangle = c_+ e^{-i\phi_+} |+\rangle + c_- e^{-i\phi_-} |-\rangle$. This system does NOT imply a given spin has probabilities c_+^2 and c_-^2 of being in state $|+\rangle$ and $|-\rangle$ respectively. There are really no cone pictures to describe this state. The polarization picture works better as shown in Figure 3.21, depicting a system of spins nearly polarized in z.

The *System 2* spins are described by a linear superposition of states as opposed to the **statistical mixture** of states in System 1. Unfortunately, in addition to being highly improbable (such a system could only be approached as the temperature dropped to absolute zero), this type of system is also inadequate to describe actual MR experiments.

In a real NMR experiment, we deal with a statistical mixture of spins, each of which is described by a linear superposition of states. An example system is shown in Figure 3.22, with N spins with wavefunctions $|\psi_i\rangle = c_{+_i} |+\rangle + c_{-_i} e^{-i\phi_-} |-\rangle$ where $i = 1, ..., N$, and c_{+_i}, c_{-_i} are complex constants subject to the constraint $c_{+_i} c_{+_i}^* + c_{-_i} c_{-_i}^* = 1$. At typical magnetic field strengths and temperatures, spins are polarized almost isotopically in space, with the term "almost" referring to a slight preference for the $+z$ component (~ 10 parts-per-million [ppm] for ^{1}H nuclei, $B_0 = 3$ T, and $T = 37\,^\circ$C).

3.5 MR in Liouville Space

The quantum mechanical derivation in Hilbert space showed that $\overline{\langle \hat{\mu}_x \rangle}$, $\overline{\langle \hat{\mu}_y \rangle}$, and $\overline{\langle \hat{\mu}_z \rangle}$ faithfully reproduce the classically derived behavior of M_x, M_y, and M_z (*e.g.*, Larmor precession and Rf excitation). This approach was rigorous but provided very limited intuition. In this section, we will show that the Liouville space description of MR is mathematically easier and retains the intuition associated with the classical vector formulation.

3.5.1 Statistical Mixture of Quantum States

Section 3.4.3 provided a rather clumsy treatment of an ensemble of spins, which we now wish to rectify. In a typical experiment, the number of nuclear spins, N, can be very large, *e.g.*, 10^{23}, whose complete quantum state is described by the wavefunction (or state vector):

$$|\psi\rangle = \sum_{i=1}^{N} a_i |\psi_i\rangle, \tag{3.76}$$

with the equation even longer if there are interactions among spins:

$$|\psi\rangle = \sum_{i=1}^{N} a_i |\psi_i\rangle + \sum_{i=1}^{N}\sum_{j=1}^{N} a_i^* a_j |\psi_i, \psi_j\rangle \dots. \tag{3.77}$$

In practice, the state of the system is almost never perfectly determined (*i.e.*, not all $|\psi_i\rangle$ precisely known). Rather, we typically have only a statistical model for the state of the system. How then can we best incorporate the partial information we have about a system to make optimal predictions and calculations?

A system described by a statistical mixture of states is *not* the same as a system whose state vector $|\psi\rangle$ is a ***linear superposition*** of states: $|\psi\rangle = \sum_i a_i |\psi_i\rangle$ with known a_is. Rather, given a system such that the state is $|\psi_1\rangle$ with probability p_1 or $|\psi_2\rangle$ with probability p_2 ($\sum_i p_i = 1$), this system is said to consist of a ***statistical mixture*** of states. Note, the $|\psi_i\rangle$s need not be orthogonal but can be assumed to be normalized.

In analyzing such a system, probability enters at **two** fundamentally different levels. First, as stated, the system state vector is not perfectly well-known, rather we only have a statistical model for being in a given state. Second, even if the state vectors *were* perfectly well-known, the probabilistic predictions arising from the QM postulates regarding the measurement process still apply.

As demonstrated in Eq. (3.52), measurements of a system described by a ***linear superposition*** of states,

$$|\psi\rangle = \sum_{i=1}^{N} a_i |\psi_i\rangle, \tag{3.78}$$

involve computing an expression such as $\langle\psi|\hat{O}|\psi\rangle$ which contains not only the terms $|a_1|^2, |a_2|^2, \dots$ but also cross terms of the form $(a_i^* a_j)$, which encompass important interference effects. Compare this with a system described by a statistical mixture of states. Namely, letting p_i be the probability of the system being in state $|\psi_i\rangle$, one might consider describing this system by a single "average" state vector:

$$|\psi\rangle = \sum_{i=1}^{N} p_i |\psi_i\rangle, \tag{3.79}$$

Measurements would then take the form:

$$\overline{\langle\hat{O}\rangle} = \sum_i p_i \langle\psi_i|\hat{O}|\psi_i\rangle. \tag{3.80}$$

However, Eq. (3.80) contains no $a_i^* a_j$ cross terms! In general, there is **no** "average state vector" that can adequately describe a system in a statistical mixture of states. Fortunately it turns out there **is** an "average operator" known as the density operator.

3.5.2 The Density Operator

We shall start by defining the density operator for a system in a ***pure state***. Namely, a system such that all $p_i = 0$ except one. Let the system wavefunction be:

$$|\psi(t)\rangle = \sum_i a_i(t)|u_i\rangle, \tag{3.81}$$

where $\{|u_i\rangle\}$ forms an orthonormal basis. The time evolution of $|\psi(t)\rangle$ is given by the Schrödinger's equation, and the expected value of an observable $\hat{A}$ is

$$\langle \hat{A}\rangle(t) = \langle\psi(t)|\hat{A}|\psi(t)\rangle = \sum_{i,j} a_i^*(t)a_j(t)A_{ij}, \tag{3.82}$$

Here, we also note that $a_i^*(t)a_j(t) = \langle u_j|\psi(t)\rangle\langle\psi(t)|u_i\rangle$, and A_{ij} are matrix elements of $\hat{A}$ in the $\{|u_i\rangle\}$ basis. Namely,

$$A_{ij} = \langle u_i|\hat{A}|u_j\rangle. \tag{3.83}$$

Let the operator $\hat{\rho}_\psi(t) = |\psi(t)\rangle\langle\psi(t)|$ be defined as the density operator for this system. For $\hat{\rho}_\psi(t)$ to be a fully adequate description of such a system, we need to (1) show the expected values for any observable $\hat{A}$, $\langle\hat{A}\rangle$, can be computed from $\hat{\rho}_\psi(t)$, and (2) find how $\hat{\rho}_\psi(t)$ evolves in time. First, using the above expressions for A_{ij} and $a_i^*(t)a_j(t)$,

$$\langle\hat{A}\rangle(t) = \sum_{i,j}\langle u_j|\hat{\rho}_\psi(t)|u_i\rangle\langle u_i|\hat{A}|u_j\rangle$$

$$= \sum_j \langle u_j|\hat{\rho}_\psi(t)\hat{A}|u_j\rangle$$

$$= \mathrm{Tr}(\hat{\rho}_\psi(t)\hat{A}) \tag{3.84}$$

where Tr denotes the trace. To find the time evolution of $\hat{\rho}_\psi(t)$, use the chain rule to take the derivative with respect to t.

$$\frac{\partial}{\partial t}\hat{\rho}_\psi(t) = \frac{\partial}{\partial t}(|\psi(t)\rangle\langle\psi(t)|)$$

$$= \frac{\partial}{\partial t}|\psi(t)\rangle\langle\psi(t)| + |\psi(t)\rangle\frac{\partial}{\partial t}\langle\psi(t)|$$

$$= -i\hat{H}|\psi\rangle\langle\psi| + i|\psi\rangle\langle\psi|\hat{H}, \tag{3.85}$$

where we made use of Schrödinger's equation to evaluate the derivatives along with the fact that $\hat{H}$ is Hermitian. Writing this in the form of a commutator,

$$\frac{\partial}{\partial t}\hat{\rho}_\psi(t) = -i[\hat{H}, |\psi(t)\rangle\langle\psi(t)|] = -i[\hat{H}, \hat{\rho}_\psi(t)]. \tag{3.86}$$

In superoperator notation,

$$\frac{\partial}{\partial t}\hat{\rho}_\psi = -i\hat{\hat{H}}\hat{\rho}_\psi. \tag{3.87}$$

Let us now find the density operator for the system described by a statistical mixture of states. That is a system whose wavefunction is given by $|\psi_i\rangle$ with probability p_i and $\sum_i p_i = 1$. Let a_i be an eigenvalue of $\hat{A}$ with associated eigenket $|u_i\rangle$. If the state vector was $|\psi_n\rangle$, then the probability of observing a_i is $P_n(a_i) = \langle\psi_n|\hat{P}_i|\psi_n\rangle = \mathrm{Tr}(\hat{\rho}_n\hat{P}_i)$, where $\hat{P}_i$ is the projection operator $|u_i\rangle\langle u_i|$. For the general case,

$$P(a_i) = \sum_n p_n P_n(a_i) = \sum_n p_n \mathrm{Tr}(\hat{\rho}_n\hat{P}_i) = \mathrm{Tr}\left(\sum_n p_n\hat{\rho}_n\hat{P}_i\right) = \mathrm{Tr}(\hat{\rho}\hat{P}_i), \tag{3.88}$$

where,

$$\hat{\rho} = \sum_n p_n \hat{\rho}_n \tag{3.89}$$

is, by definition, the density operator for the system. From this, it directly follows that the ensemble average of any observable $\hat{A}$ can be found by simply multiplying by $\hat{\rho}$ and taking the trace:

$$\overline{\langle \hat{A} \rangle} = \mathrm{Tr}(\hat{\rho}\hat{A}). \tag{3.90}$$

It is also worth noting that $\hat{\rho}$ is Hermitian. Finally, the time evolution of $\hat{\rho}$ is given by:

$$\frac{\partial}{\partial t}\hat{\rho} = -i\hat{\hat{H}}\hat{\rho}. \tag{3.91}$$

Equation (3.91) is known as the **Liouville–von Neuman equation** and replaces Schrödinger's equation when we are working in Liouville rather than Hilbert space.

3.5.3 The Spin-lattice Disconnect

One final step is needed before getting to the form of the Liouville equation of primary use in MR. The complete QM description of a molecule involves lots of terms in the Hamiltonian (such as those corresponding to nuclear spin, molecular motion, and electron–nucleus interactions), and can be written as:

$$\hat{H} = \hat{H}_l + \hat{H}_s + \hat{H}_i, \tag{3.92}$$

where $\hat{H}_s$ contains energy terms only related to a nuclear spin, $\hat{H}_l$ contains energy terms (such as vibrations and molecular motions) related to the surrounding environment (also called the lattice), and $\hat{H}_i$ representing interaction terms linking the nuclear spins and their surrounding lattice. We shall initially assume only weak interactions between nuclear spins and the lattice. In which case, the Hamiltonian can be approximated by:

$$\hat{H} \approx \hat{H}_l + \hat{H}_s. \tag{3.93}$$

Equation (3.93) is known as the spin-lattice disconnect (an assumption to be revisited when we get to relaxation theory).

Under this approximation, it suffices to solve the Liouville equation for the lattice,

$$\frac{\partial}{\partial t}\hat{\rho}_l = -i\hat{\hat{H}}_l\hat{\rho}_l, \tag{3.94}$$

independently from that associated with spin,

$$\frac{\partial}{\partial t}\hat{\sigma} = -i\hat{\hat{H}}_s\hat{\sigma}. \tag{3.95}$$

Here, we have introduced $\hat{\sigma}$ as the most commonly used notation for the spin density operator in MR. We shall initially just focus on solving Eq. (3.95), and, for any spin operator, physical quantities can be calculated via

$$\overline{\langle \hat{A}_s \rangle} = \mathrm{Tr}(\hat{\sigma}\hat{A}_s). \tag{3.96}$$

3.5.4 Hilbert Space versus Liouville Space

Figure 3.23 provides a table comparing the mathematical equations for various QM properties of a spin system as described in either Hilbert or Liouville space. In real MR experiments, our goal will

Hilbert Space versus Liouville Space

QM property	Hilbert space	Liouville space
System:	$\lvert\psi(t)\rangle$ (metric = inner product)	$\hat{\sigma}(t)$ (metric = trace)
Time evolution:	$\dfrac{\partial}{\partial t}\lvert\psi\rangle = -i\hat{H}\lvert\psi\rangle$	$\dfrac{\partial}{\partial t}\hat{\sigma} = -i\hat{\hat{H}}\hat{\sigma}$
Time independent $\hat{H}$:	$\lvert\psi(t)\rangle = \underbrace{e^{-i\hat{H}t}}\lvert\psi(0)\rangle$ rotation in ket space	$\hat{\sigma}(t) = \underbrace{e^{-i\hat{\hat{H}}t}}\hat{\sigma}(0)$ rotation in operator space
Observables:		
– Pure state	$\langle\hat{A}\rangle = \langle\psi\lvert\hat{A}\rvert\psi\rangle$	$\langle\hat{A}\rangle = \mathrm{Tr}\,(\hat{\sigma}_\psi\hat{A})$
– Statistical ensemble	computable but ugly	$\overline{\langle\hat{A}\rangle} = \mathrm{Tr}\,(\hat{\sigma}\hat{A})$

Figure 3.23 A comparison between Hilbert space versus Liouville space.

be to calculate physical quantities, such as x, y, or z magnetization, each of which will correspond to the ensemble average of the associated operator.

3.5.5 Observations About the Spin Density Operator

At this point, it is very helpful to make some important observations about the spin density operator $\hat{\sigma}$. Just as any ket in Hilbert space can be expressed as linear combination of basis kets, any operator in Liouville space can be written as a linear combination of basis operators.

Let $\hat{\sigma} = \sum_n b_n \hat{B}_n$, where $\hat{B}_n$ are an orthonormal set of Hermitian basis operators, *i.e.*,

$$(\hat{B}_n\lvert\hat{B}_m) = \mathrm{Tr}\left(\hat{B}_n^\dagger\hat{B}_m\right) = \mathrm{Tr}(\hat{B}_n\hat{B}_m) = \delta_{nm}. \tag{3.97}$$

and

$$\overline{\langle\hat{B}_n\rangle} = \mathrm{Tr}(\hat{\sigma}\hat{B}_n) = \sum_m b_m\,\mathrm{Tr}(\hat{B}_m\hat{B}_n) = b_n. \tag{3.98}$$

Hence,

$$\hat{\sigma} = \sum_n \overline{\langle\hat{B}_n\rangle}\hat{B}_n. \tag{3.99}$$

Therefore, the coefficients for any such expansion in orthonormal basis operators directly give the ensemble averages of the expected values of the respective observables!

In the case of noninteracting spin-½ particles, consider the operators $\hat{I}_x, \hat{I}_y, \hat{I}_z$, and $\hat{E}$, where $\hat{E}$ is the identify operator, and

$$\left.\begin{array}{l} \mathrm{Tr}(\hat{I}_p\hat{I}_q) = \delta_{pq} \\ \mathrm{Tr}(\hat{I}_p\hat{E}) = 0 \end{array}\right\} \text{ for } p, q \in \{x, y, z\}. \tag{3.100}$$

Here, Liouville space is four-dimensional and $\{\hat{E}, \hat{I}_x, \hat{I}_y, \hat{I}_z\}$ constitutes an orthogonal Hermitian basis set. Hence, for any expression of the spin density operatory, $\hat{\sigma}$, in terms of $\{\hat{E}, \hat{I}_x, \hat{I}_y, \hat{I}_z\}$, the expansion coefficients will directly yield the ensemble average of the expected values of the respective spin operators.

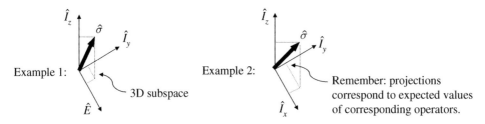

Figure 3.24 Two graphical examples depicting density operators as "vectors" in Liouville space. The basis set used in these representations is $\{\hat{E}, \hat{I}_x, \hat{I}_y, \hat{I}_z\}$, and the diagrams are drawn for two different 3D subspaces of the underlying 4D Liouville space.

QUESTION: To what physical quantity does $\overline{\langle \hat{I}_x \rangle}$ correspond?

As shown in Figure 3.24, this formulation provides a very nice geometric picture where the spin density operator, $\hat{\sigma}$, is a "vector" in Liouville space (also called "operator" or "coherence" space), and the key idea is that $\hat{\sigma}$ simply rotates around in Liouville space.

Let us now look a little closer at the spin density operator and using the eigenkets of $\hat{H}$ as a basis is a good place to start. Then, spin density matrix has elements:

$$\sigma_{i,j}(0) = \overline{a_i a_j e^{-i(\phi_i - \phi_j)}}, \tag{3.101}$$

and has the general form:

$$\underline{\sigma} = \begin{bmatrix} P_1 & \sigma_{1,2} & \cdots & \sigma_{1,n} \\ \sigma_{2,1} & P_2 & \ddots & \vdots \\ \vdots & \ddots & \ddots & \sigma_{n-1,n} \\ \sigma_{n,1} & \cdots & \sigma_{n,n-1} & P_n \end{bmatrix}, \tag{3.102}$$

where diagonal elements are called **populations** and off-diagonal elements are typically referred to as **coherences**.

We can also find the value of $\hat{\sigma}$ at thermal equilibrium (assumed to be at time $t = 0$). Using the Boltzmann distribution one can verify

$$\hat{\sigma}(0) = \frac{1}{Z} e^{-\hbar \hat{H}_0 / kT} \tag{3.103}$$

where $\hat{H}_0$ is the Hamiltonian at $t = 0$ and $Z = \mathrm{Tr}\left(e^{-\hbar \hat{H}_0 / kT}\right)$. As before, using the high temperature approximation yields:

$$\hat{\sigma}(0) \approx \frac{1}{2}\left(\hat{E} - \frac{\hbar}{kT}\hat{H}_0\right) \tag{3.104}$$

for spin-$\frac{1}{2}$ nuclei.

Piecewise constant $\hat{H}$:

$$\begin{array}{ccccc} & \hat{H}_1 & \hat{H}_2 & \hat{H}_3 & \cdots \\ \hline 0 & \leftarrow t_1 \rightarrow & \leftarrow t_2 \rightarrow & \leftarrow t_3 \rightarrow & t \end{array}$$

$$\hat{\sigma}(t_1) \qquad\qquad \hat{\sigma}(t_1 + t_2)$$

Solution: $\hat{\sigma}(0) \xrightarrow{\hat{H}_1 t_1} e^{-i\hat{H}_1 t_1}\,\hat{\sigma}(0)\,e^{i\hat{H}_1 t_1} \xrightarrow{\hat{H}_2 t_2} e^{-i\hat{H}_2 t_2}\,e^{-i\hat{H}_1 t_1}\hat{\sigma}(0)\,e^{i\hat{H}_1 t_1}\,e^{i\hat{H}_2 t_2} \xrightarrow{\hat{H}_3 t_3} \cdots$$

denotes "evolves under" $\hat{H}_1$ for time t_1

Numerical simulations most often use the matrix version of this formulation.

Figure 3.25 Solving the Liouville von Neuman equation when the Hamiltonian has a piecewise-constant time dependence.

3.5.6 Solving the Liouville von Neuman Equation

Let $\hat{\sigma}$ be the spin density operator for a system consisting of a statistical ensemble of states. The time evolution of $\hat{\sigma}$ is given by the Liouville von Neuman equation

$$\frac{\partial}{\partial t}\hat{\sigma}(t) = -i\hat{\hat{H}}(t)\hat{\sigma}(t), \tag{3.105}$$

and there are two cases of primary interest.

Case 1: $\hat{H}$ is independent of time. Here, the solution is straightforward and given by simple integration. Namely, in both superoperator and operator notation,

$$\hat{\sigma}(t) = e^{-i\hat{\hat{H}}t}\,\hat{\sigma}(0) = e^{-i\hat{H}t}\hat{\sigma}(0)e^{i\hat{H}t}. \tag{3.106}$$

Case 2: $\hat{H}$ is piecewise constant over time, that is $\hat{H} = \hat{H}_1$ for the initial time interval, $\hat{H} = \hat{H}_2$, for the second time interval. As graphically depicted in Figure 3.25, this case is solved by repeatedly computing the time evolution of $\hat{\sigma}$ during each successive time interval.

We are now ready to solve a representative MR experiment, namely a simple 90° excitation (see Figure 3.26). Ignoring relaxation, what are $\hat{\sigma}(t)$, $M_x(t)$, $M_y(t)$, and $M_z(t)$ for the time intervals t_1, t_2, and t_3?

First, a few details. Many, if not most, MRS texts define the Larmor frequency as $\omega_0 = -\gamma B_0$. However, for this text, we are going to define $\omega_0 = \gamma B_0$, which is consistent with most MRI texts and Bloch's original 1946 paper (Bloch et al. 1946). Defining ω_0 as $+\gamma B_0$ leads to the sign conventions shown in Figure 3.27. Namely, evolution of $\hat{\sigma}$ during periods of both free precession and Rf excitation correspond to *left-handed* rotations.

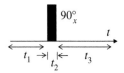

Figure 3.26 A simple pulse-and-acquire NMR experiment.

Assume we start in thermal equilibrium with B_0 (assumed to be in the z direction) the only applied magnetic field, $\hat{\sigma}(0) = \frac{1}{2}\left(\hat{E} - \frac{\hbar}{kT}\hat{H}_0\right)$. However, the identity operator, $\hat{E}$, is invariant under rotations (can you prove this using Eq. (3.50) or the more general statement that if $\hat{A}$ and $\hat{B}$ commute, then $e^{i\theta\hat{A}}\hat{B} = \hat{B}$?) and orthogonal to $\hat{I}_x$, $\hat{I}_y$, and $\hat{I}_z$. Therefore, we can ignore this term and just work in a 3D subspace. Hence, without loss of generality, we will set:

$$\hat{\sigma}(0) = C\hat{I}_z \text{ where } C = \frac{\gamma\hbar B_0}{2kt}. \tag{3.107}$$

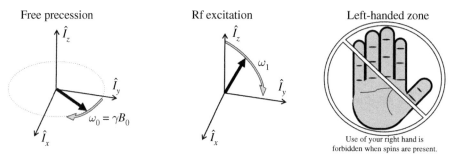

Free precession   Rf excitation   Left-handed zone

Use of your right hand is
forbidden when spins are present.

Figure 3.27 Defining $\omega_0 \equiv \gamma B_0$ leads to left-handed rotations in Liouville space for both periods of free precession and Rf excitation.

At thermal equilibrium, Eq. (3.98) yields $\hat{\sigma}_0 = 0 \cdot \hat{I}_x + 0 \cdot \hat{I}_y + C\hat{I}_z$. Thus $M_x = M_y = 0$, and

$$M_z = \rho\gamma\hbar\overline{\langle\hat{I}_z\rangle} = \rho\gamma\hbar\frac{1}{2}C = \rho\frac{\gamma^2\hbar^2 B_0}{4kT}, \tag{3.108}$$

where ρ denotes spins/volume and the factor of ½ comes from normalization value for $\mathrm{Tr}\left(\hat{I}_z^2\right)$. During the time interval t_1, $\hat{H}_1 = -\gamma B_0\hat{I}_z$, and

$$\hat{\sigma}(t) = e^{-i\hat{H}t}\hat{\sigma}(0) = Ce^{-i\gamma B_0\hat{I}_z t}\hat{I}_z = C\hat{I}_z. \tag{3.109}$$

Noting that $e^{-i\gamma B_0\hat{I}_z t}$ is simply a rotation about the $\hat{I}_z$ axis, the system remains unchanged, as expected for equilibrium conditions with:

$$M_z = \rho\frac{\gamma^2\hbar^2 B_0}{4kT} = M_0 \quad \text{for } t_0 < t < t_1. \tag{3.110}$$
$$M_x = M_y = 0$$

During time interval t_2, there is a second magnetic field corresponding to the Rf excitation pulse. Hence,

$$\hat{H}(t) = -\gamma B_0\hat{I}_z - \gamma B_1(\hat{I}_x\cos\omega t + \hat{I}_y\sin\omega t). \tag{3.111}$$

$\hat{H}$ is now time varying, and we can solve this by switching to the rotating frame $\hat{H}' = e^{-i\omega\hat{I}_z t}\hat{H}$ and $\hat{\sigma}' = e^{-i\omega\hat{I}_z t}\hat{\sigma}$ (hereafter, we will do almost everything in the rotating frame). The Liouville von Neuman equation becomes $\frac{\partial}{\partial t}\hat{\sigma}' = -i\hat{H}'\hat{\sigma}'$, where $\hat{H}' = -(\omega_0 - \omega)\hat{I}_z - \omega_1\hat{I}_x$ and $\gamma B_1 = \omega_1$. Assuming an Rf excitation pulse close to on-resonance, *i.e.*, $\omega_0 - \omega \ll \omega_1$,

$$\hat{\sigma}'(t) = Ce^{-i\hat{H}'t}\hat{I}_z = Ce^{-i\omega_1\hat{I}_x t}\hat{I}_z = C(\hat{I}_z\cos(\omega_1 t) + \hat{I}_y\sin(\omega_1 t)), \tag{3.112}$$

where we have used that fact that $\hat{I}_x$, $\hat{I}_y$, and $\hat{I}_z$ cyclically commute. Hence, in the rotating frame, the Hamiltonian, $\hat{H}_2 \approx -\omega_1\hat{I}_x$, causes the spin density operator to rotate about $\hat{I}_x$ axis. For a $90°_x$ pulse ending at time $t = t_1 + t_2$, $\hat{\sigma}'(t_1 + t_2) = C\hat{I}_y$, $M_z = M_x = 0$, and $M_y = M_0$.

For time interval t_3, let us consider the case of being slightly off-resonance (on-resonance case being trivial). During this period of free precession (Rf turned off), $\hat{H}' = -(\omega_0 - \omega)\hat{I}_z = -\Omega\hat{I}_z$, and $\hat{\sigma}'(t) = Ce^{-i\Omega t \hat{I}_z}\hat{I}_y = C(\hat{I}_y\cos(\Omega t) + i\hat{I}_x\sin(\Omega t))$. This is simply Larmor precession with $M_z = 0$, $M_x = M_0\sin(\Omega t)$, and $M_y = M_0\cos(\Omega t)$ for $t > t_1 + t_2$.

3.6 Summary

Figure 3.28 shows a direct comparison between this simple experiment as drawn in physical versus Liouville space. Analogous to classical case, the QM behavior of this system can be solved using an intuitive vector model and associated left-handed rotations.

QUESTION: For independent spin ½ particles, Liouville space is 4D. What happened to the fourth dimension in Figure 3.28?

Do not underestimate the intuitive power of using your left (or right) hand to compute rotations – many physics problems have been solved with just such handwaving! Magnetization components can be readout via their respective projections along the associated axes with $M_x = \gamma\hbar\langle\hat{I}_x\rangle$, $M_y = \gamma\hbar\langle\hat{I}_y\rangle$, and $M_z = \gamma\hbar\langle\hat{I}_z\rangle$. However, this brings up an important question. Have we really made any progress other than gaining the personal satisfaction of knowing how the familiar classical analysis of MR can be derived from the postulates of QM (see Figure 3.29)?

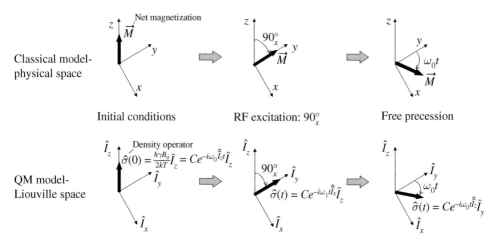

Figure 3.28 Graphical depictions of a simple MR experiment drawn as a classical magnetization vector, $\vec{M}$, rotating in physical space versus a spin density operator, $\hat{\sigma}$, rotating in Liouville space. Note, here the $\hat{\sigma}$ is expressed in superoperator notation.

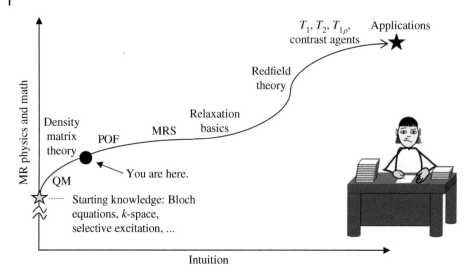

Figure 3.29 The roadmap at the end Chapter 3.

Exercises

E.3.1 Invariance of the Trace
Prove that the trace of the matrix representation of an operator is independent of the corresponding basis.

E.3.2 Hermitian Operators
a) Prove that the eigenvalues of a Hermitian operator are real.
b) Prove that the eigenkets corresponding to nondegenerate eigenvalues of a Hermitian operator are orthogonal.

E.3.3 Wavefunctions
Let $|\phi(t)\rangle$ be a normalized ket satisfying the following differential equation,

$$\frac{d\,|\phi(t)\rangle}{dt} = \hat{A}\,|\phi(t)\rangle.$$

Solve for $|\phi(t)\rangle$.

E.3.4 Dirac Notation
Using Dirac notation, let $|f\rangle$ and $|g\rangle$ be kets in a Hilbert space and $\hat{A}$ an operator in that space. For each of the following expressions, state whether the result is a ket, a bra, operator, or scalar. (a) $\langle f\,|\,g\rangle$, (b) $\langle f|\,\hat{A}$, (c) $|f\rangle\langle g|$, (d) $A|f\rangle\langle g|$, (e) $\langle f|\,\hat{A}^{\dagger}$, and (f) $\langle f|\hat{A}|g\rangle$.

E.3.5 Projection Operators
Let $|\alpha\rangle$ and $|\beta\rangle$ form a complete orthonormal basis set for a two-dimensional system. Consider the wavefunction $|\psi\rangle = \lambda_1|\alpha\rangle + \lambda_2|\beta\rangle$ where $|\lambda_1|^2 + |\lambda_2|^2 = 1$, and its corresponding projection operator $\hat{P}_\psi = |\psi\rangle\langle\psi|$. The information content of the system can be measured by the quantity $\mathrm{Tr}\left(\hat{P}_\psi^2\right)$.

a) Let $\lambda_2 = 0$ (or equivalently $\lambda_1 = 0$). Find the eigenkets, eigenvalues, trace of $\hat{P}_{\psi_1}$, and the trace of $\hat{P}_{\psi}^2$.

b) Let $\lambda_1 \neq 0$ and $\lambda_2 \neq 0$. Namely, $|\psi_2\rangle = \lambda_1 |\alpha\rangle + \lambda_2 |\beta\rangle$. Find the trace of $\hat{P}_{\psi}^2$.

E.3.6 Superoperators

Given operators $\hat{A}$, $\hat{B}$, and $\hat{C}$, prove the following. (a) If $\hat{\hat{A}}\hat{B} = 0$, then $e^{\hat{A}+\hat{B}} = e^{\hat{A}}e^{\hat{B}} = e^{\hat{B}}e^{\hat{A}}$, (b) $e^{\hat{A}}\hat{B} = e^{\hat{A}}\hat{B}e^{-\hat{A}}$, and (c) If $\hat{\hat{A}}\hat{B} = 0$, then $e^{\hat{A}+\hat{B}}\hat{C} = e^{\hat{A}}e^{\hat{B}}\hat{C} = e^{\hat{B}}e^{\hat{A}}\hat{C}$.

E.3.7 The Pauli Matrices

Let $|\psi_{x_i}\rangle$, $|\psi_{y_i}\rangle$, and $|\psi_{z_i}\rangle$, for $i = 1, 2$, be the eigenkets for $\hat{I}_x$, $\hat{I}_y$, and $\hat{I}_z$. The Pauli matrices, expressed in the $|\psi_{z_i}\rangle$ basis, are given by:

$$I_x = \frac{1}{\sqrt{2}}\begin{bmatrix} 0 & 1 \\ 1 & 0 \end{bmatrix}, \; I_y = \frac{1}{\sqrt{2}}\begin{bmatrix} 0 & -i \\ i & 0 \end{bmatrix} \text{ and } I_z = \frac{1}{\sqrt{2}}\begin{bmatrix} 1 & 0 \\ 0 & -1 \end{bmatrix}$$

a) Find the corresponding eigenvectors and eigenvalues for I_x, I_y, and I_z.

b) Show by direct matrix calculation that

$$\sum_{i=1}^{2} \left|\psi_{x_i}\right\rangle\left\langle\psi_{x_i}\right| = \sum_{i=1}^{2} \left|\psi_{y_i}\right\rangle\left\langle\psi_{y_i}\right| = \sum_{i=1}^{2} \left|\psi_{z_i}\right\rangle\left\langle\psi_{z_i}\right| = \hat{E},$$

where $\hat{E}$ is the identity operator.

c) Find $[\hat{I}_z, \hat{I}_x]$, $[\hat{I}_z, \hat{I}_y]$, and $[\hat{I}_x, \hat{I}_y]$.

d) Express $\hat{I}_x$, $\hat{I}_y$, and $\hat{I}_z$ in the $|\psi_{y_i}\rangle$ basis.

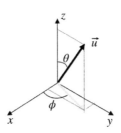

E.3.8 The Stern–Gerlach Experiment

The x, y, and z components of the intrinsic angular momentum (spin) of a spin-½ particle are given by the cyclically commuting operators: $\hat{I}_x$, $\hat{I}_y$, and $\hat{I}_z$, each with eigenvalues $\pm\frac{1}{2}$. Consider an arbitrary unit vector $\vec{u}$.

a) What are the eigenvalues of $\hat{I}_u$, the operator corresponding to the spin in the $\vec{u}$ direction?

b) What is the matrix form of $\hat{I}_u$ expressed in the $\{|+\rangle, |-\rangle\}$ basis, where $|+\rangle$ and $|-\rangle$ are the eigenkets of $\hat{I}_z$ corresponding to eigenvalues $+½$ and $-½$ respectively?

c) The general form of a ket for single spin-1/2 particle is $|\psi\rangle = \alpha|+\rangle + \beta|-\rangle$, where $|\alpha|^2 + |\beta|^2 = 1$. Can you find α and β such that $|\phi\rangle$ represents an unpolarized spin. Namely, $\langle\hat{I}_x\rangle = \langle\hat{I}_y\rangle = \langle\hat{I}_z\rangle = 0$?

d) Consider a collection of spin-½ nuclei each described by the same ket $\left|s_p\right\rangle = \frac{1}{\sqrt{3}}\left(|s_x\rangle + |s_y\rangle\right)$, where $|s_x\rangle$ and $|s_y\rangle$ are spin states oriented along the $+x$ and $+y$ directions, respectively. This system is said to be in a "pure" state. Show $|s_p\rangle$ is normalized. Find $\langle\widehat{M}_x\rangle$, $\langle\widehat{M}_y\rangle$, and $\langle\widehat{M}_z\rangle$.

e) Silver atoms leaving the oven in the Stern–Gerlach experiment can be polarized in any direction. It can be shown that the state of a spin polarized in an arbitrary direction is given by:

$$|\psi\rangle = \cos\frac{\theta}{2}\,|+\rangle + \sin\frac{\theta}{2}e^{i\phi}\,|-\rangle .$$

For a statistical ensemble of spins where all directions $\vec{u}(\theta,\phi)$ equally likely show $\overline{\langle\widehat{M}_x\rangle} = \overline{\langle\widehat{M}_y\rangle} = \overline{\langle\widehat{M}_z\rangle} = 0$ (which represents an unpolarized spin system).

E.3.9 Bloch Equations Revisited

Consider a spin-½ particle with wavefunction $\psi(t)$ and associated magnetic moment operators $\widehat{\mu}_x$, $\widehat{\mu}_y$, and $\widehat{\mu}_z$ (written more compactly as $\widehat{\vec{\mu}} = \gamma\hbar\widehat{\vec{I}}$).

a) While the operator $\widehat{\vec{\mu}}$ is independent of time, the expected value $\langle\widehat{\vec{\mu}}\rangle$ is, in general, time varying. Show

$$i\frac{d}{dt}\langle\widehat{\vec{\mu}}\rangle(t) = \langle[\widehat{\vec{\mu}},\widehat{H}]\rangle.$$

where $\widehat{H}$ is the Hamiltonian operator of the system.

b) When placed in a magnetic field $\vec{B}(t)$, $\hbar\widehat{H}(t) = -\widehat{\vec{\mu}}\Delta\vec{B}(t)$. Show

$$\frac{d}{dt}\langle\widehat{\vec{\mu}}\rangle(t) = \gamma\langle\widehat{\vec{\mu}}\rangle(t) \times \vec{B}(t).$$

(Hint: Use the commutator relationships for $\widehat{I}_x$, $\widehat{I}_y$, and $\widehat{I}_z$).

Hence, the expected value of $\widehat{\vec{\mu}}$ obeys the classical Bloch equations (ignoring relaxation)!

E.3.10 Basis Set in Liouville Space

Prove that if the kets $|f_i\rangle$ form an orthonormal basis for an n-dimensional Hilbert space, then the transition operators $\widehat{T}_{ij} = |f_i\rangle\langle f_j|$ constitute an orthonormal basis in the corresponding Liouville space.

E.3.11 Pure Versus Mixed States

Consider the kets corresponding to the two collections of spin-1/2 nuclei:

$$\text{System 1: } |s_1\rangle = \frac{1}{\sqrt{2}}(|+\rangle + |-\rangle)$$

$$\text{System 2: } |s_2\rangle = \begin{cases} |+\rangle \text{ with probability } 0.5 \\ |-\rangle \text{ with probability } 0.5 \end{cases}$$

where $|+\rangle$ and $|-\rangle$ are spin states oriented along the $+z$ and $-z$ directions.

a) For each system, calculate the corresponding density matrix $\underline{\sigma}$ in the $\{|+\rangle, |-\rangle\}$ basis. Which systems are in a pure versus mixed state?

b) For each system, calculate $\text{Tr}\{\underline{\sigma}^2\}$. If $\text{Tr}\{\underline{\sigma}^2\}$ is considered as a measure of the information content of the system, what do these results say about systems in pure versus mixed states?

c) For each system, compute $\langle \widehat{M}_x \rangle$, $\langle \widehat{M}_y \rangle$, and $\langle \widehat{M}_z \rangle$.

E.3.12 Transition Operators

Using the $\{|+\rangle, |-\rangle\}$ basis, the transition operators for a spin-$\frac{1}{2}$ system are defined as $\widehat{T}_{++} = |+\rangle\langle+|$, $\widehat{T}_{+-} = |+\rangle\langle-|$, $\widehat{T}_{-+} = |-\rangle\langle+|$, and $\widehat{T}_{--} = |-\rangle\langle-|$. Find the relationship between the Liouville basis sets $\{\widehat{T}_{++}, \widehat{T}_{+-}, \widehat{T}_{-+}, \widehat{T}_{--}\}$ and $\{\widehat{I}^2, \widehat{I}_x, \widehat{I}_y, \widehat{I}_z\}$. Note, $\widehat{I}^2 = \widehat{I}_x\widehat{I}_x + \widehat{I}_y\widehat{I}_y + \widehat{I}_z\widehat{I}_z$.

Historical Notes

Sir William Rowan Hamilton, born on August 3/4, 1805, in Dublin, Ireland, was a remarkable mathematician whose contributions spanned various fields, including optics, dynamics, algebra, and QM. His most significant achievement was the discovery of quaternions, an algebraic system that later played a crucial role in QM and NMR.

From a young age, Hamilton showed exceptional talent and linguistic aptitude. Under the guidance of his uncle, James Hamilton, an Anglican priest, he quickly mastered multiple languages, such as Latin, Greek, Hebrew, Arabic, Sanskrit, Persian, Syriac, French, and Italian, surpassing his peers in knowledge.

In 1823, Hamilton entered Trinity College, Dublin, where he excelled in mathematics, physics, and classics. Throughout his undergraduate years, he delved into his own mathematical investigations and showed a keen interest in optics. In 1827, despite still being an undergraduate, he had a substantial paper on optics accepted for publication by the Royal Irish Academy, marking the beginning of his academic recognition. Remarkably, that same year, Hamilton was appointed as the professor of astronomy at Trinity College and became the Royal Astronomer of Ireland. He settled at the Dunsink Observatory, just outside Dublin, where he continued groundbreaking research in mathematics and astronomy.

However, it was during the last 22 years of his life that Hamilton dedicated himself to developing the theory of quaternions and related systems. Quaternions extended the concept of complex numbers to four dimensions, providing a powerful tool for solving problems in 3D geometry. His seminal papers on quaternions laid the foundation for many fundamental concepts and results in vector analysis. In 1853, he published "Lectures on Quaternions" to disseminate his findings, although it did not gain widespread recognition among mathematicians and physicists.

Unfortunately, his magnum opus, "Elements of Quaternions," remained unfinished at the time of his death on September 2, 1865. Nonetheless, his groundbreaking contributions influenced further developments in mathematics and physics, becoming particularly relevant in the emerging field of QM. Hamilton's work on quaternions and extending complex numbers to higher dimensions found unexpected applications in the field of NMR.

(Source: Photo from https://commons.wikimedia.org/wiki/File:Sir_William_Rowan_Hamilton, _head-and-shoulders_portrait,_facing_slightly_right_LCCN90713420.jpg, public domain)

Paul Adrien Dirac, born on August 8, 1902, in Bristol, Gloucestershire, England, was an eminent English theoretical physicist who left an indelible mark in the fields of QM and quantum electrodynamics. His pioneering work in theoretical physics was characterized by extraordinary mathematical abilities and a preference for solitary contemplation.

From a young age, Dirac displayed exceptional mathematical talents, but he showed little interest in literature and art. A reserved individual with limited social and emotional skills, he often found it challenging to engage in small talk. Nevertheless, his physics papers and books were renowned for their literary excellence, featuring impeccable mathematical expressions and precision in wording.

One of Dirac's most monumental contributions came in 1927 with his quantum theory of radiation, widely regarded as the true beginning of quantum electrodynamics. In this groundbreaking theory, he introduced innovative methods to quantize electromagnetic waves and formulated what later became known as "second quantization." This revolutionary approach facilitated the description of a single quantum particle to be extended into a formalism applicable to systems comprising multiple particles.

In 1928, Dirac achieved what is arguably his most significant accomplishment, the development of the relativistic wave equation for the electron. This revolutionary equation elegantly described the behavior of electrons traveling at relativistic speeds, seamlessly integrating QM with special relativity. The profound implications of this equation paved the way for successfully unifying QM and special relativity, representing a crucial milestone in the evolution of quantum field theory.

Among Dirac's notable contributions was the prediction of the existence of antiparticles, which was later experimentally confirmed. For his remarkable achievements, he was honored with the 1933 Nobel Prize in Physics, which he shared with the Austrian physicist Erwin Schrödinger for their respective and foundational contributions to the development of QM.

(Source: Photo from https://commons.wikimedia.org/wiki/File:Paul_Dirac,_1933.jpg, public domain.)

David Hilbert, a distinguished German mathematician, born on January 23, 1862, in Königsberg, Prussia (now Kaliningrad, Russia), left an indelible mark in the field of mathematics. Renowned for his groundbreaking work in reducing geometry to a series of axioms and for his substantial contributions to establishing the formalistic foundations of mathematics, Hilbert's influence extended beyond pure mathematics into the realm of mathematical physics.

Hilbert's academic journey commenced at the University of Königsberg, where he completed his PhD in 1885 and ultimately became a full professor in 1893.

In 1895, Hilbert's career took a significant turn as he accepted a prestigious professorship in mathematics at the University of Göttingen, where he would spend the rest of his professional life. His presence at Göttingen further elevated the university's reputation in physics, as his profound interest in mathematical physics brought forth noteworthy contributions to the field.

Collaborating with his friend and colleague, Hermann Minkowski, Hilbert ventured into exploring novel applications of mathematics in physics, thereby solidifying Göttingen's status as a prominent center for scientific inquiry. Notably, this period saw the arrival of several eminent physicists at the university, including Nobel laureates Max von Laue, James Franck, and Werner Heisenberg, all of whom were associated with Göttingen during Hilbert's lifetime.

Hilbert created algorithms for space-filling curves in higher dimensional spaces. This extended the methods of vector algebra and calculus to spaces with any finite or infinite number of dimensions. Hilbert Spaces provided the basis for significant contributions to physics.

Among Hilbert's enduring legacies in mathematics, his formulation of "Hilbert's Problems" stands as one of the most notable. In the year 1900, during the International Mathematical Congress in Paris, he presented a list of 23 research problems that he believed would shape the course of mathematical exploration in the 20th century. Over time, many of these problems were successfully solved, with each solution representing a significant milestone in the field. However, it is worth mentioning that one of the challenges on his list, which involves the partial resolution of the Riemann hypothesis, remains an essential open question in number theory to this day.

(Source: Photo from https://commons.wikimedia.org/wiki/File:Hilbert.jpg, public domain.)

Joseph Liouville (1809–1882) was a brilliant French mathematician renowned for his significant contributions to analysis, differential geometry, and number theory. Born on March 24, 1809, in Saint-Omer, France, Liouville displayed a profound interest in mathematics from an early age. His groundbreaking work across diverse mathematical fields earned him a place among the most influential mathematicians of the 19th century.

Initially focusing on electrodynamics and the theory of heat, Liouville's most notable achievement came in the early 1830s when he established the first comprehensive theory of fractional calculus. This theory extended the meaning of differential and integral operators, opening new avenues for exploring and applying fractional calculus in various scientific disciplines.

Between 1832 and 1833, Liouville conducted research on integration in finite terms, seeking to determine if algebraic functions could have integrals expressed in finite or elementary terms. His investigations significantly advanced the understanding of integrals and their properties, leading to important developments in mathematical analysis.

In collaboration with Charles-François Sturm, a devoted friend and mathematician, Liouville published a series of articles between 1836 and 1837, giving rise to a new subject in mathematical analysis known as Sturm–Liouville theory. This theory underwent further generalization and rigorization in the late 19th century and found applications in 20th century mathematical physics and the theory of integral equations.

In 1844, Liouville achieved another groundbreaking feat by proving the existence of transcendental numbers. These numbers, distinct from algebraic numbers that are roots of algebraic equations with rational coefficients, revolutionized number theory and had significant implications in various mathematical contexts.

In the field of analysis, Liouville made pioneering advancements in the theory of doubly periodic functions, deriving this theory from general theorems, including his own, in the realm of analytic functions of a complex variable. His work enriched the understanding of functions with periodic properties and their practical applications.

Liouville's dedication to number theory resulted in over 200 publications, mostly in the form of concise notes. While some of his works were initially published without explicit proofs, later mathematicians provided rigorous justifications for his results. His vast collection of publications, totaling about 400 memoirs, articles, and notes, illustrates the breadth of his mathematical inquiries and the depth of his contributions.

(Source: Photo from https://commons.wikimedia.org/wiki/File:Joseph_liouville.jpeg, public domain.)

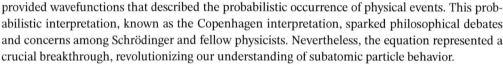

Erwin Schrödinger, an Austrian theoretical physicist, was born on August 12, 1887, in Vienna. He made groundbreaking contributions to the wave theory of matter and the development of quantum mechanics, leaving a lasting impact in the field of physics.

In 1926, Schrödinger published influential papers that laid the foundation for quantum wave mechanics. These papers introduced the Schrödinger equation, a partial differential equation that became the cornerstone of QM. Building on Louis de Broglie's proposal that matter particles have both particle and wave-like properties, Schrödinger's wave equation elegantly described particle behavior as waves in specific situations, offering profound insights into the quantum nature of matter.

Unlike classical mechanics, where Newton's equations predict the future state of a system with certainty, Schrödinger's equation provided wavefunctions that described the probabilistic occurrence of physical events. This probabilistic interpretation, known as the Copenhagen interpretation, sparked philosophical debates and concerns among Schrödinger and fellow physicists. Nevertheless, the equation represented a crucial breakthrough, revolutionizing our understanding of subatomic particle behavior.

One of Schrödinger's most famous challenges to the probabilistic interpretation of QM was the thought experiment "Schrödinger's cat," proposed in 1935. This scenario involves placing a cat inside a sealed box with a vial of poisonous gas and a radioactive atom with a 50% chance of decaying within an hour. According to quantum theory, until the box is opened and the atom's wavefunction collapses, the cat exists in a superposition of two states: alive and dead. Schrödinger considered this absurd and used the thought experiment to highlight apparent paradoxes arising from the interpretation of QM.

(Source: Photo from https://commons.wikimedia.org/wiki/File:Erwin_Schr%C3%B6dinger_-_Narodowe_Archiwum_Cyfrowe_(1-E-939).jpg, public domain.)

Otto Stern, a renowned scientist and Nobel laureate in Physics, made pioneering contributions to molecular beam technology and NMR. Born on February 17, 1888, in Sohrau, Germany (now Zory, Poland), Stern's scientific journey began with theoretical studies in statistical thermodynamics, laying the groundwork for his groundbreaking discoveries later in life.

In 1914, Stern became a lecturer in theoretical physics at the University of Frankfurt, where he explored fundamental properties of matter through theoretical research. Later, in 1923, he became a professor of physical chemistry at the University of Hamburg. It was during this time that Stern, in collaboration with Walther Gerlach, conducted a historic experiment that revolutionized our understanding of QM and atomic behavior.

The famous molecular-beam experiment, designed by Stern and Gerlach in the early 1920s at the University of Hamburg, involved directing a beam of silver atoms through a nonuniform magnetic field onto a glass plate. The surprising outcome was that instead of broadening into a continuous band, the beam split into two distinct beams. This discovery confirmed the space quantization theory, which stated that atoms could only align in specific directions in a magnetic field, contrary to classical physics' predictions. This groundbreaking experiment came to be known as the Stern–Gerlach experiment, cementing Stern's reputation as a brilliant experimental physicist.

However, as the Nazis came to power in Germany in 1933, Stern faced persecution due to his Jewish heritage and was forced to leave his homeland. Seeking refuge in the United States, he became a research professor of physics at the Carnegie Institute of Technology in Pittsburgh. During his time in the U.S., Stern continued his groundbreaking research, leading to another significant achievement in the field of NMR.

Using his expertise in molecular beam technology, Stern measured the magnetic moment of the proton in 1933. The magnetic moment represents a subatomic particle's magnetic property and had been a subject of intense theoretical interest. Stern's meticulous experiments revealed that the magnetic moment of the proton was about 2.5 times greater than the previously theoretical value. This groundbreaking finding deepened our understanding of atomic and subatomic structures, opening new possibilities in the study of nuclear properties.

For his outstanding contributions to molecular beams and NMR, Otto Stern was awarded the prestigious Nobel Prize in Physics in 1943.

(Source: Photo from https://commons.wikimedia.org/wiki/File:Otto_Stern.jpg, public domain.)

George Uhlenbeck and **Samuel Goudsmit**, Dutch–American physicists, made groundbreaking contributions to QM, particularly in the understanding of nuclear and electron spin. Their work significantly advanced modern physics.

Uhlenbeck and Goudsmit, studied physics and mathematics under the guidance of Paul Ehrenfest at the University of Leiden; a collaboration that would have a profound impact on QM. In 1925, while both were still graduate students, Goudsmit and Uhlenbeck made their groundbreaking contribution to understanding electron spin. They were inspired by Wolfgang Pauli's exclusion principle, which introduced four quantum numbers to describe an electron's properties. However, the fourth quantum number lacked a clear physical interpretation.

Drawing on Paul Dirac's work, which depicted electrons as rotating spheres, Goudsmit and Uhlenbeck proposed a resolution to this mystery. They envisioned the electron not as a point particle but as a small, rotating sphere, introducing the concept of "electron spin." This idea revolutionized the classical understanding of electrons as point-like particles moving in orbits around the nucleus.

Initially, hesitant to publish their speculative findings, Goudsmit and Uhlenbeck received encouragement from Paul Ehrenfest and discussed their idea with Hendrik Lorentz. Eventually, they submitted a short note to the physics research journal Naturwissenschaften.

Their work on electron spin and intrinsic angular momentum opened new research avenues in QM and shed light on the magnetic properties of electrons. The associated magnetic moment was found to be 2, a value confirmed experimentally and known as the "gyromagnetic ratio." This contribution was crucial for explaining atomic spectra's fine structure and laid the foundation for future developments in quantum electrodynamics.

(Source: Photo of George Uhlenbeck, Hendrik Kramers, and Samuel Goudsmit around 1928 in Ann Arbor, MI; from https://commons.wikimedia.org/wiki/File:George_Uhlenbeck_(left)_and_ Samuel_Goudsmit_(right)_along_with_Hendrik_Kramers_(middle)_in_Ann_Arbor_c._1928.jpg, public domain.)

4

Nuclear Spins

4.1 Review of the Spin Density Operator and the Hamiltonian

In the previous chapter, we derived the spin density operator, $\hat{\sigma}(t)$, and its time evolution as a metric to characterize the state of a spin system. For a system in a statistical mixture of states, the ensemble average of any observable $\hat{A}$ is easily found using the trace operation, $\langle \hat{A} \rangle = \text{Tr}(\hat{\sigma}\hat{A})$. The most important quantities for MR are $M_x = \rho\gamma\hbar\langle\hat{I}_x\rangle = \rho\gamma\hbar\text{Tr}(\hat{\sigma}\hat{I}_x)$, $M_y = \rho\gamma\hbar\text{Tr}(\hat{\sigma}\hat{I}_y)$, and $M_z = \rho\gamma\hbar\text{Tr}(\hat{\sigma}\hat{I}_z)$.

QUESTION: Is not the trace a matrix rather than an operator operation?

The time evolution of $\hat{\sigma}$ is given by the Liouville-von Neuman equation (Equation (3.105)). If the Hamiltonian $\hat{H}$, the operator corresponding to the energy of the spin system, is time independent, $\hat{\sigma}$ simply rotates around in operator space, $\hat{\sigma}(t) = e^{-i\hat{H}t}\hat{\sigma}(t)e^{i\hat{H}t} = e^{-i\hat{H}t}\hat{\sigma}(0)$. The key is to find the appropriate Hamiltonian for a given experiment.

QUESTION: Around what axis does $\hat{\sigma}$ rotate?

In general, $\hat{H}$ is the sum of different terms representing different physical interactions. That is, $\hat{H} = \hat{H}_1 + \hat{H}_2 + \hat{H}_3 + \cdots$, with some examples being interactions between nuclei and applied magnetic fields, nearby electrons, or other nuclei. There are at least three conditions under which calculations simplify: (1) $\hat{H}_i$ are time independent, (2) $\hat{H}_i$ depend on the spatial orientation of the respective molecule and average to zero with rapid molecular tumbling (a random process distinct from a systematic rotation), or (3) $\hat{H}_i$ and $\hat{H}_j$ terms commute. For the third case, rotations

Fundamentals of In Vivo Magnetic Resonance: Spin Physics, Relaxation Theory, and Contrast Mechanisms, First Edition. Daniel M. Spielman and Keshav Datta.
© 2024 John Wiley & Sons, Inc. Published 2024 by John Wiley & Sons, Inc.
Companion website: www.wiley.com/go/Spielman

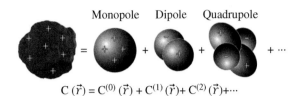

$$C(\vec{r}) = C^{(0)}(\vec{r}) + C^{(1)}(\vec{r}) + C^{(2)}(\vec{r}) + \cdots$$

Figure 4.1 The charge distribution of nucleus can be expressed as a sum of multipoles.

described by $e^{-i\hat{\hat{H}}t}\hat{\sigma}$ can be computed in any order. Namely,

$$[\hat{H}_i, \hat{H}_j] = 0 \implies e^{-i(\hat{H}_i + \hat{H}_j)t} = e^{-i\hat{H}_i t}e^{-i\hat{H}_j t} = e^{-i\hat{H}_j t}e^{-i\hat{H}_i t} \tag{4.1}$$

In terms of how a spin with a magnetic moment can interact with its environment, one can think of an atomic nucleus as a lumpy magnet with non-uniform positive electric charge. The nuclear spin Hamiltonian contains terms which describe the orientation dependence of the nuclear energy. The nuclear magnetic moment interacts with magnetic fields, and nuclear electric charge interacts with electric fields. Hence, $\hat{H} = \hat{H}_{elec} + \hat{H}_{mag}$.

To evaluate the effects of electric fields, the nuclear electric charge distributions can be expressed as a sum of multipole components $C(\vec{r}) = C^{(0)}(\vec{r}) + C^{(1)}(\vec{r}) + C^{(2)}(\vec{r}) + \cdots$ (see Figure 4.1). However, based on symmetry arguments relating to the shape of a nucleus and its spin, $C^{(n)} = 0$ for $n > 2I$, and, within experimental error, odd terms disappear (though this last fact is not at all obvious (Levitt 2008)). Hence, there are no electrical energy terms that depend on orientation or internal nuclear structure. That is to say, all spin-$\frac{1}{2}$ nuclei behave exactly like point charges!

The nuclear magnetic moment arises primarily from the dipole moment (the quadrupole moment can cause some small shifts in the hyperfine structure (Levitt 2008)). As previously discussed, $\hat{H}_{mag} = -\vec{\mu} \cdot \vec{B} = -\gamma \hbar \hat{\vec{I}} \cdot \vec{B}$, where $\vec{\mu}$ is the magnetic dipole moment and $\vec{B}$ is the local magnetic field. Figure 4.2 schematically depicts the relative contributions of the various terms in the nuclear spin Hamiltonian. We will now discuss these in more detail.

QUESTION: Is tissue a solid or a liquid?

4.2 External Interactions

The interaction energy between a magnetic field, $\vec{B}$, and the magnetic moment, $\vec{\mu} = \gamma \hbar \vec{I}$, is given by the **Zeeman Hamiltonian**. Classically, the $\mathcal{E} = -\hbar \gamma \vec{I} \cdot \vec{B}$. Given we have defined the MR Hamiltonian as $\mathcal{E}/\hbar$, QM notation yields:

$$\hat{H}_Z = -\gamma \hat{\vec{I}} \cdot \vec{B}, \tag{4.2}$$

Hence, a large static magnetic field in the z direction, $\vec{B} = B_0\vec{z}$ results in the Hamiltonian

$$\hat{H}_{static} = -\gamma B_0 \hat{I}_z = -\omega_0 \hat{I}_z. \tag{4.3}$$

If we were to add an Rf excitation field of magnitude B_1 rotating in the $x-y$ plane,

$$\vec{B}_{Rf} = B_1(\cos \omega t \vec{x} - \sin \omega t \vec{y}), \tag{4.4}$$

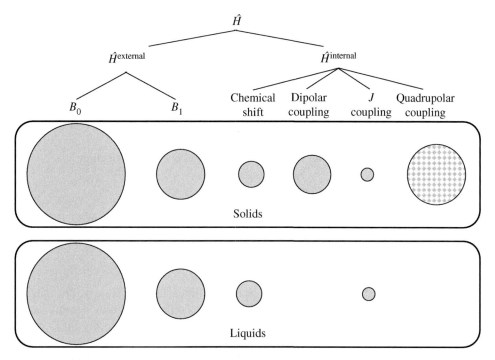

Figure 4.2 The dominant external and internal nuclear spin Hamiltonian contributors and their relative sizes. Note, the dipolar and quadrupolar (for nuclei with spin > ½.) coupling terms disappear for liquids with isotropic molecular tumbling; however, both terms contribute to relaxation processes. Source: Adapted with permission from Levitt (2008), Fig. 7–63, p. 217.

the full Hamiltonian for external magnetic fields becomes:

$$\hat{H}_{ext} = \hat{H}_{static} + \hat{H}_{Rf} = -\gamma B_0 \hat{I}_z - \gamma B_1 (\hat{I}_x \cos \omega t - \hat{I}_y \sin \omega t). \tag{4.5}$$

In the rotating frame near resonance ($\omega \approx \omega_0$), the Hamiltonian reduces to

$$\hat{H}_{ext} \approx -\omega_1 \hat{I}_x. \tag{4.6}$$

Assuming ^{1}H imaging on a typical 3 T clinical scanner, $\omega_0 = 128$ MHz, and peak values of ω_1 are generally much smaller, around 1 kHz.

4.3 Internal Interactions

In general, the internal nuclear spin Hamiltonian is quite complicated, and we will make two simplifying approximations. First, the large B_0 field typically dominates internal spin interactions. We can thus make use of what is known as a ***secular approximation***. Consider a large vector subject to a small perturbation. The perturbation can be divided into the sum of a parallel and a perpendicular component. The secular approximation is to ignore perturbations perpendicular to the large main vector, and this approximation is used in multiple scientific and engineering applications (see Figure 4.3).

The second simplification arises from molecular motion. With rapid molecular tumbling, the spatially dependent interaction terms fluctuate

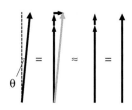

Figure 4.3 The secular approximation for a vector is valid for small values of θ.

with time and can be replaced by their motion-averaged values. Furthermore, terms with time averages of zero can be dropped. Note, the discarded (for now) internal spin Hamiltonian terms are responsible for spin relaxation and will be revisited in Chapter 7. Specifically, when molecular orientations depend on time, the secular Hamiltonian terms can be written as $\widehat{H}^{int}(\theta(t), \phi(t))$, where θ and ϕ are the standard angles used in spherical coordinates. These terms can be replaced by their time averages:

$$\overline{\widehat{H}}_{int} = \frac{1}{\tau} \int_0^\tau \widehat{H}_{int}(\theta(t), \phi(t))dt. \tag{4.7}$$

Assuming ergodicity, time averages can be replaced by averages over space, yielding:

$$\overline{\widehat{H}}_{int} = \frac{1}{4\pi} \int_0^{2\pi} \int_0^\pi \widehat{H}_{int}(\theta, \phi)p(\theta, \phi) \sin\theta d\theta d\phi. \tag{4.8}$$

where $p(\theta, \phi)$ = probability density for molecule having orientation ϕ and θ. With isotropic tumbling,

$$\overline{\widehat{H}}_{int,iso} = \frac{1}{4\pi} \int_0^{2\pi} \int_0^\pi \widehat{H}_{int}(\theta, \phi) \sin\theta d\theta d\phi. \tag{4.9}$$

On the timescale of a typical NMR experiment, molecules in a liquid largely diffuse within a small spherical volume a few tens of microns in diameter (known as a diffusion sphere). Because each spin effectively experiences every possible spatial orientation with respect to every other spin, intermolecular interactions within the diffusion sphere averages out to zero.

We will now examine some of the primary interactions between nuclei and nearby electrons. Any material placed in a magnetic field is magnetized to some degree and modifies the applied field. As shown in Figure 4.4, electrons in an atom circulate about B_0, generating a magnetic moment opposing the applied magnetic field. The global effect, known as bulk susceptibility, results in a reduction of the field inside the material such that:

$$B_0^s = (1 - \chi)B_0, \tag{4.10}$$

where χ is the ***magnetic susceptibility*** constant. Materials with magnetic susceptibilities <1 are called diamagnetic, versus >1 for paramagnetic materials. For water, $\chi_{water} \approx -9.1$ ppm. Hereafter, we will use "B_0" to refer to the internal field (to be revisited when we talk about field inhomogeneities and shimming).

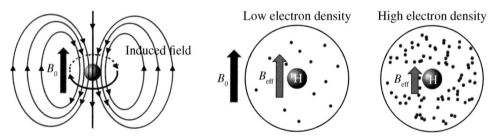

Figure 4.4 Electron shielding. Electrons in an atom circulate about B_0, generating a magnetic moment opposing the applied magnetic field. As a result, nuclei experience a reduced effective field, B_{eff}, the strength of which is inversely proportional to the local electron density.

4.3.1 Chemical Shift

As compared to the global effects, different atoms locally experience different electron cloud densities (Ramsey 1950) (see Figure 4.4). This is typically expressed in terms of a shielding constant, σ. Do not confuse it with the spin density operator! Namely,

$$B = B_0(1 - \sigma). \tag{4.11}$$

The formal correction for chemical shielding is:

$$\hat{H}_Z = -\gamma \vec{\hat{I}} \cdot (1 - \underline{\sigma})\vec{B} \tag{4.12}$$

where $\underline{\sigma} = 3 \times 3$ shielding tensor and corresponds to the field created in the ith direction when applying an external field $\vec{B}$ in the jth direction for $\{i, j = x, y, z\}$. *In vivo*, rapid molecular tumbling largely averages out the non-isotropic components (we note that anisotropic components do contribute to spin relaxation), leaving,

$$\sigma = \sigma_{iso} = \mathrm{Tr}(\underline{\sigma}/3). \tag{4.13}$$

Hence, for $\vec{B} = [0, 0, B_0]$ and including chemical shift, the Zeeman Hamiltonian becomes:

$$\hat{H}_Z = -\gamma(1 - \sigma)B_0\hat{I}_z = -\omega\hat{I}_z. \tag{4.14}$$

Figure 4.5 depicts the alterations in spin energy due to chemical shift as well as representative examples for ^1H nuclei in water versus fat molecules. Oxygen is highly electrophilic, drawing the electrons away from the ^1H nuclei in a water molecule, reducing the amount of shielding, and shifting the corresponding ^1H spectrum. In contrast, for the representative lipid molecule, carbon is less electrophilic, resulting in increased shielding of the attached ^1H nuclei. For historical reasons, NMR spectroscopists plot spectra with resonant frequency increasing to the left.

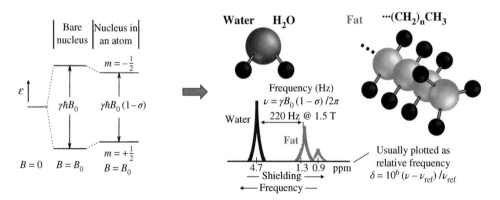

Figure 4.5 Chemical shift. The left diagram showing shifts in energy due to electron shielding. Examples, water and fat molecules and the corresponding ^1H NMR spectra (right). First complete theory of chemical shift was developed by Norman Ramsey in 1950.

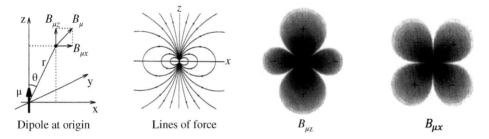

Figure 4.6 A magnetic dipole with the corresponding lines of force and cross-sectional images of the z and x components of the resulting field.

4.3.2 Dipolar Coupling

Nuclei with spin $\geq \frac{1}{2}$ behave like tiny magnetic dipoles, with the resulting dipole magnetic field (see Figure 4.6) falling off as r^3. In cartesian coordinates,

$$B_x = \left(\frac{\mu_0}{4\pi}\right)\left(\frac{\mu}{r^3}\right)(3\sin\theta\cos\theta\cos\phi)$$

$$B_y = \left(\frac{\mu_0}{4\pi}\right)\left(\frac{\mu}{r^3}\right)(3\sin\theta\cos\theta\sin\phi)$$

$$B_z = \left(\frac{\mu_0}{4\pi}\right)\left(\frac{\mu}{r^3}\right)(3\cos^2\theta - 1), \tag{4.15}$$

where μ_0 is the permeability of free space (see Figure 4.6).

The dipole fields from nearby spins interact, an effect known as ***dipolar coupling***. Given the rapid fall off with distance, this is largely an intramolecular effect, and we will further simplify this by just considering pairs of coupled spins labeled I and S. The complete dipolar coupling Hamiltonian is then given by:

$$\widehat{H}_D = -\frac{\mu_0\gamma_I\gamma_S}{4\pi r^3}\hbar\left(\hat{\vec{I}}\cdot\hat{\vec{S}} - \frac{3}{r^2}(\hat{\vec{I}}\cdot\vec{r})(\hat{\vec{S}}\cdot\vec{r})\right), \tag{4.16}$$

where $\vec{r}$ is a vector from spin I to spin S. In the presence of a large field B_0, the secular approximation becomes:

$$\widehat{H}_D = d(3\hat{I}_z\hat{S}_z - \hat{\vec{I}}\cdot\hat{\vec{S}}), \tag{4.17}$$

where,

$$d = -\frac{\mu_0\gamma_I\gamma_S}{4\pi r^3}\hbar(3\cos^2\theta - 1) \tag{4.18}$$

and θ is the angle between B_0 and the vector connecting spins I and S.

The dependence on the spatial angle θ is very important, as, with molecular tumbling, the dipolar coupling Hamiltonian becomes time variant! An example of water molecule tumbling in a magnetic field is shown in Figure 4.7. Defining the origin to be the right-hand ^{1}H nuclei, the magnetic field seen by its coupled partner is B_0 plus a smaller random function of time. With isotropic tumbling (e.g., as seen in liquids), $\overline{\widehat{H}_D} = 0$, and can be ignored for now, but will later play a major role in relaxation. Typical values for dipolar coupling are 10–15 kHz.

4.3.3 J Coupling

The most obvious magnetic interaction between neighboring nuclei is their mutual dipole coupling. However, as discussed in the previous section, this anisotropic interaction averages out to zero for

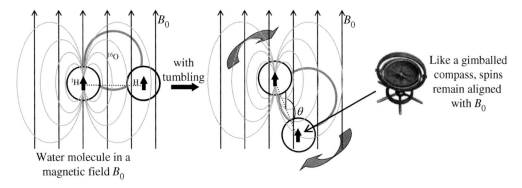

Figure 4.7 A water molecule undergoing molecular tumbling in a magnetic field.

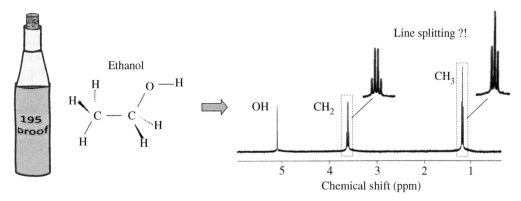

Figure 4.8 ^{1}H NMR spectrum of ethanol.

freely tumbling molecules. Consider the ^{1}H NMR spectrum of ethanol as shown in Figure 4.8. Due to local symmetry, the three ^{1}H nuclei bonded to the first carbon experience identical chemical shifts, a phenomenon known as ***equivalent spins***. Similarly, the two ^{1}Hs bound to the second carbon are also equivalent to each other but resonate at a slightly shifted frequency due to asymmetries in the overall ethanol molecule. However, the spectra for each of these two groups of equivalent spins show a distinct structure indicative of splitting of the associated energy levels.

QUESTION: For the ^{1}H spectrum of ethanol shown in Figure 4.8, why is the peak from ^{1}H bound to the oxygen not also split?

This splitting is due to a process known as ***J coupling*** (also called ***scalar coupling*** among other names), which contributes an additional term to the spin Hamiltonian. This is an interaction between nuclei involved in a chemical bond, and the physical mechanism responsible for J coupling can be explained as follows. Consider a J-coupled pair of spins I and S. When sharing a chemical bond, the I spin "senses" the polarization of the S spin. Specifically, the electrons participating in the chemical bond and their associated spins influence the energy of the S spin depending on the polarization of the *I* spin. At *very* small distances (comparable to the nuclear radius), the dipolar interaction between a bonded electron and a nucleus is replaced by an *isotropic* interaction called ***Fermi contact interaction***. The electron's magnetic dipole moment creates a nonzero magnetic field at the nucleus. However, consistent with the Pauli exclusion principle, any two electrons involved in a common chemical bond must be polarized in opposite directions. This results in the paired electron creating a non-zero magnetic field at the position of the second nucleus. Consequently, there exists an energy difference for bonded nuclei whose spin states are parallel versus antiparallel (see Figure 4.9). This *J* coupling effect can also extend over more than a single chemical bond. The ethanol molecule is a good example, where the two groups of ^{1}H nuclei are not directly bonded to each but rather bonded via the intermediary carbons.

The associated *J* coupling interaction energy is proportional to $-\gamma_e\gamma_n\hat{\vec{I}} \cdot \hat{\vec{S}}$, and the corresponding Hamiltonian is:

$$\hat{H}_J = 2\pi J\hat{\vec{I}} \cdot \hat{\vec{S}} = 2\pi J(\hat{I}_x\hat{S}_x + \hat{I}_y\hat{S}_y + \hat{I}_z\hat{S}_z), \tag{4.19}$$

where *J* is the *J* coupling constant usually in the range of 1–150 Hz. In this expression, we find some unfamiliar operators, namely $\hat{I}_x\hat{S}_x$, $\hat{I}_y\hat{S}_y$, and $\hat{I}_z\hat{S}_z$ (remember products of operators are themselves operators). In this case, these new terms represent ***product operators***, which will be discussed in detail in Chapter 5. Figure 4.10 shows the energy splitting and corresponding spectra for a theoretical *J*-coupled pair of spins. This figure also includes an alternative diagram (to prove useful in later discussions of relaxation) depicting the different energy levels with arrows representing transitions between states (also known as single quantum coherences).

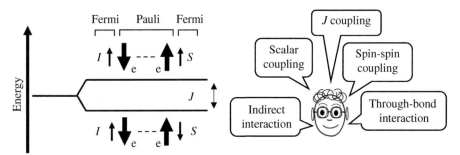

Figure 4.9 Energy diagram for *J* coupling along with the multiple names describing the phenomenon.

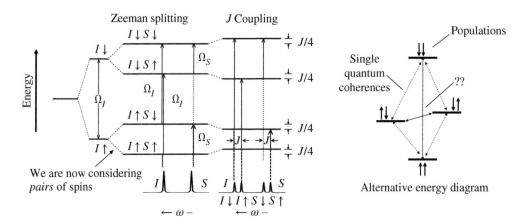

Figure 4.10 Energy diagrams and corresponding spectra for a pair of spins, *I* and *S*, in the presence of both Zeeman splitting and J coupling.

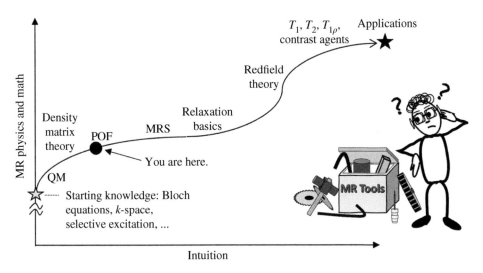

Figure 4.11 The roadmap at the end Chapter 4.

4.4 Summary

In summary, the nuclear spin Hamiltonian is given by $\hat{H} = \hat{H}_Z + \hat{H}_D + \hat{H}_J + \hat{H}_Q$. For now, we can ignore the dipolar coupling term (averages to zero with molecular tumbling) and the quadrupolar term, $\hat{H}_Q$, which only exists for nuclei with spin $> \frac{1}{2}$. Written for a frame of reference rotating at frequency ω, the resulting Hamiltonian is,

$$\hat{H} = -\Omega_I \hat{I}_z - \Omega_S \hat{S}_z + 2\pi J(\hat{I}_x \hat{S}_x + \hat{I}_y \hat{S}_y + \hat{I}_z \hat{S}_z),\tag{4.20}$$

where Ω_I and Ω_S are rotating frame resonant frequencies of the respective spins given by $\Omega_I = \gamma(1 - \sigma_I)B_0 - \omega = \omega_I - \omega$ and $\Omega_S = \gamma(1 - \sigma_S)B_0 - \omega = \omega_S - \omega$. During Rf excitation (assuming B_1 along the x axis), $\hat{H} = -\Omega_I \hat{I}_z - \Omega_S \hat{S}_z + 2\pi J(\hat{I}_x \hat{S}_x + \hat{I}_y \hat{S}_y + \hat{I}_z \hat{S}_z) - \omega_1^I \hat{I}_x - \omega_1^S \hat{S}_x$, which reduces to:

$$\hat{H} \approx -\omega_1^I \hat{I}_x - \omega_1^S \hat{S}_x,\tag{4.21}$$

for $\omega_1^I \gg \Omega_I$, $\omega_1^S \gg \Omega_S$, and $\omega_1^I, \omega_1^S \gg 2\pi J$. This is another example of using a secular approximation, and the above equations hold for both homonuclear and heteronuclear cases.

In the next chapter, we will further explore the use of product operators (see Figure 4.11). The goal will be to provide a more practical and intuitive approach for analyzing common MR pulse sequences, particularly for molecules with J coupling.

Exercises

E.4.1 The Spin Lattice Disconnect

In general, the complete Hamiltonian for a nuclear spin system is $\hat{H} = \hat{H}_l + \hat{H}_s + \hat{H}_i$, where these components represent terms only involving the lattice, terms only involving spin (*e.g.*, magnetic moment), and interaction terms coupling spin and lattice variables. If we assume that the interaction term is small, *i.e.*, spin variables have no effect on lattice variables and vice versa, then $\hat{H} \cong \hat{H}_l + \hat{H}_s$. Given $\hat{\rho}_l$ and $\hat{\sigma}$ are the density operators for lattice and spin systems, respectively, find an equation for the density operator of the full system.

E.4.2 **Polarization and Boltzmann distribution**

Fill in the missing entries in the table below that lists the thermal equilibrium polarization (%) obtained by systems of ^{1}H nuclei, ^{13}C nuclei, and unpaired electrons e^-, placed in a 3 T magnetic field under a variety of different temperatures.

	Polarization (%)		
Temp	^{1}H	^{13}C	e^-
310 K	?	?	?
4 K	?	?	?
1 K	?	?	?
0.017 K	25	?	?

What conclusions can you draw regarding the feasibility of using low temperatures to increase the SNR for *in vivo* studies?

E.4.3 **Two-spin Operators**

For a single spin-½ nucleus, matrices corresponding to the spin operators are the identity matrix (here denoted by $\underline{E}$) plus the 2×2 Pauli matrices:

$$\underline{I}_x = \frac{1}{\sqrt{2}}\begin{bmatrix} 0 & 1 \\ 1 & 0 \end{bmatrix}, \quad \underline{I}_y = \frac{1}{\sqrt{2}}\begin{bmatrix} 0 & -i \\ i & 0 \end{bmatrix}, \quad \underline{I}_z = \frac{1}{\sqrt{2}}\begin{bmatrix} 1 & 0 \\ 0 & -1 \end{bmatrix}, \quad \text{and} \quad \underline{E} = \frac{1}{\sqrt{2}}\begin{bmatrix} 1 & 0 \\ 0 & 1 \end{bmatrix}.$$

If we now consider *pairs* of spins, e.g., I and S, the two-spin operators needed to describe the system correspond to 4×4 matrices and are calculated using the **matrix direct product** denoted by $\otimes$ (also known as the Kronecker product, tensor product, or direct product). Namely,

$$\begin{bmatrix} a & b \\ c & d \end{bmatrix} \otimes \begin{bmatrix} \alpha & \beta \\ \gamma & \delta \end{bmatrix} = \begin{bmatrix} a\begin{bmatrix} \alpha & \beta \\ \gamma & \delta \end{bmatrix} & b\begin{bmatrix} \alpha & \beta \\ \gamma & \delta \end{bmatrix} \\ c\begin{bmatrix} \alpha & \beta \\ \gamma & \delta \end{bmatrix} & d\begin{bmatrix} \alpha & \beta \\ \gamma & \delta \end{bmatrix} \end{bmatrix} = \begin{bmatrix} a\alpha & a\beta & b\alpha & b\beta \\ a\gamma & a\delta & b\gamma & b\delta \\ c\alpha & c\beta & d\alpha & d\beta \\ c\gamma & c\delta & d\gamma & d\delta \end{bmatrix}$$

a) Find the matrices corresponding to the two-spin operators $\underline{I}_x$, $\underline{S}_x$, $2\underline{I}_z\underline{S}_z$, and $2\underline{I}_y\underline{S}_z$.

b) Show $2\hat{I}_z\hat{S}_z$, $\hat{I}_x$, and $2\hat{I}_y\hat{S}_z$ cyclically commute.

E.4.4 **The Steady-state Density Matrix**

Given a time independent Hamiltonian, $\hat{H}$ with nondegenerate eigenvalues e_i and eigenkets $|i\rangle$, i.e., $\hat{H}|i\rangle = e_i|i\rangle$,

a) Show that in steady-state conditions, the diagonal elements of σ_0, expressed in terms of the eigenkets of $\hat{H}$, are constants and the off-diagonal elements are zero.

b) For a two-spin system of spin-½ nuclei I and S, the steady-state density operator, $\hat{\sigma}_0$, can be expressed as a linear combination of the following four product operators: $\hat{E}$, $\hat{I}_z$, $\hat{S}_z$, and $2\hat{I}_z\hat{S}_z$. Why do we not need any of the other 12 two-spin product operators?

E.4.5 J Coupling

Consider a homonuclear two-spin system with a J coupling of $J = 16$ Hz, $T_2 = 100$ ms, and chemical shift difference between the two spins equal to $\Delta\Omega$. Using the full density matrix, simulate (e.g., using Matlab) the series of spectra generated by the 90°-acquire sequence with $\Delta\Omega/J$ ranging from 0 to 10. Account for the effect of T_2 by assuming the signal following the 90° excitation is weighted by a factor of e^{-t/T_2}.

E.4.6 Dipolar Coupling

Consider a homonuclear two-spin system from an anisotropic material with $J = 0$ but a residual dipole coupling of $d = 16$ Hz. Assume a $T_2 = 100$ ms and a chemical shift difference between the two spins equal to $\Delta\Omega$. Using the full density matrix, simulate (e.g., using Matlab) the series of spectra generated by the 90°-acquire sequence with $\Delta\Omega/d$ ranging from 0 to 10. Use the secular approximation to the dipolar coupling spin Hamiltonian. Account for the effect of T_2 by assuming the signal following the 90° excitation is weighted by a factor of e^{-t/T_2}. Comment on the differences between the results found with J coupling (E.4.5) versus dipolar coupling.

E.4.7 Secular Approximation

In a uniform magnetic field B_0 (assumed to be in the $+z$ direction), the Hamiltonian for a J-coupled two-spin system of spin-½ nuclei can be written in the rotating frame as $\hat{H}_0 = \hat{A} + \hat{B}$, where $\hat{A} = -\Omega_I\hat{I}_z - \Omega_S\hat{S}_z$ and $\hat{B} = 2\pi J(\hat{I}_x\hat{S}_x + \hat{I}_y\hat{S}_y + \hat{I}_z\hat{S}_z)$ for $\Omega_I = \gamma(1 - \sigma_I)B_0 - \omega_0$ and $\Omega_S = \gamma(1 - \sigma_S)B_0 - \omega_0$.
a) Find a secular approximation for the Hamiltonian, i.e., $\hat{H}_0 \cong \hat{A} + \hat{B}^s$.
b) Under what conditions is the secular approximation valid? How does this compare to the strong versus weak coupling approximation?

E.4.8 Eigenoperators

Just as operators have associated eigenkets, superoperators have associated eigenoperators. Let $\hat{\hat{A}}$ be a commutation superoperator. Then, if $\hat{\hat{A}}\hat{B} = \lambda\hat{B}$ where λ is a scalar, $\hat{B}$ is an eigenoperator of $\hat{\hat{A}}$ with corresponding eigenvalue λ. Consider the Hamiltonian $\hat{H}_0$ with eignenkets $|m\rangle$, i.e., $\hat{H}_0|m\rangle = E_m|m\rangle$ for $m = 1, \cdots, N$.
a) Show that the transition operators, defined as $\hat{T}_{nm} = |n\rangle\langle m|$, are eigenoperators of $\hat{\hat{H}}_0$. What are their corresponding eigenvalues?
b) Let and $\hat{H}_0 = -\omega_I\hat{I}_z - \omega_S\hat{S}_z$ with $\hat{I}_\pm \equiv \hat{I}_x \pm i\hat{I}_y$ and $\hat{S}_\pm \equiv \hat{S}_x \pm i\hat{S}_y$. Show that $2\hat{I}_z\hat{S}_z$, $\hat{I}_\pm\hat{S}_\pm$, $\hat{I}_\pm\hat{S}_\mp$, $2\hat{I}_\pm\hat{S}_z$, and $2\hat{I}_z\hat{S}_\pm$ are all eigenoperators of $\hat{\hat{H}}_0$ and complete the following table.

Eigenoperator	Eigenvalue	Corresponding transition operator (s)
$2\hat{I}_z\hat{S}_z$	0	$\hat{T}_{11} - \hat{T}_{22} - \hat{T}_{33} + \hat{T}_{44}$
$\hat{I}_+\hat{S}_+$	?	?
$\hat{I}_-\hat{S}_-$	?	?
$\hat{I}_+\hat{S}_-$	?	?
$\hat{I}_-\hat{S}_+$	?	?
$\hat{I}_+\hat{S}_z$	?	?
$\hat{I}_-\hat{S}_z$	?	?
$\hat{I}_z\hat{S}_+$	?	?
$\hat{I}_z\hat{S}_-$	?	?

Historical Notes

Norman Foster Ramsey Jr. (August 27, 1915 – November 4, 2011) was an influential American physicist known for his groundbreaking contributions to the field. Born in Washington, D.C., Ramsey developed a keen interest in science and mathematics from an early age, setting the stage for his remarkable career in academia and research.

Ramsey pursued his higher education at Columbia University, earning his bachelor's degree in 1935 and a Ph.D. in physics in 1940. Under the guidance of Nobel laureate Isidor I. Rabi, he conducted doctoral research on molecular beams for studying nuclear magnetic moments. This work became the foundation for his future endeavors in nuclear and atomic physics.

During World War II, Ramsey joined the renowned Massachusetts Institute of Technology's (MIT) Radiation Laboratory, where he made significant advances in radar technology. His exceptional service in the war effort earned him the Presidential Medal for Merit in 1946. After the war, Ramsey returned to academia and joined Harvard University's faculty in 1947. It was during this time that he made one of his most significant contributions to physics – the development of the separated oscillatory fields method, also known as the Ramsey method.

In 1949, Ramsey and his research group achieved a breakthrough by creating the first successful atomic hydrogen maser. This invention revolutionized timekeeping and enabled precise atomic clocks, becoming a vital tool in global navigation systems like the global positioning system (GPS). Subsequently, in 1950, he developed the separated oscillatory fields technique, which allowed accurate measurements of atomic and nuclear energy levels, further enhancing the precision of atomic clocks and finding applications in various fields. In recognition of his pioneering work, Norman Ramsey was awarded the Nobel Prize in Physics in 1989.

Throughout his career, Ramsey remained at the forefront of experimental physics, conducting research on diverse topics, including nuclear magnetic resonance, molecular beams, and fundamental symmetries in particle physics. His work in precision spectroscopy played a crucial role in the development of quantum field theory and contributed to fundamental tests of quantum electrodynamics (QED).

Despite primarily being an experimental physicist, Ramsey also published several theoretical papers, covering subjects like parity and time reversal symmetry, the first successful theory of the NMR chemical shifts, theories of nuclear interactions in molecules, and the theory of thermodynamics and statistical mechanics at negative absolute temperatures.

(Source: Photo from https://upload.wikimedia.org/wikipedia/commons/0/05/HD.4G.049_%2810537741015%29_crop.jpg, public domain.)

Pieter Zeeman, a renowned Dutch physicist, was born on May 25, 1865, in Zonnemaire, Netherlands. Throughout his illustrious career, he made significant contributions to the field of physics, most notably for his groundbreaking discovery of the Zeeman effect, which ultimately led to him being awarded the Nobel Prize in Physics in 1902 alongside his mentor, Hendrik A. Lorentz.

Zeeman's academic journey commenced at the University of Leiden, where he became a devoted student under the tutelage of the esteemed physicist Hendrik A. Lorentz. Under Lorentz's mentorship, Zeeman honed his skills and deepened his understanding of physics. Upon completing his studies, he returned to Leiden in 1890, this time as a lecturer.

In 1896, Lorentz encouraged Zeeman to investigate the effect of magnetic fields on a source of light. This suggestion proved to be a pivotal moment in Zeeman's career. During his experiments, he made a remarkable observation – when a source of light was subjected to a magnetic field, the spectral lines it emitted split into several components. This groundbreaking phenomenon, later known as the Zeeman effect, provided crucial evidence for the quantization of atomic energy levels and became a significant discovery in the study of atomic and molecular physics.

The discovery of the Zeeman effect brought international recognition to Pieter Zeeman and Hendrik A. Lorentz, leading to their prestigious Nobel Prize in Physics in 1902. Their collaborative work shed light on the interaction between magnetic fields and light, opening up new avenues for research in spectroscopy and quantum mechanics.

In 1900, owing to his remarkable achievements, Pieter Zeeman was appointed as a professor of physics at the University of Amsterdam. Later, in 1908, he assumed the position of director of the University's Physical Institute, where he continued to delve into various aspects of theoretical and experimental physics.

Remaining dedicated to his research, Zeeman focused on studying the propagation of light in moving media, such as water, quartz, and flint. His investigations in this area contributed to the understanding of how light behaves under different conditions, thereby having practical applications in fields like optics and meteorology.

(Source: Photo from https://commons.wikimedia.org/wiki/File:Pieter_Zeeman.jpg, public domain.)

Enrico Fermi, a renowned Italian-born American scientist, was born on September 29, 1901, in Rome, Italy. He became a key figure in shaping the nuclear age, making significant contributions to physics and nuclear science. His work in mathematical statistics, nuclear transformations, and nuclear fission research earned him international acclaim and numerous accolades.

Fermi demonstrated exceptional intellectual abilities early in his academic journey. He earned his physics doctorate from the University of Pisa in 1922, at just 21 years old, gaining prominence for his research in quantum mechanics and statistical mechanics. In 1926, he became a professor of theoretical physics at the University of Rome, where he continued to produce groundbreaking work.

In 1938, Fermi received the Nobel Prize in Physics for his discoveries regarding new radioactive elements produced by neutron irradiation and his theory of beta decay. However, due to increasing discrimination in Italy as fascism rose and his Jewish descent, Fermi emigrated to the United States with his family in 1938 and became a naturalized citizen in 1944.

In the United States, Fermi joined the Manhattan Project, a top-secret initiative to develop an atomic bomb during World War II, where he played a crucial role as an associate director of the Los Alamos laboratory. On July 16, 1945, Fermi witnessed the world's first successful detonation of a nuclear device near Alamogordo, New Mexico, and ingeniously calculated the bomb's explosive energy using slips of paper blown from the vertical.

Despite his involvement in atomic weapons development, Fermi expressed ethical concerns. In 1949, he opposed the development of a more powerful thermonuclear bomb, fearing immense destruction. His legacy extended beyond nuclear science, as he was renowned as an exceptional teacher, influencing generations of scientists.

In his later years, Fermi pondered the "Fermi paradox," speculating on the absence of detectable extraterrestrial civilizations despite the vastness of the universe. He voiced concerns that advanced civilizations might be self-destructive, possibly through nuclear annihilation, reflecting his awareness of nuclear weaponry's catastrophic implications.

Enrico Fermi's scientific achievements earned him lasting recognition. The U.S. Department of Energy established the Enrico Fermi Award, and Fermilab, the National Accelerator Laboratory in Illinois, was named after him. Element 100 in the periodic table, fermium, was also named in his honor. Fermi's pioneering spirit and commitment to science continue to inspire and shape the world of physics and nuclear research to this day.

(Source: Photo from https://commons.wikimedia.org/wiki/File:Enrico_Fermi_1943-49_ (cropped).jpg, public domain).

5

Product Operator Formalism

5.1 The Density Operator, Populations, and Coherences

5.1.1 Spin Systems and Associated Density Operators

Before proceeding further, it is helpful to gain a little more insight and intuition regarding the spin density operator and phase coherences. First, consider $\hat{\sigma}$ for a single-spin system, that is, a system consisting of many isolated spins. At thermal equilibrium, $\hat{H}_0 = -\omega_0 \hat{I}_z$, and using the high-temperature approximation and ignoring the $\hat{E}$ component,

$$\hat{\sigma}_0 = \frac{\hbar \gamma B_0}{4kT} \hat{I}_z. \tag{5.1}$$

For this system, the most convenient basis set is the eigenkets of $\hat{H}_0$: $\{|+\rangle, |-\rangle\}$. The wavefunction for a given spin is simply $|\psi\rangle = a_+|+\rangle + a_-|-\rangle$, with a_+ and a_- arbitrary complex scalars. The matrix representation of the density operator is

$$\underline{\sigma} = \begin{bmatrix} \overline{|a_+|^2} & \overline{a_+ a_-^*} \\ \overline{a_+^* a_-} & \overline{|a_-|^2} \end{bmatrix} \implies \begin{array}{c} \\ |+\rangle \\ |-\rangle \end{array} \begin{array}{cc} |+\rangle & |-\rangle \\ \begin{bmatrix} L & I \\ I & L \end{bmatrix} \end{array}, \tag{5.2}$$

where the right-hand expression highlights the density operator as measuring coherences between eigenstates listed across the top and along the left-hand side. In this case, the entries labeled "L" are referred to as "longitudinal magnetization," "spin population," or "single-spin order." The "I" entries correspond to "transverse magnetization" and are zero unless there exists some phase coherence between states $|+\rangle$ and $|-\rangle$. Such phase coherences are called ***coherent superposition of quantum states*** or, more commonly, simply ***coherences***.

For this single-spin system, $\underline{\sigma}$ could be expressed as

$$\underline{\sigma} = L_1 \begin{bmatrix} 1 & 0 \\ 0 & 0 \end{bmatrix} + I_1 \begin{bmatrix} 0 & 1 \\ 0 & 0 \end{bmatrix} + I_2 \begin{bmatrix} 0 & 0 \\ 1 & 0 \end{bmatrix} + L_2 \begin{bmatrix} 0 & 0 \\ 0 & 1 \end{bmatrix}. \tag{5.3}$$

In operator notation, $\hat{\sigma} = L_1 \hat{T}_{11} + I_1 \hat{T}_{12} + I_2 \hat{T}_{21} + L_2 \hat{T}_{22}$, where the $\hat{T}_{ij}$ form an orthonormal basis set called ***transition operators***. Alternatively, the density operator could be expanded as:

$$\underline{\sigma} = a_1 \frac{1}{\sqrt{2}} \begin{bmatrix} 1 & 0 \\ 0 & 1 \end{bmatrix} + a_2 \frac{1}{\sqrt{2}} \begin{bmatrix} 0 & 1 \\ 1 & 0 \end{bmatrix} + a_3 \frac{1}{\sqrt{2}} \begin{bmatrix} 0 & -i \\ i & 0 \end{bmatrix} + a_4 \frac{1}{\sqrt{2}} \begin{bmatrix} 1 & 0 \\ 0 & -1 \end{bmatrix}, \tag{5.4}$$

which, in operator notation, is $\hat{\sigma} = a_1 \hat{E} + a_2 \hat{I}_x + a_3 \hat{I}_y + a_4 \hat{I}_z$. The set of operators $\{\hat{E}, \hat{I}_x, \hat{I}_y, \hat{I}_z\}$ is called the ***product operator*** basis set (although the reason for the word "product" will not become clear until we consider two-spin systems).

Fundamentals of In Vivo Magnetic Resonance: Spin Physics, Relaxation Theory, and Contrast Mechanisms, First Edition.
Daniel M. Spielman and Keshav Datta.
© 2024 John Wiley & Sons, Inc. Published 2024 by John Wiley & Sons, Inc.
Companion website: www.wiley.com/go/Spielman

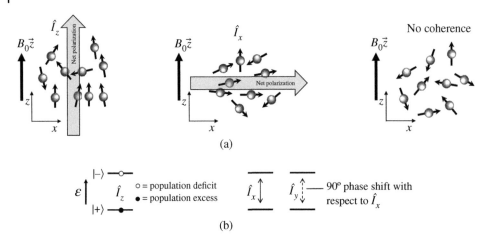

(a)

(b)

Figure 5.1 A system of single spins. (a) Polarization diagrams for systems with a net $\hat{I}_z$ coherence, a net $\hat{I}_x$ coherence, and no net coherence. (b) Corresponding energy diagrams for $\hat{I}_z, \hat{I}_x,$ and $\hat{I}_y$ coherences. Note that spin polarizations generally point along all three spatial axes; 2D representations in the $y = 0$ plane are used here only for easier visualization.

Let us first try and visualize a spin system with phase coherences. This is most readily accomplished using the "polarization" picture of spins introduced in Chapter 3, and Figure 5.1 illustrates single-spin systems with various coherences. A system in an $\hat{I}_z$ coherence is a collection of spins such that there is a net tendency for the spin polarizations to be aligned in z, but no net tendency to be aligned in x or y.

It is important to note that there are many different configurations of a given system that will yield the same value for a particular coherence. From an MR experimental perspective, these systems are effectively indistinguishable. All will yield the same results for measurements of the physical quantities we really care about, such as M_x, M_y, and M_z, without having to keep track of individual spins. This represents a huge advantage of using the density operator, in this case just a 2×2 matrix, in calculations.

We can now consider a collection of many nuclei each of which is paired with a partner. In the case of the pairing being due to chemical bonds, this would be known as J-coupled two-spin system. The eigenkets of the Hamiltonian $\hat{H}_0$ are now $\{|++\rangle, |+-\rangle, |-+\rangle, |--\rangle\}$, and, using this basis set, the density matrix can be written in the form

$$
\underline{\sigma} \Longrightarrow
\begin{array}{c}
\\
|++\rangle \\
|+-\rangle \\
|-+\rangle \\
|--\rangle
\end{array}
\begin{array}{c}
|++\rangle \; |+-\rangle \; |-+\rangle \; |--\rangle \\
\begin{bmatrix}
L & S & I & M \\
S & L & M & I \\
I & M & L & S \\
M & I & S & L
\end{bmatrix}
\end{array},
\tag{5.5}
$$

where the corresponding energy diagram is shown in Figure 5.2.

The density matrix now has four subspaces with the labels corresponding to $L =$ longitudinal populations, $S =$ transverse S coherences, $I =$ transverse I coherences, and $M =$ multiple quantum (double and zero) coherences. These multiple quantum coherences are new and will be discussed in more detail later.

As with the single-spin case, the product operators form a very convenient basis set. For a two-spin system, the Hilbert space is 4D and the corresponding Liouville space is 16 dimensional,

Figure 5.2 Energy diagram for 2-spin system.

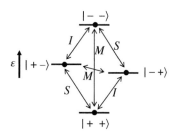

resulting in 16 product operators for the basis set. There is the identity operator $\frac{1}{2}\hat{E}$ (the factor of ½ being used for normalization in this two-spin system) and six familiar terms representing ***in-phase quantum coherences*** $\hat{I}_x, \hat{I}_y, \hat{I}_z, \hat{S}_x, \hat{S}_y$, and $\hat{S}_z$. There are also four unfamiliar terms, $2\hat{I}_x\hat{S}_z, 2\hat{I}_y\hat{S}_z, 2\hat{I}_z\hat{S}_x$, and $2\hat{I}_z\hat{S}_y$, known as ***anti-phase quantum coherences***. Finally, the terms $2\hat{I}_z\hat{S}_z$ (called ***longitudinal two-spin order***) $2\hat{I}_x\hat{S}_x, 2\hat{I}_y\hat{S}_y, 2\hat{I}_x\hat{S}_y$, and $2\hat{I}_y\hat{S}_x$ (representing linear combinations of double- and zero-quantum (DQ and ZQ) coherences) are left for later discussion. For two-spin product operators, the factor of 2 is needed for normalization.

Figure 5.3 illustrates physical pictures corresponding to several of these coherences. For example, in an $I-S$ spin system having only a $2\hat{I}_x\hat{S}_z$ anti-phase coherence, there would be no net tendency of the polarizations of the I or S spins to be aligned in x, y, or z. However, if a given I spin polarization has an $\pm x$ component, its paired S spin would have an increased probability of having a $\pm z$ component.

Each of these phase coherences represents possible physical configurations for a given system, and a system can, of course, exhibit multiple coherences simultaneously. Some coherences generate M_x or M_y magnetization that we can detect with an MRI scanner; others generate no measurable signals but are just as real!

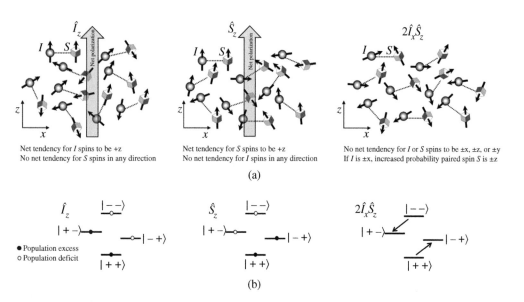

Figure 5.3 Pictorial representations of three example two-spin coherence operators as depicted by (a) polarization plots and (b) energy diagrams. For practice, try drawing the corresponding pictures for $2\hat{I}_x\hat{S}_x$ as compared to $\hat{I}_x + \hat{S}_z$.

QUESTION: How many operators are needed to span Liouville space for a 2-spin system? What about a 3-spin system?

Let us now consider the matrix representations of the product operators. For a single-spin system, $|\psi\rangle = a_+|+\rangle + a_-|+\rangle$, the matrices for the corresponding product operators are just the familiar 2×2 Pauli matrices given in Eq. (3.59). For a two-spin system, $|\psi\rangle = a_{++}|++\rangle + a_{+-}|+-\rangle + a_{-+}|-+\rangle + a_{--}|--\rangle$, the 4×4 product operators are computed using a tensor product (also called the Kronecker or direct matrix product). For example, the two-spin matrix for I_x is given by

$$I_x = \frac{1}{\sqrt{2}}\begin{bmatrix} 0 & 1 \\ 1 & 0 \end{bmatrix} \otimes \frac{1}{\sqrt{2}}\begin{bmatrix} 1 & 0 \\ 0 & 1 \end{bmatrix} = \frac{1}{2}\begin{bmatrix} 0 & 0 & 1 & 0 \\ 0 & 0 & 0 & 1 \\ 1 & 0 & 0 & 0 \\ 0 & 1 & 0 & 0 \end{bmatrix}, \tag{5.6}$$

where $\frac{1}{\sqrt{2}}\begin{bmatrix} 0 & 1 \\ 1 & 0 \end{bmatrix}$ and $\frac{1}{\sqrt{2}}\begin{bmatrix} 1 & 0 \\ 0 & 1 \end{bmatrix}$ are the normalized single-spin I_x and identity matrices in the two-spin case. Examples for S_x and $2I_xS_z$ are given as follows:

$$S_x = \frac{1}{\sqrt{2}}\begin{bmatrix} 1 & 0 \\ 0 & 1 \end{bmatrix} \otimes \frac{1}{\sqrt{2}}\begin{bmatrix} 0 & 1 \\ 1 & 0 \end{bmatrix} = \frac{1}{2}\begin{bmatrix} 0 & 1 & 0 & 0 \\ 1 & 0 & 0 & 0 \\ 0 & 0 & 0 & 1 \\ 0 & 0 & 1 & 0 \end{bmatrix}, \tag{5.7}$$

and

$$2I_xS_z = \frac{1}{\sqrt{2}}\begin{bmatrix} 0 & 1 \\ 1 & 0 \end{bmatrix} \otimes \frac{1}{\sqrt{2}}\begin{bmatrix} 1 & 0 \\ 0 & -1 \end{bmatrix} = \frac{1}{2}\begin{bmatrix} 0 & 0 & 1 & 0 \\ 0 & 0 & 0 & -1 \\ 1 & 0 & 0 & 0 \\ 0 & -1 & 0 & 0 \end{bmatrix}. \tag{5.8}$$

See Figure 5.4 for the matrix form of all 2-spin operators

$$I_x = \frac{1}{2}\begin{bmatrix} 0 & 0 & 1 & 0 \\ 0 & 0 & 0 & 1 \\ 1 & 0 & 0 & 0 \\ 0 & 1 & 0 & 0 \end{bmatrix} \quad S_x = \frac{1}{2}\begin{bmatrix} 0 & 1 & 0 & 0 \\ 1 & 0 & 0 & 0 \\ 0 & 0 & 0 & 1 \\ 0 & 0 & 1 & 0 \end{bmatrix} \quad 2I_xS_x = \frac{1}{2}\begin{bmatrix} 0 & 0 & 0 & 1 \\ 0 & 0 & 1 & 0 \\ 0 & 1 & 0 & 0 \\ 1 & 0 & 0 & 0 \end{bmatrix} \quad 2I_yS_x = \frac{i}{2}\begin{bmatrix} 0 & 0 & 0 & -1 \\ 0 & 0 & -1 & 0 \\ 0 & 1 & 0 & 0 \\ 1 & 0 & 0 & 0 \end{bmatrix}$$

$$I_y = \frac{i}{2}\begin{bmatrix} 0 & 0 & -1 & 0 \\ 0 & 0 & 0 & -1 \\ 1 & 0 & 0 & 0 \\ 0 & 1 & 0 & 0 \end{bmatrix} \quad S_y = \frac{i}{2}\begin{bmatrix} 0 & -1 & 0 & 0 \\ 1 & 0 & 0 & 0 \\ 0 & 0 & 0 & -1 \\ 0 & 0 & 1 & 0 \end{bmatrix} \quad 2I_xS_y = \frac{i}{2}\begin{bmatrix} 0 & 0 & 0 & -1 \\ 0 & 0 & 1 & 0 \\ 0 & -1 & 0 & 0 \\ 1 & 0 & 0 & 0 \end{bmatrix} \quad 2I_yS_y = \frac{1}{2}\begin{bmatrix} 0 & 0 & 0 & -1 \\ 0 & 0 & 1 & 0 \\ 0 & 1 & 0 & 0 \\ -1 & 0 & 0 & 0 \end{bmatrix}$$

$$I_z = \frac{1}{2}\begin{bmatrix} 1 & 0 & 0 & 0 \\ 0 & 1 & 0 & 0 \\ 0 & 0 & -1 & 0 \\ 0 & 0 & 0 & -1 \end{bmatrix} \quad S_z = \frac{1}{2}\begin{bmatrix} 1 & 0 & 0 & 0 \\ 0 & -1 & 0 & 0 \\ 0 & 0 & 1 & 0 \\ 0 & 0 & 0 & -1 \end{bmatrix} \quad 2I_xS_z = \frac{1}{2}\begin{bmatrix} 0 & 0 & 1 & 0 \\ 0 & 0 & 0 & -1 \\ 1 & 0 & 0 & 0 \\ 0 & -1 & 0 & 0 \end{bmatrix} \quad 2I_yS_z = \frac{i}{2}\begin{bmatrix} 0 & 0 & -1 & 0 \\ 0 & 0 & 0 & 1 \\ 1 & 0 & 0 & 0 \\ 0 & -1 & 0 & 0 \end{bmatrix}$$

$$\frac{1}{2}E = \frac{1}{2}\begin{bmatrix} 1 & 0 & 0 & 0 \\ 0 & 1 & 0 & 0 \\ 0 & 0 & 1 & 0 \\ 0 & 0 & 0 & 1 \end{bmatrix} \quad 2I_zS_x = \frac{1}{2}\begin{bmatrix} 0 & 1 & 0 & 0 \\ 1 & 0 & 0 & 0 \\ 0 & 0 & 0 & -1 \\ 0 & 0 & -1 & 0 \end{bmatrix} \quad 2I_zS_y = \frac{i}{2}\begin{bmatrix} 0 & -1 & 0 & 0 \\ 1 & 0 & 0 & 0 \\ 0 & 0 & 0 & 1 \\ 0 & 0 & -1 & 0 \end{bmatrix} \quad 2I_zS_z = \frac{1}{2}\begin{bmatrix} 1 & 0 & 0 & 0 \\ 0 & -1 & 0 & 0 \\ 0 & 0 & -1 & 0 \\ 0 & 0 & 0 & 1 \end{bmatrix}$$

Figure 5.4 Matrix representations of all 16 normalized 2-spin product operators as expressed in the $\{|++\rangle, |+-\rangle, |-+\rangle, |--\rangle\}$ basis.

5.1.2 Density Matrix Calculations

We will now find the x, y, and z magnetizations corresponding to the simple MR experiment shown in Figure 5.5. These computations will initially be performed using matrix calculations and then revisited using superoperator notation, which will prove easier for many of the systems we deal with *in vivo*. This is a 2-spin system and, for simplicity, let us solve for the homonuclear case $\gamma_I = \gamma_S = \gamma$. For $t < 0$ and using the Boltzmann distribution

Figure 5.5 A pulse-and-acquire MR experiment.

$$\underline{\sigma} = \frac{\hbar \gamma B_0}{4kT} \left(\underline{I}_z + \underline{S}_z \right), \tag{5.9}$$

where the $\underline{E}$ term has been ignored (why?). For the next time interval, $0 \leq t < t_1$, we need to first find the Hamiltonian, which during the Rf pulse is well approximated by

$$\underline{H} \approx -\omega_1 \left(\underline{I}_x + \underline{S}_x \right). \tag{5.10}$$

Next, the time evolution of the density operator is

$$\underline{\sigma}(t_1) = e^{-i\underline{H}t_1} \underline{\sigma}(0) e^{i\underline{H}t_1}, \tag{5.11}$$

for $\omega_1 t_1 = \pi/2$. After some algebra, we arrive at

$$\underline{\sigma}(t_1) = e^{-i\underline{H}t_1} \underline{\sigma}(0) e^{i\underline{H}t_1} = \frac{\hbar \gamma B_0}{4kT} \frac{1}{2} \begin{bmatrix} 0 & -i & -i & 0 \\ i & 0 & 0 & -i \\ i & 0 & 0 & -i \\ 0 & i & i & 0 \end{bmatrix} = \frac{\hbar \gamma B_0}{4kT} \left(\underline{I}_y + \underline{S}_y \right). \tag{5.12}$$

As expected, a 90° Rf pulse along the x-axis generates coherences corresponding to M_y magnetization.

For $t > t_1$, there is no Rf pulse, and the spin system evolves under the Hamiltonian that only includes Zeeman and J coupling terms. Namely,

$$\underline{H} = -\Omega_I \underline{I}_z - \Omega_S \underline{S}_z + 2\pi J \left(\underline{I}_x \underline{S}_x + \underline{I}_y \underline{S}_y + \underline{I}_z \underline{S}_z \right). \tag{5.13}$$

After even more algebra, the Hamiltonian is

$$\underline{H} = \frac{1}{2} \begin{bmatrix} -\Omega_I - \Omega_S + \pi J & 0 & 0 & 0 \\ 0 & -\Omega_I + \Omega_S - \pi J & 2\pi J & 0 \\ 0 & 2\pi J & \Omega_I - \Omega_S - \pi J & 0 \\ 0 & 0 & 0 & \Omega_I + \Omega_S + \pi J \end{bmatrix}. \tag{5.14}$$

Let's now consider the special case known as **weak coupling** where the difference in chemical shift between the I and S spins is much larger than the J coupling constant, $|\Omega_I - \Omega_S| \gg 2\pi J$. Again, this is an example of a secular approximation, which yields

$$\underline{H} = \frac{1}{2} \begin{bmatrix} -\Omega_I - \Omega_S + \pi J & 0 & 0 & 0 \\ 0 & -\Omega_I + \Omega_S - \pi J & 0 & 0 \\ 0 & 0 & \Omega_I - \Omega_S - \pi J & 0 \\ 0 & 0 & 0 & \Omega_I + \Omega_S + \pi J \end{bmatrix}. \tag{5.15}$$

Written as a linear combination of product operators, $\underline{H} = -\Omega_I \underline{I}_z - \Omega_S \underline{S}_z + 2\pi J \underline{I}_z \underline{S}_z$. Compare this matrix with the energy diagram shown in Figure 5.3. Remember, when a matrix is expressed in diagonal form, the diagonal elements are the corresponding eigenvalues, and the eigenvalues of $\hat{H}$ correspond to the possible energies of the system.

We can now compute $\underline{\sigma}(t) = e^{-i\underline{H}t} \underline{\sigma}(t_1) e^{i\underline{H}t}$, which is easy given $\underline{H}$ is diagonal.

A bit more algebra yields

$$\underline{\sigma}(t) = e^{-i\underline{H}t}\underline{\sigma}(t_1)e^{i\underline{H}t} = \frac{\hbar\gamma B_0}{8kT}\begin{bmatrix} 0 & e^{i(\Omega_S-\pi Jt)t} & e^{i(\Omega_I-\pi Jt)t} & 0 \\ e^{i(\Omega_S-\pi Jt)t} & 0 & 0 & e^{i(\Omega_I+\pi Jt)t} \\ e^{i(\Omega_I-\pi Jt)t} & 0 & 0 & e^{i(\Omega_S+\pi Jt)t} \\ 0 & e^{i(\Omega_I+\pi Jt)t} & e^{i(\Omega_S+\pi Jt)t} & 0 \end{bmatrix}. \tag{5.16}$$

From this, we can directly compute the I-spin transverse magnetization (the S-spin terms are similar). Namely, as shown in Figure 5.6,

$$\overline{\langle \hat{I}_x \rangle} = \mathrm{Tr}\left(\underline{\sigma I_x}\right) = \frac{\hbar\gamma B_0}{4kT}\sin \Omega_I t \cos \pi Jt, \tag{5.17}$$

and

$$\overline{\langle \hat{I}_y \rangle} = \mathrm{Tr}\left(\underline{\sigma I_y}\right) = \frac{\hbar\gamma B_0}{4kT}\cos \Omega_I t \cos \pi Jt. \tag{5.18}$$

From Figure 5.6, it is apparent something strange is happening to our x and y magnetization. The detected M_{xy} magnetization seems to appear and disappear due to the $\cos\pi Jt$ modulation. Where does the magnetization go (hint: it does **not** go into an in-phase S coherence)? Let us compute the antiphase coherences over this time interval.

$$\overline{\langle 2\hat{I}_x\hat{S}_z \rangle} = 2\mathrm{Tr}\left(\underline{\sigma I_x S_z}\right) = \frac{\hbar\gamma B_0}{4kT}\cos \Omega_I t \sin \pi Jt, \tag{5.19}$$

and

$$\overline{\langle 2\hat{I}_y\hat{S}_z \rangle} = 2\mathrm{Tr}\left(\underline{\sigma I_y S_z}\right) = \frac{\hbar\gamma B_0}{4kT}\sin \Omega_I t \sin \pi Jt. \tag{5.20}$$

These are the missing $\sin\pi Jt$ modulated components!

Remember, a typical MR spectrometer or scanner has receiver coils that measure voltages induced by time-varying magnetic fields in the x and y directions. There is, for example, no way to directly detect z magnetization (other than perhaps via the use of some sort of SQUID device measurement, which would be particularly challenging in the presence of the very large $B_0\vec{z}$ field). How do we know a nonzero z magnetization for our spin system exists? We convert the z coherence into an x or y coherence via an Rf pulse, which can then be detected by the receiver coils.

Does antiphase magnetization exist? Absolutely, but it is not directly detectable by any Rf coil. To measure antiphase magnetization (or, in fact any of the other unfamiliar product operator coherences), these coherences will have to be converted into measurable x and y components. This happens via the combination of Rf pulses and the passage of time. Hence, in the simple example discussed above, the magnetization, or more formally the spin phase coherences, does not disappear, but rather oscillates between detectable and undetectable components! Both the in-phase and anti-phase coherences are real, but only the in-phase terms generate voltages in our receiver coils.

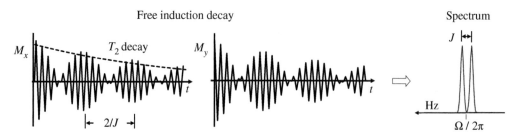

Figure 5.6 Detected x and y magnetization components and corresponding spectrum for the I spin from a homonuclear I – S spin system following the application of a nonselective 90° Rf pulse along the x-axis. A transverse magnetization decay rate of T_2 is assumed.

Let's finish this section with a brief discussion of the concepts of equivalent spins and strong coupling. The complete Hamiltonian for $t > t_1$ in this experiment is given by Eq. (5.14). Simplification occurs for two important cases. The first is the weak coupling case, $|\Omega_I - \Omega_S| \gg 2\pi J$, described above. The second is when the I and S spins have the same chemical shift, namely, $\Omega_I = \Omega_S$. In this case, the spins are said to be equivalent, and the effects of J coupling will disappear entirely, resulting in a spectrum with only a single peak. The weak coupling assumption is always valid for heteronuclear I–S spin systems, e.g. coupled proton and carbon nuclei. However, in the homonuclear case, we are sometimes forced to deal with the more complicated **strong coupling** condition. Figure 5.7 illustrates the observed spectral changes as a function

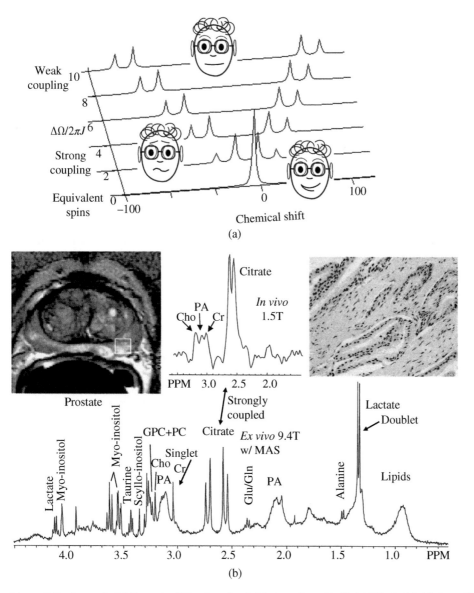

Figure 5.7 Example 1.5 T spectra (Why does the field strength matter?). (a) Effect of field strength on J-coupled spectra. (b) *In vivo* and *ex vivo* spectra from a human prostate (MAS, Magic Angle Spinning, a technique widely used in solid-state NMR to reduce the effects of residual dipolar coupling). Source: (b) Adapted with permission from Swanson et al. (2003).

of the strength of coupling from a *J*-coupled spin pair along with *in-vivo* and *ex-vivo* tissue examples.

5.2 POF for Single-Spin Coherence Space

Let's now develop a graphical and more intuitive analysis known as the Product Operator Formalism (POF) (Sorensen et al. 1983). Consider first a single-spin system. Starting with Eqs. (5.3) and (5.4), the spin density operator can be expressed as "vector" in a 4D Liouville space also called "coherence space" (see Figure 5.8).

We have derived that, given the density operator $\hat{\sigma}$, physical quantities can be computed using $\overline{\langle \hat{A} \rangle} = \text{Tr}(\hat{\sigma}\hat{A})$, and the time evolution of $\hat{\sigma}$ is governed by the Liouville-von Neuman equation. Given a Hamiltonian $\hat{H}$, independent of time, yields, in superoperator notation,

$$\hat{\sigma}(t) = e^{-i\hat{\hat{H}}t}\hat{\sigma}(0). \tag{5.21}$$

For $\hat{H}$ piecewise constant in time, where we previously wrote expressions such as $\hat{\sigma}(0) \xrightarrow{\hat{H}_1}$ $\xrightarrow{\hat{H}_2} \xrightarrow{\hat{H}_3} \hat{\sigma}(t_1 + t_2 + t_3)$ for the time evolution of the density operator, we can equivalently write

$$\hat{\sigma}(0) \xrightarrow{\hat{H}_1(\omega_1 t_1)} \xrightarrow{\hat{H}_2(\omega_2 t_2)} \xrightarrow{\hat{H}_3(\omega_3 t_3)} \hat{\sigma}(t_1 + t_2 + t_3) \dots, \tag{5.22}$$

with the interpretation that during the *i*th time interval, the spin density operator rotates around $\hat{H}_i$ through an angle $\omega_i t_i$ (note, technically the minus sign in the Liouville-von Neuman equation says the rotation is about $-\hat{H}$). Hence, borrowing terminology from real estate, MR is all about rotations, rotations, rotations!

In Eq. (3.50) we showed that, for cyclically commuting operators, the equation $e^{i\theta\hat{A}}\hat{B} = \hat{B}\cos\theta + i[\hat{A}, \hat{B}]\sin\theta$ describes a rotation about the $\hat{A}$ axis in $\hat{B} \times [\hat{A}, \hat{B}]$ plane. Here we note that the operators $\hat{I}_x, \hat{I}_y,$ and $\hat{I}_z$ cyclically commute,

$$[\hat{I}_x, \hat{I}_y] = i\hat{I}_z, [\hat{I}_y, \hat{I}_z] = i\hat{I}_x, \text{ and } [\hat{I}_z, \hat{I}_x] = i\hat{I}_y. \tag{5.23}$$

Then, ignoring relaxation, Eq. (5.23) implies that all possible rotations of $\hat{\sigma}$ during a typical MR experiment can be fully described by the following equation:

$$\hat{I}_p \xrightarrow{\hat{I}_q(\theta)} \begin{cases} \hat{I}_p & \text{if } [\hat{I}_q, \hat{I}_p] = 0 \\ \hat{I}_p \cos\theta + i[\hat{I}_q, \hat{I}_p]\sin\theta & \text{if } [\hat{I}_q, \hat{I}_p] \neq 0 \end{cases}, \tag{5.24}$$

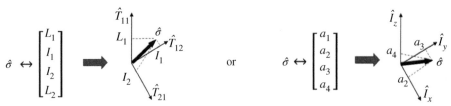

| "vector" expressed in transition operator basis | A 3D subspace of coherence space | | "vector" expressed in product operator basis | A 3D subspace of coherence space |

Figure 5.8 The spin density operator as a vector in coherence space is described by two different sets of basis operators. For the analysis that follows, the product operators will be the most convenient basis.

for $p, q \in \{x, y, z\}$ with the minus sign coming from the spin density operator rotation being about the $-\hat{H}$ axis.

Let us consider a 90° excitation about the x-axis. The Hamiltonian is $\hat{H} \approx -\omega_1 \hat{I}_x$ with the rotation angle $\omega_1 t = \pi/2$. From which, Eq. (5.24) yields the following time evolutions for various starting spin coherences,

$$\hat{I}_z \xrightarrow{\hat{I}_x(\pi/2)} \hat{I}_y, \quad \hat{I}_x \xrightarrow{\hat{I}_x(\pi/2)} \hat{I}_x, \quad \text{and} \quad \hat{I}_y \xrightarrow{\hat{I}_x(\pi/2)} -\hat{I}_z. \tag{5.25}$$

Similarly, a 90° excitation about the y-axis would produce the following:

$$\hat{I}_z \xrightarrow{\hat{I}_y(\pi/2)} -\hat{I}_x, \quad \hat{I}_x \xrightarrow{\hat{I}_y(\pi/2)} \hat{I}_z, \quad \text{and} \quad \hat{I}_y \xrightarrow{\hat{I}_y(\pi/2)} \hat{I}_y. \tag{5.26}$$

As a second example, consider the evolution of the spin density operator during free precession where, in the absence of any Rf, the Hamiltonian is just the chemical shift Zeeman term $\hat{H} = -\Omega \hat{I}_z$. Then,

$$\hat{I}_x \xrightarrow{\hat{I}_z(\Omega t)} \hat{I}_x \cos \Omega t - \hat{I}_y \sin \Omega t$$

$$\hat{I}_y \xrightarrow{\hat{I}_z(\Omega t)} \hat{I}_y \cos \Omega t + \hat{I}_x \sin \Omega t$$

$$\hat{I}_z \xrightarrow{\hat{I}_z(\Omega t)} \hat{I}_z. \tag{5.27}$$

Alternatively, one could just use the rotation diagrams shown in Figure 5.9.

A slightly more complicated example is the spin echo sequence shown in Figure 5.10. Here there are two Rf pulses, the first about the x-axis and the second along y. Starting with an initial thermal equilibrium Hamiltonian proportional to $\hat{I}_z$, we now need to compute the following rotations, $\hat{\sigma}(0) \xrightarrow{\hat{H}_1} \hat{\sigma}(0^+) \xrightarrow{\hat{H}_2} \hat{\sigma}(\tau) \xrightarrow{\hat{H}_3} \hat{\sigma}(\tau^+) \xrightarrow{\hat{H}_4} \hat{\sigma}(2\tau)$, in order to find the coherences that exist at the echo time $t = 2\tau$. After the two 90° pulses, the coherences are

$$\hat{I}_z \xrightarrow{\hat{I}_x(\pi/2)} \hat{I}_y \xrightarrow{\hat{I}_z(\Omega t)} \hat{I}_y \cos \Omega \tau + \hat{I}_x \sin \Omega \tau \xrightarrow{\hat{I}_y(\pi)} \hat{I}_y \cos \Omega \tau - \hat{I}_x \sin \Omega \tau. \tag{5.28}$$

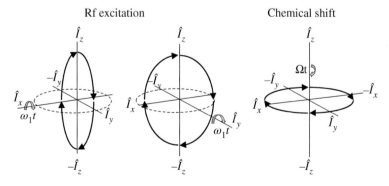

Figure 5.9 Coherence space rotations for Rf excitation and chemical shift of a single-spin system. Note that all rotations are left-handed.

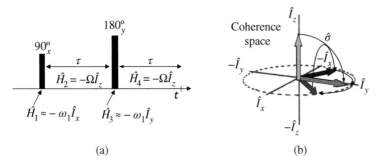

(a)   (b)

Figure 5.10 A spin-echo pulse sequence with (a) corresponding Hamiltonians and (b) evolution of the spin density operator $\hat{\sigma}(t)$.

Continuing with the evolution under $\hat{H}_4$ yields the expected single $\hat{I}_y$ coherence (see Figure 5.10b).

$$\xrightarrow{\hat{I}_z(\Omega t)}(\hat{I}_y\cos\Omega\tau + \hat{I}_x\sin\Omega\tau)\cos\Omega\tau - (\hat{I}_x\cos\Omega\tau - \hat{I}_y\sin\Omega\tau)\sin\Omega\tau$$

$$= \hat{I}_y\cos^2\Omega\tau + \hat{I}_y\sin^2\Omega\tau = \hat{I}_y. \tag{5.29}$$

QUESTION: Don't we really care about the physical quantities $\overline{\langle\hat{I}_x\rangle}$, $\overline{\langle\hat{I}_y\rangle}$, and $\overline{\langle\hat{I}_z\rangle}$ rather than the operators themselves?

5.3 POF for Two-Spin Coherence Space

Now on to the more interesting case of a two-spin system. As before, the product operators form an orthonormal basis in coherence space, only now coherence space is 16-dimensional, *i.e.*, $\hat{\sigma} = \sum_j a_j \hat{C}_j$ where $\hat{C}_j \in \left\{\frac{1}{2}\hat{E}, \hat{I}_x, \hat{S}_x, \hat{I}_y, \hat{S}_y, \dots, 2\hat{I}_z\hat{S}_z\right\}$. Remembering that for any physical quantity C, $\overline{\langle\hat{C}\rangle} = \mathrm{Tr}(\hat{\sigma}\hat{C})$. The density operator, again, simply rotates around in this higher dimensional space where the coefficients are the expected values of the corresponding operators.

$$\hat{\sigma} = \sum_j \overline{\langle\hat{C}_j\rangle}\hat{C}_j \Longleftrightarrow \begin{bmatrix} \frac{1}{2}\overline{\langle\hat{E}\rangle} \\ \overline{\langle\hat{I}_x\rangle} \\ \overline{\langle\hat{S}_x\rangle} \\ \vdots \\ \overline{\langle 2\hat{I}_z\hat{S}_z\rangle} \end{bmatrix} \tag{5.30}$$

Given that coherence space is 16-dimensional for a two-spin system, life can get rather complicated (not to mention for higher-order spin systems). Fortunately, we will only need to deal with sequential rotations in 3-dimensional subspaces. In particular, for a given rotation about an axis, for example $\hat{A}$, we need only be concerned with the 3D subspace defined by the axes $\hat{A}$, $\hat{B}$, and $[\hat{A}, \hat{B}]$.

Again, start with our favorite relation for cyclically commuting operators, now given in superoperator notation, $\hat{\hat{A}}\hat{\hat{A}}\hat{B} = \hat{B}$. Examination of Figure 5.11 listing the two-spin product operator commutators reveals that *all* two-spin product operators either commute or commute cyclically! Hence, ignoring relaxation, all possible rotations of $\hat{\sigma}$ can be fully described by the following equation:

$$\hat{C}_p \xrightarrow{\ \hat{C}_q(\theta)\ } \begin{cases} \hat{C}_p & \text{if } [\hat{C}_q, \hat{C}_p] = 0 \\ \hat{C}_p \cos\theta + i[\hat{C}_q, \hat{C}_p]\sin\theta & \text{if } [\hat{C}_q, \hat{C}_p] \neq 0 \end{cases} \tag{5.31}$$

where $\hat{C}_p$ and $\hat{C}_q$ are any pair of product operators. Hence, for any rotation $e^{i\theta\hat{A}}\hat{B}$, we need only work in the 3D subspace defined by the axes $\hat{A}$, $\hat{B}$, and $[\hat{A}, \hat{B}]$.

Now let us look at the three basic rotations for Rf excitation, chemical shift, and the new rotation for J coupling. Start by noting that, for Hamiltonians containing multiple terms, the rotations can be performed independently (and in any order) so long as the operators involved commute with each other. Namely, if $[\hat{A}, \hat{B}] = 0$, then $e^{\hat{A}+\hat{B}} = e^{\hat{A}}e^{\hat{B}} = e^{\hat{B}}e^{\hat{A}}$ (see Exercise E.3.6).

For Rf excitations, the rotation axes are essentially the same as before for both the I and S spins. In particular, we note that $[\hat{I}_x, \hat{S}_x] = 0$, hence for a rotation due to, for example, a nonselective Rf pulse along the x-axis corresponding to the Hamiltonian $\hat{H} = -\omega_1^I \hat{I}_x - \omega_1^S \hat{S}_x$, the rotations about $\hat{I}_x$ and $\hat{S}_x$, can be performed as either first a rotation about $\hat{I}_x$ followed by a rotation about $\hat{S}_x$ or in the opposite order. Here are some examples for Rf excitation of a two-spin system.

$$\hat{I}_z \xrightarrow{\ \hat{S}_x(\pi/2)\ } \hat{I}_z \xrightarrow{\ \hat{I}_x(\pi/2)\ } \hat{I}_y$$

$$2\hat{I}_z\hat{S}_z \xrightarrow{\ \hat{S}_x(\pi/2)\ } 2\hat{I}_z\hat{S}_y$$

$$2\hat{I}_x\hat{S}_z \xrightarrow{\ \hat{S}_x(\pi/2)\ } 2\hat{I}_x\hat{S}_y \xrightarrow{\ \hat{I}_y(\pi/2)\ } 2\hat{I}_z\hat{S}_y. \tag{5.32}$$

	$\hat{I}_x$	$\hat{I}_y$	$\hat{I}_z$	$\hat{S}_x$	$\hat{S}_y$	$\hat{S}_z$	$2\hat{I}_x\hat{S}_x$	$2\hat{I}_x\hat{S}_y$	$2\hat{I}_x\hat{S}_z$	$2\hat{I}_y\hat{S}_x$	$2\hat{I}_y\hat{S}_y$	$2\hat{I}_y\hat{S}_z$	$2\hat{I}_z\hat{S}_x$	$2\hat{I}_z\hat{S}_y$	$2\hat{I}_z\hat{S}_z$
$\hat{I}_x$	0	$i\hat{I}_z$	$-i\hat{I}_y$	0	0	0	0	0	0	$i2\hat{I}_z\hat{S}_x$	$i2\hat{I}_z\hat{S}_y$	$i2\hat{I}_z\hat{S}_z$	$-i2\hat{I}_y\hat{S}_x$	$-i2\hat{I}_y\hat{S}_y$	$-i2\hat{I}_y\hat{S}_z$
$\hat{I}_y$	$-i\hat{I}_z$	0	$i\hat{I}_x$	0	0	0	$-i2\hat{I}_z\hat{S}_x$	$-i2\hat{I}_z\hat{S}_y$	$-i2\hat{I}_z\hat{S}_z$	0	0	0	$i2\hat{I}_x\hat{S}_x$	$i2\hat{I}_x\hat{S}_y$	$i2\hat{I}_x\hat{S}_z$
$\hat{I}_z$	$i\hat{I}_y$	$-i\hat{I}_x$	0	0	0	0	$i2\hat{I}_y\hat{S}_x$	$i2\hat{I}_y\hat{S}_y$	$i2\hat{I}_y\hat{S}_z$	$-i2\hat{I}_x\hat{S}_x$	$-i2\hat{I}_x\hat{S}_y$	$-i2\hat{I}_x\hat{S}_z$	0	0	0
$\hat{S}_x$	0	0	0	0	$i\hat{S}_z$	$-i\hat{S}_y$	0	$i2\hat{I}_x\hat{S}_z$	$-i2\hat{I}_x\hat{S}_y$	0	$i2\hat{I}_y\hat{S}_z$	$-i2\hat{I}_y\hat{S}_y$	0	$i2\hat{I}_z\hat{S}_z$	$-i2\hat{I}_z\hat{S}_y$
$\hat{S}_y$	0	0	0	$-i\hat{S}_z$	0	$i\hat{S}_x$	$-i2\hat{I}_x\hat{S}_z$	0	$i2\hat{I}_x\hat{S}_x$	$-i2\hat{I}_y\hat{S}_z$	0	$i2\hat{I}_y\hat{S}_x$	$-i2\hat{I}_z\hat{S}_z$	0	$i2\hat{I}_z\hat{S}_x$
$\hat{S}_z$	0	0	0	$i\hat{S}_y$	$-i\hat{S}_x$	0	$i2\hat{I}_x\hat{S}_y$	$-i2\hat{I}_x\hat{S}_x$	0	$i2\hat{I}_y\hat{S}_y$	$-i2\hat{I}_y\hat{S}_x$	0	$i2\hat{I}_z\hat{S}_y$	$-i2\hat{I}_z\hat{S}_x$	0
$2\hat{I}_x\hat{S}_x$	0	$i2\hat{I}_z\hat{S}_x$	$-i2\hat{I}_y\hat{S}_x$	0	$i2\hat{I}_x\hat{S}_z$	$-i2\hat{I}_x\hat{S}_y$	0	$i\hat{S}_z$	$-i\hat{S}_y$	$i\hat{I}_z$	0	0	$-i\hat{I}_y$	0	0
$2\hat{I}_x\hat{S}_y$	0	$i2\hat{I}_z\hat{S}_y$	$-i2\hat{I}_y\hat{S}_y$	$-i2\hat{I}_x\hat{S}_z$	0	$i2\hat{I}_x\hat{S}_x$	$-i\hat{S}_z$	0	$i\hat{S}_x$	0	$i\hat{I}_z$	0	0	$-i\hat{I}_y$	0
$2\hat{I}_x\hat{S}_z$	0	$i2\hat{I}_z\hat{S}_z$	$-i2\hat{I}_y\hat{S}_z$	$i2\hat{I}_x\hat{S}_y$	$-i2\hat{I}_x\hat{S}_x$	0	$i\hat{S}_y$	$-i\hat{S}_x$	0	0	0	$i\hat{I}_z$	0	0	$-i\hat{I}_y$
$2\hat{I}_y\hat{S}_x$	$-i2\hat{I}_z\hat{S}_x$	0	$i2\hat{I}_x\hat{S}_x$	0	$i2\hat{I}_y\hat{S}_z$	$-i2\hat{I}_y\hat{S}_y$	$-i\hat{I}_z$	0	0	0	$i\hat{S}_z$	$-i\hat{S}_y$	$i\hat{I}_x$	0	0
$2\hat{I}_y\hat{S}_y$	$-i2\hat{I}_z\hat{S}_y$	0	$i2\hat{I}_x\hat{S}_y$	$-i2\hat{I}_y\hat{S}_z$	0	$i2\hat{I}_y\hat{S}_x$	0	$-i\hat{I}_z$	0	$-i\hat{S}_z$	0	$i\hat{S}_x$	0	$i\hat{I}_x$	0
$2\hat{I}_y\hat{S}_z$	$-i2\hat{I}_z\hat{S}_z$	0	$i2\hat{I}_x\hat{S}_z$	$i2\hat{I}_y\hat{S}_y$	$-i2\hat{I}_y\hat{S}_x$	0	0	0	$-i\hat{I}_z$	$i\hat{S}_y$	$-i\hat{S}_x$	0	0	0	$i\hat{I}_x$
$2\hat{I}_z\hat{S}_x$	$i2\hat{I}_y\hat{S}_x$	$-i2\hat{I}_x\hat{S}_x$	0	0	$i2\hat{I}_z\hat{S}_z$	$-i2\hat{I}_z\hat{S}_y$	$i\hat{I}_y$	0	0	$-i\hat{I}_x$	0	0	0	$i\hat{S}_z$	$-i\hat{S}_y$
$2\hat{I}_z\hat{S}_y$	$i2\hat{I}_y\hat{S}_y$	$-i2\hat{I}_x\hat{S}_y$	0	$-i2\hat{I}_z\hat{S}_z$	0	$i2\hat{I}_z\hat{S}_x$	0	$i\hat{I}_y$	0	0	$-i\hat{I}_x$	0	$-i\hat{S}_z$	0	$i\hat{S}_x$
$2\hat{I}_z\hat{S}_z$	$i2\hat{I}_y\hat{S}_z$	$-i2\hat{I}_x\hat{S}_z$	0	$i2\hat{I}_z\hat{S}_y$	$-i2\hat{I}_z\hat{S}_x$	0	0	0	$i\hat{I}_y$	0	0	$-i\hat{I}_x$	$i\hat{S}_y$	$-i\hat{S}_x$	0

Figure 5.11 Commutators for all 16 two-spin product operators. The table entries are the commutation between the row and column operators, i.e. $[C_i, C_j]$.

When there is no Rf pulse, the Hamiltonian for a two-spin system has the form $\hat{H} = -\Omega_I \hat{I}_z - \Omega_S \hat{S}_z + 2\pi J \hat{I}_z \hat{S}_z$. This equation corresponds to the weak coupling case, where the POF is the most useful. This is due to the fact that $[\hat{I}_z, \hat{S}_z] = 0$, $[\hat{I}_z, 2\hat{I}_z\hat{S}_z] = 0$, and $[\hat{S}_z, 2\hat{I}_z\hat{S}_z] = 0$, hence, these rotations can be performed in any order. For evolution under chemical shift, the $\{\hat{I}_x, \hat{I}_y, \hat{I}_z\}$ and $\{\hat{S}_x, \hat{S}_y, \hat{S}_z\}$ coherences behave pretty much as expected. Product operators, such as $2\hat{I}_x\hat{S}_x$, behave similarly. $\hat{I}_z$ rotations affect the I spin, and $\hat{S}_z$ rotations affect the S spin. For example, under $\hat{H} = -\Omega_I \hat{I}_z$, $2\hat{I}_x\hat{S}_p$, rotation occurs in the $2\hat{I}_x\hat{S}_p \times 2\hat{I}_y\hat{S}_p$ plane where p is x, y, or z. Some examples are given below.

$$\hat{I}_x \xrightarrow{\hat{I}_z(\Omega_I t)} \hat{I}_x \cos \Omega_I t - \hat{I}_y \sin \Omega_I t$$

$$2\hat{I}_x\hat{S}_x \xrightarrow{\hat{I}_z(\Omega_I t)} 2\hat{I}_x\hat{S}_x \cos \Omega_I t - 2\hat{I}_y\hat{S}_x \sin \Omega_I t. \tag{5.33}$$

J coupling involves rotations of the spin density operator around the $2\hat{I}_z\hat{S}_z$ axis in coherence space. Transverse magnetization, for example, oscillates between in-phase and antiphase coherences. Namely,

$$\hat{I}_x \xrightarrow{2\hat{I}_z\hat{S}_z(\pi J t)} \hat{I}_x \cos \pi J t + 2\hat{I}_y\hat{S}_z \sin \pi J t$$

$$\hat{I}_y \xrightarrow{2\hat{I}_z\hat{S}_z(\pi J t)} \hat{I}_y \cos \pi J t - 2\hat{I}_x\hat{S}_z \sin \pi J t. \tag{5.34}$$

 NOTA BENE: For experts only: show that rotating around the tilted axis $2\left(\hat{I}_x\hat{I}_y - \hat{I}_y\hat{I}_x\right)$ is advantageous for analyzing strong coupling effects with product operators.

Additional examples for Rf excitation, chemical shift, and J coupling rotations are shown in Figure 5.12, and using these diagrams is generally much easier than direct matrix computations. There is one caveat, unlike Rf and chemical shift, J coupling involves a right-handed rotation.

Consider a simple MR experiment consisting of a nonselective 90° pulse along x followed by a readout. Namely, $\hat{\sigma}(0) \xrightarrow{\hat{I}_x(\pi/2)} \xrightarrow{\hat{S}_x(\pi/2)} \xrightarrow{\hat{I}_z(\Omega_I t)} \xrightarrow{\hat{S}_z(\Omega_S t)} \xrightarrow{2\hat{I}_z\hat{S}_z(\pi J t/2)} \hat{\sigma}(t)$. Starting with the I spin and ignoring commuting operators, the excitation pulse generates $\hat{I}_z \xrightarrow{\hat{I}_x(\pi/2)} \hat{I}_y$. Evolving over chemical shift produces $\hat{I}_y \xrightarrow{\hat{I}_z(\Omega_I t)} \hat{I}_y \cos \Omega_I t + \hat{I}_x \sin \Omega_I t$, and finally, the J coupling evolution results in a final set of coherences $\xrightarrow{2\hat{I}_z\hat{S}_z(\pi J t/2)} (\hat{I}_y \cos \Omega_I t + \hat{I}_x \sin \Omega_I t) \cos \pi J t + (2\hat{I}_x\hat{S}_z \sin \Omega_I t - 2\hat{I}_y\hat{S}_z \cos \Omega_I t) \sin \pi J t$. The coefficient of the $\cos\pi J t$ term is referred to as "in-phase" or "observable" magnetization with the $\sin\pi J t$ term coefficient called "anti-phase" or "unobservable". There are analogous equations for the S spin. Keeping track of the constants needed to calculate the final magnetization yields the spectrum shown in Figure 5.13.

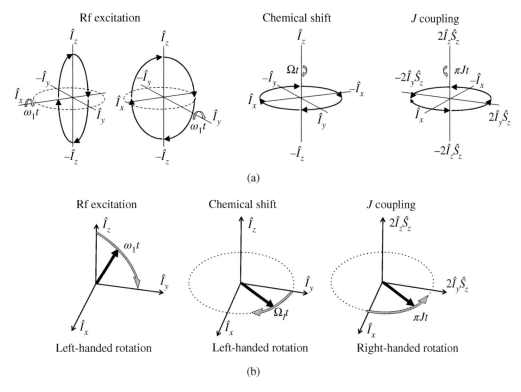

Figure 5.12 POF rotation diagrams. (a) Examples of 3D-subspaces for Rf, chemical shift, and J coupling rotation in coherence space, and (b) sign conventions for the rotations.

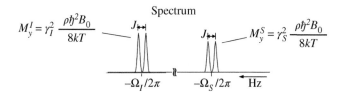

Figure 5.13 Spectrum resulting from a nonselective 90° Rf pulse applied to a two-spin system starting in thermal equilibrium.

Figure 5.14 A spin-echo pulse sequence with a selective 180°.

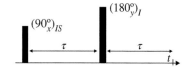

Let's now consider a spin echo pulse sequence where the first 90° is nonselective and the 180° refocusing pulse is selective only on the I spin (see Figure 5.14). Continuing the coherences for the I spin prior to the 180 are: $(\hat{I}_y \cos \Omega_I \tau + \hat{I}_x \sin \Omega_I \tau) \cos \pi J \tau + (2\hat{I}_y \hat{S}_z \sin \Omega_I \tau - 2\hat{I}_x \hat{S}_z \cos \Omega_I \tau) \sin \pi J \tau$. Applying the 180° Rf pulse selection on the I spin yields

$$\xrightarrow{\hat{I}_y(\pi)} (\hat{I}_y \cos \Omega_I \tau - \hat{I}_x \sin \Omega_I \tau) \cos \pi J \tau + (2\hat{I}_y \hat{S}_z \sin \Omega_I \tau + 2\hat{I}_x \hat{S}_z \cos \Omega_I \tau) \sin \pi J \tau. \qquad (5.35)$$

Computing the rotations for chemical shift and J coupling over the second time interval τ results (after some algebra that we will clean up in the next section) in the following coherences at the echo time $t = 2\tau$,

$$\xrightarrow{\hat{I}_z(\Omega_I t)} \xrightarrow{2\hat{I}_z\hat{S}_z(\pi Jt)} \hat{I}_y \cos \pi Jt - 2\hat{I}_x\hat{S}_z \sin \pi Jt. \tag{5.36}$$

About which there are two noteworthy points. Equation (5.36) contains no Ω_I, hence chemical shift for the I spin is refocused. However, the J coupling evolution continues over the entire 2τ time period. Computing the S spin coherences in this example is left as an exercise for the reader.

5.4 Branch Diagrams

The calculations to derive Eq. (5.36) involved lots of algebra with multiple sine and cosine functions. One simpler approach is the use of rotation matrices. For example, consider $\hat{\sigma}$ as a 16-dimensional vector in coherence space,

$$\vec{\sigma} = \begin{bmatrix} \dfrac{1}{\langle \hat{I}_x \rangle} \\ \vdots \end{bmatrix}, \tag{5.37}$$

but note that rotations take place in subspaces. For example, evolution due to chemical shift and J coupling could be expressed as $\vec{\sigma}(t) = \underline{\Omega}_I(t)\underline{J}(t)\vec{\sigma}(0)$. The full matrix equations are

$$\begin{bmatrix} \hat{I}_x(t) \\ \hat{I}_y(t) \\ 2\hat{I}_x\hat{S}_z(t) \\ 2\hat{I}_y\hat{S}_z(t) \end{bmatrix} = \begin{bmatrix} c_I & -s_I & 0 & 0 \\ s_I & c_I & 0 & 0 \\ 0 & 0 & c_I & -s_I \\ 0 & 0 & s_I & c_I \end{bmatrix} \begin{bmatrix} c_I' & 0 & 0 & -s_I' \\ 0 & c_I' & s_I' & 0 \\ 0 & -s_I' & c_I' & 0 \\ s_I' & 0 & 0 & c_I' \end{bmatrix} \begin{bmatrix} \hat{I}_x(0) \\ \hat{I}_y(0) \\ 2\hat{I}_x\hat{S}_z(0) \\ 2\hat{I}_y\hat{S}_z(0) \end{bmatrix}, \tag{5.38}$$

where $c_I = \cos \Omega_I t$, $c_I' = \cos \pi Jt$, $s_I = \sin \Omega_I t$, and $s_I' = \sin \pi Jt$. Such rotation matrix calculations can be very helpful for arbitrary rotations but are not the most efficient for describing mutually commuting rotations. Instead, for weakly coupled spin systems, there is a neat graphical approach to avoid these computations altogether, known as ***branch diagrams***.

A more compact notation for Eq. (5.31) is

$$\hat{C}_p \xrightarrow{\hat{C}_q(\theta)} \begin{cases} \hat{C}_p & \Longleftarrow \text{``cosine'' branch} \\ i[\hat{C}_q, \hat{C}_p] & \Longleftarrow \text{``sine'' branch} \end{cases}, \tag{5.39}$$

where the upper and lower equations are referred to as the "cosine" and "sine" branches, respectively. As an example, an excitation of an $\hat{I}_z$ coherence by an Rf pulse along the x-axis followed by J coupling evolution is written as

$$\hat{I}_z \xrightarrow{\hat{I}_x(\theta)} \begin{cases} \hat{I}_z \\ \hat{I}_y \end{cases}, \quad \text{and} \quad \hat{I}_y \xrightarrow{2\hat{I}_z\hat{S}_z(\pi Jt)} \begin{cases} \hat{I}_y \\ -2\hat{I}_x\hat{S}_z \end{cases}. \tag{5.40}$$

Branches are simply dropped altogether when there is no rotation at all (*i.e.*, operators commute) or when equal to zero, as in the examples $2\hat{I}_y\hat{S}_x \xrightarrow{\hat{I}_y(\theta)} 2\hat{I}_y\hat{S}_x$ and $2\hat{I}_z\hat{S}_z \xrightarrow{\hat{I}_x(\pi/2)\hat{S}_x(\pi/2)} -2\hat{I}_z\hat{S}_y$.

Let us now construct the branch diagram for the simple case of the received signal following an Rf pulse of angle α exciting both the I and S spins. For the I spin, we get the following branch diagram.

$$\hat{I}_z \xrightarrow{\hat{I}_x(\alpha)} \begin{cases} \hat{I}_z \xrightarrow{\hat{S}_x(\alpha)} \hat{I}_z \xrightarrow{\hat{S}_z(\Omega_S t)} \hat{I}_z \xrightarrow{\hat{I}_z(\Omega_I t)} \hat{I}_z \xrightarrow{2\hat{I}_z\hat{S}_z(\pi Jt)} \hat{I}_z \\[2em] \hat{I}_y \xrightarrow{\hat{S}_x(\alpha)} \hat{I}_y \xrightarrow{\hat{S}_z(\Omega_S t)} \hat{I}_y \xrightarrow{\hat{I}_z(\Omega_I t)} \begin{cases} \hat{I}_y \xrightarrow{2\hat{I}_z\hat{S}_z(\pi Jt)} \begin{cases} \hat{I}_y \\ -2\hat{I}_x\hat{S}_z \Longleftarrow \end{cases} \\[1.5em] \hat{I}_x \xrightarrow{2\hat{I}_z\hat{S}_z(\pi Jt)} \begin{cases} \hat{I}_x \\ 2\hat{I}_y\hat{S}_z \end{cases} \end{cases} \end{cases}$$

The coherences of interest can now just be "read off" directly from the diagram. For example, the coefficients for the $\hat{I}_x$ coherence are found by moving from right to left and, whenever a branch occurs, either adding a cos(angle of rotation) if on the upper branch or a sin(angle of rotation) if on the lower branch. Hence in this example, we get $\Longrightarrow \hat{I}_x \cos \pi Jt \sin \Omega_I t \sin \alpha$.

NOTA BENE: Try generating branch diagrams for different permutations of the operators. Just be sure to only permute commuting operators! Verify that you get the same final coherences.

The POF and associated branch diagram essentially eliminate the need to write down and solve long equations. In fact, they are so useful that many MR texts just start with writing down the product operators, $\hat{I}_x, \hat{S}_x, \hat{I}_y, \hat{S}_y, \ldots, 2\hat{I}_z\hat{S}_z$, and giving the associated rotation diagrams. If you want, you can now forget all about quantum mechanics and just systematically follow the algebraic rules to find out what happens in any number of MR pulse sequences by using the orthogonal expansion property of $\hat{\sigma}$ to write down M_x, M_y, and M_z. However, skipping all the quantum mechanics and associated physics does leave quantities such as anti-phase magnetization and multiple-quantum coherences as rather mysterious entities. Furthermore, having a solid understanding of the quantum mechanical operators and how to manipulate them provides a solid foundation for the delve into relaxation theory to come.

Spin echoes are an important component of many MR pulse sequences (see Figure 5.15), so let us take a closer look at the associated branch diagrams. Consider the magnetization right after a $90°$ Rf pulse and, for now, just focus on chemical shift (*i.e.*, ignore J coupling). The corresponding

Figure 5.15 A spin-echo pulse sequence.

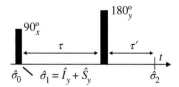

branch diagram is

$$\hat{I}_y \xrightarrow{\hat{I}_z(\Omega_I\tau)} \begin{cases} \hat{I}_y \xrightarrow{\hat{I}_y(\pi)} \hat{I}_y \xrightarrow{\hat{I}_z(\Omega_I\tau')} \begin{cases} \hat{I}_y \\ \hat{I}_x \end{cases} \\ \hat{I}_x \xrightarrow{\hat{I}_y(\pi)} -\hat{I}_x \xrightarrow{\hat{I}_z(\Omega_I\tau')} \begin{cases} -\hat{I}_x \\ \hat{I}_y \end{cases} \end{cases}.$$

Reading off the two terms for each of the $\hat{I}_y$ and $\hat{I}_x$ coherences yields $\hat{I}_y \to \hat{I}_y(\cos\Omega_I\tau'\cos\Omega_I\tau + \sin\Omega_I\tau'\sin\Omega_I\tau) + \hat{I}_x(\sin\Omega_I\tau'\cos\Omega_I\tau - \cos\Omega_I\tau'\sin\Omega_I\tau)$, which can be simplified to $\hat{I}_y\cos\Omega_I(\tau' - \tau) + \hat{I}_x\sin\Omega_I(\tau' - \tau)$. Hence, for this case, the branch diagram for chemical shift during a spin echo (Eq. (5.39)) can be replaced by the simpler

$$\hat{I}_y \xrightarrow{\hat{I}_y(\pi)} \hat{I}_y \xrightarrow{\hat{I}_z(\Omega_I(\tau'-\tau))} \begin{cases} \hat{I}_y \\ \hat{I}_x \end{cases}.$$

Now let us ignore chemical shift, and just consider the effects of J coupling. We will look at three cases. Case 1: the 180° is selective for the I spin. Here, we get the following branch diagram, which is then simplified using a similar trick. Namely,

$$\hat{I}_y \xrightarrow{2\hat{I}_z\hat{S}_z(\pi J\tau)} \begin{cases} \hat{I}_y \xrightarrow{\hat{I}_y(\pi)} \hat{I}_y \xrightarrow{2\hat{I}_z\hat{S}_z(\pi J\tau')} \begin{cases} \hat{I}_y \\ -2\hat{I}_x\hat{S}_z \end{cases} \\ -2\hat{I}_x\hat{S}_z \xrightarrow{\hat{I}_y(\pi)} 2\hat{I}_x\hat{S}_z \xrightarrow{2\hat{I}_z\hat{S}_z(\pi J\tau')} \begin{cases} 2\hat{I}_x\hat{S}_z \\ \hat{I}_y \end{cases} \end{cases}$$

simplifies to

$$\hat{I}_y \xrightarrow{\hat{I}_y(\pi)} \hat{I}_y \xrightarrow{2\hat{I}_z\hat{S}_z(\pi J(\tau'-\tau))} \begin{cases} \hat{I}_y \\ -2\hat{I}_x\hat{S}_z \end{cases},$$

yielding $\hat{I}_y \Longrightarrow \hat{I}_y\cos\pi J(\tau' - \tau) - 2\hat{I}_x\hat{S}_z\sin\pi J(\tau' - \tau)$, and the effects of J coupling will disappear at $\tau' = \tau$. For case 2, make the 180° selective for the S spin. Here the S terms are simply swapped for those with I. Finally, consider case 3: 180° is non-selective and excites both the I and S spins. Here the branch diagram is

$$\hat{I}_y \xrightarrow{2\hat{I}_z\hat{S}_z(\pi J\tau)} \begin{cases} \hat{I}_y \xrightarrow{\hat{I}_y(\pi)} \hat{I}_y \xrightarrow{\hat{S}_y(\pi)} \hat{I}_y \xrightarrow{2\hat{I}_z\hat{S}_z(\pi J\tau')} \begin{cases} \hat{I}_y \\ -2\hat{I}_x\hat{S}_z \end{cases} \\ -2\hat{I}_x\hat{S}_z \xrightarrow{\hat{I}_y(\pi)} 2\hat{I}_x\hat{S}_z \xrightarrow{\hat{S}_y(\pi)} -2\hat{I}_x\hat{S}_z \xrightarrow{2\hat{I}_z\hat{S}_z(\pi J\tau')} \begin{cases} -2\hat{I}_x\hat{S}_z \\ -\hat{I}_y \end{cases} \end{cases}.$$

Collecting terms and simplifying, we get $\hat{I}_y \to \hat{I}_y\cos\pi J(\tau' + \tau) - 2\hat{I}_x\hat{S}_z\sin\pi J(\tau' + \tau)$, and the modulation due to J coupling continues uninterrupted!

Combining the effects of chemical shift and J coupling initially looks rather daunting with 64 branches to compute, $\hat{\sigma}_1 \xrightarrow{\hat{S}_z(\Omega_S\tau)} \xrightarrow{2\hat{I}_z\hat{S}_z(\pi J\tau)} \xrightarrow{\hat{I}_z(\Omega_I\tau)} \xrightarrow{\hat{I}_y(\pi)} \xrightarrow{\hat{I}_z(\Omega_I\tau')} \xrightarrow{\hat{S}_z(\Omega_S\tau')} \xrightarrow{2\hat{I}_z\hat{S}_z(\pi J\tau')} \hat{\sigma}_2$. But by repeatedly simplifying and rearranging commuting terms, we get the following important results. With a 180° selective for I, $\hat{\sigma}_1 \xrightarrow{\hat{I}_y(\pi)} \xrightarrow{\hat{S}_z(2\Omega_S\tau)} \hat{\sigma}_2$, and the effects of chemical shift and J coupling are refocused at the echo time $\tau' = \tau$ for spin I but chemical shift for the spin S is not. Alternatively, a

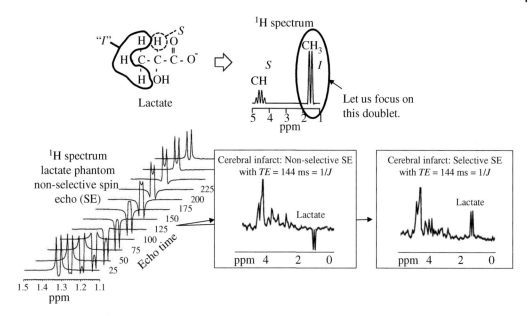

Figure 5.16 Human *in vivo* spin-echo ^{1}H MRS brain data identifying lactate, an important marker of anaerobic metabolism, following a stroke. *In vivo*, the ^{1}H NMR lactate peak is often masked by overlapping signals from lipids, a complication that can be eliminated using a combination of selective and non-selective spin-echoes. Source: Figure adapted with permission from Adalsteinsson et al. (1993).

non-selective 180° refocuses chemical shift for both spins while J coupling continues unabated. These types of manipulations of chemical shift and J coupling refocusing can be very useful *in vivo* as illustrated by the lactate ^{1}H-MRS study shown in Figure 5.16. Note, lactate is technically what is known as an AX_3 spin system (*i.e.,* a weakly coupled system with three spins having the same chemical shift coupled to a fourth nuclei). For now, we will group together the three chemically equivalent ^{1}H nuclei on CH_3 group together and analyze this as a simple *I–S* system.

5.5 Multiple Quantum Coherences and 2D NMR

Two-spin operators that we have not said much about relate to multiple quantum coherences. These are $2\hat{I}_x\hat{S}_x$, $2\hat{I}_x\hat{S}_y$, $2\hat{I}_y\hat{S}_x$, and $2\hat{I}_y\hat{S}_y$, and, similar to anti-phase magnetization, are not directly observable by our Rf receiver coils. Linear combinations of these operators are called DQ and ZQ coherences (see Figure 5.17). Under chemical shift, DQ and ZQ coherences evolve at the sum,

$$\widehat{DQ}_x = \tfrac{1}{2}(2\hat{I}_x\hat{S}_x - 2\hat{I}_y\hat{S}_y) \quad \widehat{DQ}_y = \tfrac{1}{2}(2\hat{I}_y\hat{S}_x + 2\hat{I}_x\hat{S}_y) \quad \widehat{ZQ}_x = \tfrac{1}{2}(2\hat{I}_x\hat{S}_x + 2\hat{I}_y\hat{S}_y) \quad \widehat{ZQ}_y = \tfrac{1}{2}(2\hat{I}_y\hat{S}_x - 2\hat{I}_x\hat{S}_y)$$

Figure 5.17 Product operators corresponding to double- and zero-quantum coherences and their associated energy diagrams.

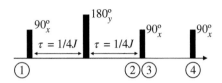

Figure 5.18 An MR pulse sequence generating double-quantum coherence.

$\Omega_I + \Omega_S$, and difference, $\Omega_I - \Omega_S$, of the chemical shifts, respectively,

$$\widehat{DQ}_x \xrightarrow{\widehat{I}_z(\Omega_I \tau)} \xrightarrow{\widehat{S}_z(\Omega_S \tau)} \widehat{DQ}_x \cos(\Omega_I + \Omega_S)t + \widehat{DQ}_y \sin(\Omega_I + \Omega_S)t$$

$$\widehat{ZQ}_y \xrightarrow{\widehat{I}_z(\Omega_I \tau)} \xrightarrow{\widehat{S}_z(\Omega_S \tau)} \widehat{ZQ}_x \cos(\Omega_I - \Omega_S)t + \widehat{ZQ}_x \sin(\Omega_I - \Omega_S)t.$$

Using the commutator tables, $[\widehat{I}_x\widehat{S}_x, \widehat{I}_z\widehat{S}_z] = [\widehat{I}_y\widehat{S}_y, \widehat{I}_z\widehat{S}_z] = 0$, and $[\widehat{I}_x\widehat{S}_y, \widehat{I}_z\widehat{S}_z] = [\widehat{I}_y\widehat{S}_x, \widehat{I}_z\widehat{S}_z] = 0$, which reflect neither DQ nor ZQ coherences evolving under J coupling (assuming weak coupling).

A simple pulse sequence that generates a DQ coherence is shown in Figure 5.18, where we have added the 180°_y Rf pulse and selected the echo times to simplify the branch diagram by refocusing chemical shift and eliminating the effects of J coupling. At the first time point, let $\hat{\sigma} = \widehat{I}_z + \widehat{S}_z$. Calculating the coherences as they evolve from time points 2 to 3 yields

I spin: $-2\widehat{I}_x\widehat{S}_z \xrightarrow{\widehat{I}_x(\pi/2)+\widehat{S}_x(\pi/2)} -2\widehat{I}_x\widehat{S}_y$

S spin: $-2\widehat{I}_z\widehat{S}_x \xrightarrow{\widehat{I}_x(\pi/2)+\widehat{S}_x(\pi/2)} -2\widehat{I}_y\widehat{S}_x.$

the sum of which is a pure $\widehat{DQ}_y$ coherence. If there were no additional Rf pulses, the $\widehat{DQ}$ would continue to evolve, but we have no Rf coils capable of measuring this coherence. Analogous to using Rf pulses to convert an undetectable $\widehat{I}_z$ to a detectable $\widehat{I}_x$ or $\widehat{I}_y$, a final 90° pulse is added as the "readout". It is left as an exercise for the reader to show that this final 90° pulse converts $\widehat{DQ}$ into anti-phase magnetization that will then evolve into a detectable signal.

Variations of this simple pulse sequence are the central building blocks for many 2D NMR sequences (van de Ven 1996). Measuring the J coupling constants and associated peak splittings allows the determination of how close two nuclei are on a given molecule, as measured by the number and types of connecting chemical bonds. A related technique, Nuclear Overhauser Effect Spectroscopy (NOESY), measures correlations between nuclei which are physically close in space regardless of whether there is a bond between them (Neuhaus and Williamson 2000). These high-dimensional NMR techniques were later utilized by Kurt Wüthrich to determine the three-dimensional structure of biological macromolecules in solution, with the first complete determination of a protein structure achieved in 1985. Such techniques have been applied *in vivo*, but to date the results have been limited by the multiple overlapping metabolite signals characteristic of these applications (see Figure 5.19).

A second important NMR method involves J coupling and is known as 2D-J (van Zijl et al. 2021). This method results in a two-dimensional dataset for which chemical shift is resolved along one axis with J coupling resolved along the second. Examples of this approach are shown in Figure 5.20.

Hence, the 2D-J method has demonstrated some limited utility in analyzing *in vivo* spectra. Additional applications include measurement of metabolite T_2 relaxation times (Adalsteinsson et al. 2004), suppression of unwanted water or lipid frequency modulation sidebands due to mechanical gradient vibrations (Hurd, Gurr, and Sailasuta 1998), and simplification of *in vivo* spectra by integrating 2D-J plots over the *J*-dimension to collapse peak splitting. The later technique is also known as TE-averaged MRS (Hancu et al. 2005; Hurd et al. 2004; Zhang et al. 2015) and an example is shown in Figure 5.21.

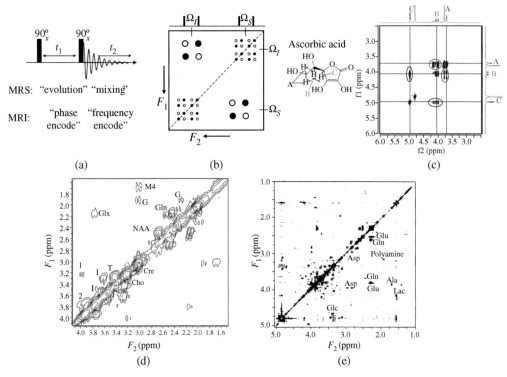

Figure 5.19 2D-NMR examples. (a) building block for many 2D NMR sequences, (b), representative COSY spectrum for coupled *I* and *S* spins, (c) COSY spectrum of ascorbic acid, an important antioxidant, (d) a representative *in vivo* 7 T rat brain 2D-NMR COSY ^{1}H spectra, and (e) *ex vivo* perchloric acid (PCA) extract of tumor tissue from a patient with an intraductal carcinoma. Source: (d,e) Adapted with permission from Welch et al. (2003) Figure 2d and Gribbestad et al. (1994) Figure 2, respectively.

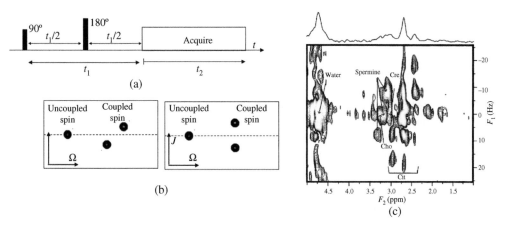

Figure 5.20 2D-J NMR. (a) Pulse sequence diagram for a generic 2D-J acquisition. During time interval t_1, chemical shift, Ω, is refocused, while *J* evolution is not; during t_2 neither chemical shift nor *J* is refocused. (b) Acquired for multiple values of t_1, a two-dimensional Fourier transform of the acquired data yields the 2D-J spectrum, which is usually processed by "tilting" by 45°. (c) Representative *in vivo* human 2D-J spectra acquired from the prostate of 60-year-old subject with benign prostatic hyperplasia (BPH). Source: (c) Adapted with permission from Yue et al. (2002) Figure 3a.

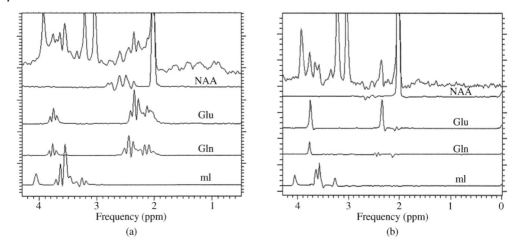

Figure 5.21 Measurement of brain glutamate using TE-averaged ^{1}H PRESS at 3 T. (a) 3 T PRESS spectrum of normal mid-parietal white matter and scaled spectra of *N*-acetyl-aspartate (NAA), glutamate (Glu), glutamine (Gln), which overlap of signals from NAA, Glu, Gln, and myo-Inositol (ml). (b) TE-averaged spectrum of mid-parietal gray matter. The peak at 2.35 ppm is exclusively from Glu. The resolved signal at 3.75 ppm is typically refer to as Glx, the sum of Glu and Gln. In conventional PRESS, Glu, Gln, and NAA are overlapped. Source: Adapted with permission from Hurd et al. (2004) Figures 2 and 5.

5.6 Polarization Transfer

The POF provides an ideal tool to analyze one of the important methods used for *in vivo* multinuclear MRS studies, namely, ***polarization transfer***. The basic idea is to study relatively insensitive nuclei by taking advantage of higher-sensitivity *J*-coupled partners. The signal-to-noise ratio (SNR) for *in vivo* MRS experiments at typical clinical field strengths is determined by the number of spins per unit volume ρ, the magnetic moment of the targeted nuclei, and the degree of polarization. From Chapter 2, the signal induced in the receiver coil is proportional to $dM/dt = \omega M$, and at typical clinical scanners where body noise dominates coil noise,

$$SNR \propto \rho \frac{\gamma^2 \hbar^2 B_0}{4kT}. \tag{5.41}$$

Hence, experiments with ^{1}H versus ^{13}C nuclei, where $\gamma_H \approx 4\gamma_C$, yield a 16-fold SNR advantage. Therefore, it is often desirable to enhance the SNR of ^{13}C signals by exploiting the *J*-coupled ^{1}H nuclei. This can be performed using polarization transfer.

Consider the pulse sequence shown in Figure 5.22a where Rf pulses are played using both the proton and carbon exciters. This is a simplified version of the pulse sequence known as ***Insensitive Nuclei Enhanced by Polarization Transfer (INEPT)*** (de Graaf et al. 2011). For now, let us assume that the chemical shift evolution during $1/2J$ is negligible (later we will add $180°$s to ensure this), and the ^{1}H and ^{13}C spins are weakly coupled. The initial density operator is

$$\widehat{\sigma}(0) = \frac{\hbar B_0}{4kT}(\gamma_I \widehat{I}_z + \gamma_S \widehat{S}_z), \tag{5.42}$$

and the Hamiltonian during $1/2J$ time interval is $\widehat{H} = 2\pi J \widehat{I}_z \widehat{S}_z$ which changes to $\widehat{H} = -\Omega_I \widehat{I}_z - \Omega_S \widehat{S}_z + 2\pi J \widehat{I}_z \widehat{S}_z$ during data acquisition. Let us construct the branch diagram.

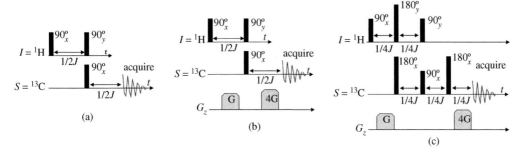

Figure 5.22 Polarization transfer. (a) Basic pulse sequence module, (b) basic pulse sequence with gradients for dephasing unwanted coherences, (c) full pulse sequence using spin-echoes to refocus chemical shift and gradients to dephase unwanted coherences.

Starting with the I spin, the coherences existing at the start of the acquisition are

$$\hat{I}_z \xrightarrow{\hat{I}_x(\pi/2)} \hat{I}_y \xrightarrow{2\hat{I}_z\hat{S}_z(\pi/2)} -2\hat{I}_x\hat{S}_z \xrightarrow{\hat{I}_y(\pi/2)} -2\hat{I}_z\hat{S}_z \xrightarrow{\hat{S}_x(\pi/2)} -2\hat{I}_z\hat{S}_y \xrightarrow{2\hat{I}_z\hat{S}_z(\pi/2)} \hat{S}_x.$$

During acquisition, we have

$$\hat{S}_x \xrightarrow{2\hat{I}_z\hat{S}_z(\pi J\tau/2)} \begin{cases} \hat{S}_x \xrightarrow{\hat{S}_z(\Omega_S t)} \begin{cases} \hat{S}_x \\ -\hat{S}_y \end{cases} \\ \\ 2\hat{I}_z\hat{S}_y \xrightarrow{\hat{S}_z(\Omega_S t)} \begin{cases} 2\hat{I}_z\hat{S}_y \\ 2\hat{I}_z\hat{S}_x \end{cases} \end{cases},$$

from which we can read-off the S spin coherences as $\hat{S}_x \cos(\Omega_S t)\cos(\pi Jt) - \hat{S}_y \sin(\Omega_S t)\cos(\pi Jt)$. Keeping track of the leading constants generates the desired J-coupled spectrum with SNR of $\rho\gamma_C\gamma_H\hbar^2 B_0/4kT$; a factor of $\gamma_H/\gamma_C \cong 4$ SNR gain over direct ^{13}C detection.

However, there is also the coherence coming directly from the S spins (as opposed to the signal transferred from the I spins). Namely,

$$\hat{S}_z \xrightarrow{\hat{S}_x(\pi J\tau/2)} \hat{S}_y \xrightarrow{2\hat{I}_z\hat{S}_z(\pi/2)} -2\hat{I}_z\hat{S}_x \xrightarrow{2\hat{I}_z\hat{S}_z(\pi Jt)} \begin{cases} -2\hat{I}_z\hat{S}_x \xrightarrow{\hat{S}_z(\Omega_S t)} \begin{cases} -2\hat{I}_z\hat{S}_x \\ 2\hat{I}_z\hat{S}_y \end{cases} \\ \\ -\hat{S}_y \xrightarrow{\hat{S}_z(\Omega_S t)} \begin{cases} -\hat{S}_y \\ -\hat{S}_x \end{cases} \end{cases},$$

which results in a second S spin coherence contribution of $-\hat{S}_x \sin(\Omega_S t)\sin(\pi Jt) - \hat{S}_y \cos(\Omega_S t)\sin(\pi Jt)$. This term produces an unwanted anti-phase doublet with $SNR = \rho\gamma_C^2\hbar^2 B_0/4kT$ (see Figure 5.23). In addition, the final spectrum will also contain contributions from any uncoupled ^{13}C spins.

We typically want to get rid of these unwanted signal contributions, and there are two approaches. The first is phase cycling whereby two acquisitions are performed for which the sign of the first I Rf pulse is inverted, *i.e.*, acquisitions using $\pm90°_x$. Subtraction of the two resulting signals retains only the desired enhanced in-phase doublet (see Figure 5.22a). Alternatively, we can use gradients (see Figure 5.22b).

Figure 5.23 An anti-phase doublet.

Consider the effect of the application of a gradient pulse on the spin density operator $\hat{\sigma}$, where for simplicity we will assume a linear gradient in the z-direction having a spatial dependent amplitude of $G_z z$ (see Figure 5.24). The gradient adds to B_0 such that $\hat{H}$ becomes a function of position. Remembering that the Rf receiver coil simply integrates all the signal over the sensitive volume. The new term in the Hamiltonian becomes

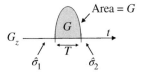

Figure 5.24 The use of gradients to dephase coherences.

$$\hat{H}_G = -\gamma z \hat{I}_z \int_0^T G_z(t)dt, \qquad (5.43)$$

which adds the following term to our branch diagram, $\hat{\sigma}_1 \xrightarrow{\hat{I}_z(zGT)} \hat{\sigma}_2$. For example, consider $\hat{\sigma}_1 = \hat{I}_y$ with the gradient area set to add π rads per unit z of phase. Then $\hat{I}_y \xrightarrow{\hat{I}_z(\pi z)} \hat{I}_y \cos \pi z + \hat{I}_x \sin \pi z$, and, to get the total coherence, we need to integrate over z. Also keep in mind that one must consider both the evolution under the gradient for both the I and S spins.

The new Hamiltonian for free precession in such a gradient-enhanced INEPT sequence is

$$\hat{H} = -\Omega_I \hat{I}_z - \Omega_S \hat{S}_z - G_z z \hat{I}_z - G_z z \hat{S}_z + 2\pi J \hat{I}_z \hat{S}_z. \qquad (5.44)$$

Analysis of the resulting branch diagram, where we start with a 90° excitation of the I spin,

$$\hat{I}_z \xrightarrow{\left(\frac{\pi}{2}\right)_x^I} \hat{I}_y \xrightarrow{-\pi J\left(\frac{1}{2J}\right)} -2\hat{I}_x\hat{S}_z \xrightarrow{\gamma_I G_1 z}$$

$$\begin{cases} -2\hat{I}_x\hat{S}_z \xrightarrow{\left(\frac{\pi}{2}\right)_y^I} -2\hat{I}_z\hat{S}_z \xrightarrow{\left(\frac{\pi}{2}\right)_x^S} -2\hat{I}_z\hat{S}_y \xrightarrow{\gamma_S G_2 z} \begin{cases} -2\hat{I}_z\hat{S}_y \xrightarrow{-\pi J\left(\frac{1}{2J}\right)} \hat{S}_x \\ -2\hat{I}_z\hat{S}_x \xrightarrow{-\pi J\left(\frac{1}{2J}\right)} -\hat{S}_y \end{cases} \\ 2\hat{I}_y\hat{S}_z \xrightarrow{\left(\frac{\pi}{2}\right)_y^I} 2\hat{I}_y\hat{S}_z \xrightarrow{\left(\frac{\pi}{2}\right)_x^S} \cdots \end{cases}$$

yields an additional modulation of the received signal by $\cos(\gamma_C G_2 z) \cos(\gamma_H G_1 z)$. The final result is

$$M_{xy}(t) \propto e^{-i\Omega_S t} \cos(\pi J t) \int_z \cos(\gamma_C G_2 z) \cos(\gamma_H G_1 z)dz. \qquad (5.45)$$

A judicious choice for the relative gradient areas is $G_2 = G_1 \gamma_H / \gamma_C \approx 4G_1$. This yields

$$M_{xy}(t) \propto e^{-i\Omega_S t} \cos(\pi J t) \int_z \cos^2(\gamma_H G_1 z)dz = \frac{1}{2} e^{-i\Omega_S t} \cos(\pi J t), \qquad (5.46)$$

a doublet with $SNR = \rho \gamma_C \gamma_H \hbar^2 B_0 / 8kT$. Hence, this sequence has a 50%yield but is a single-shot technique. It is left to the reader to show that this gradient-enhance INEPT method also filters out both the undesired anti-phase doublet as well as all contributions from uncoupled carbon spins.

A more complete gradient-enhanced INEPT pulse sequence including 180° pulses to refocus chemical shift is shown in Figure 5.22c. We further note that this gradient-enhanced INEPT pulse sequence can either be performed so that the final ^{1}H-enhanced signal is read out on the carbon channel (^{13}C having the advantage of having a larger chemical range than ^{1}H) or reversed whereby the ^{13}C signal is readout is on the ^{1}H channel, but only ^{1}H nuclei coupled to ^{13}C nuclei are detected. Both approaches yield the same $\gamma_H / \gamma_C \approx 4$ SNR gain over direct ^{13}C detection.

The same branch diagram analysis can also be extended to more complicated molecules such as those having S spins bonded to more than one I spin. In the case on an $I_2 S$ spin system, the free precession Hamiltonian becomes $\hat{H} = -\Omega_I \hat{I}_{1z} - \Omega_I \hat{I}_{2z} - \Omega_S \hat{S}_z + 2\pi J \hat{I}_{1z} \hat{S}_z + 2\pi J \hat{I}_{2z} \hat{S}_z + 2\pi J \hat{I}_{1z} \hat{I}_{2z}$, and rotations take place in 3-D subspaces of a 64-D coherence space.

 NOTA BENE: Try a CH_3 system for a 256-dimensional adventure!

5.7 Spectral Editing

Although *in vivo* ^{1}H spectra are simplified by concentration and relaxation time detection limits (to be discussed in Chapter 6), multiple overlapping peaks complicate unambiguous metabolite measurements (see Figure 5.25). Consequently, there are several techniques, known collectively as spectral editing, used to simplify *in vivo* MRS data.

5.7.1 *J*-difference Editing

One approach is to utilize *J* coupling. As an extension of the data shown in Figure 5.16, one can add spatial localization to a spin echo sequence, and, by alternating between the 180° pulse being non-selective verses selective for only the *I* or *S* spins, *in vivo* spectra can be considerably simplified. Figure 5.26 shows *in vivo* human ^{1}H brain data for which J editing was used to separate (a) the normally overlapping lactate and lipid signals in a patient with suspected stroke, and (b) the neurotransmitter GABA. The GABA editing technique, usually referred to as MEGA-PRESS (Mullins et al. 2014), has become a standard approach for measuring this primary human brain inhibitory transmitter. In general, *in vivo* *J*-editing methods have high sensitivity and are generally robust to B_0 inhomogeneities, though subtraction artifacts due to motion or other system instabilities may limit data quality.

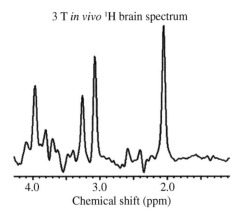
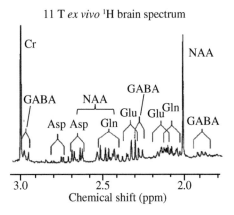

Figure 5.25 Representative *in vivo* and *ex vivo* human brain ^{1}H spectra. The metabolites identified in this example are aspartate (Asp), choline (Cho), creatine (Cre), γ-aminobutyric acid (GABA), glutathione (GSH), glutamate (Glu), glutamine (Gln), glycine (Gly), lactate (Lac), myo-Inositol (mI), and *N*-acetyl-aspartate (NAA).

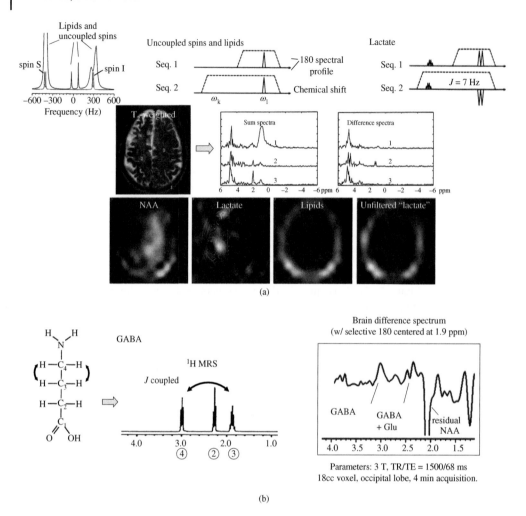

Figure 5.26 *In vivo* J editing: (a) editing lactate in a patient with MELAS (a metabolic disorder characterized by multiple strokes). (b) MEGA-PRESS editing for GABA. Source: (a) adapted with permission from Adalsteinsson et al. (1993) Figures 1, 3, and 4.

5.7.2 Multiple-quantum Filtering

A second approach for *in vivo* spectral editing is to use multiple-quantum filtering. A generic DQ filter is shown in Figure 5.27a, where the 180° Rf pulses are used to refocus chemical shift and the selection of coherences is done via the use of gradients. A branch diagram analysis can be used to show the detectable coherences at the start of data acquisition are $\frac{1}{2}(\hat{I}_y + \hat{S}_y)$, a single-shot filter with 50% yield. Figure 5.27b shows the results for *in vivo* lactate imaging using a DQ-filter for assessing metabolism in a mouse flank tumor model.

QUESTION: Referring to Figure 5.27b, aren't lipids also *J*-coupled? Shouldn't subcutaneous lipid signals be as large or larger than that due to lactate?

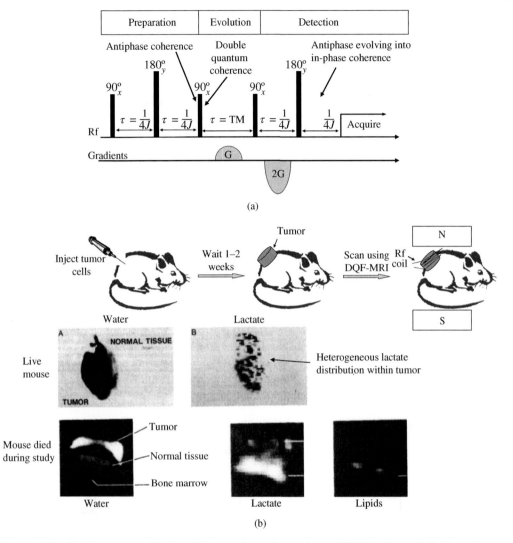

Figure 5.27 Double-quantum filtering. (a) A generic gradient-enhanced DQ filter for an *I–S* spin system, and (b) *in vivo* double-quantum filtering results in a murine tumor model. Source: (b) Adapted with permission from Hurd and Freeman (1991) Figures 10 and 11.

5.8 Summary

This chapter introduces two-spin product operators ($\hat{E}$, $\hat{I}_x$, $\hat{I}_y$, $\hat{I}_z$, $\hat{S}_x$, $\hat{S}_y$, $\hat{S}_z$, $2\hat{I}_x\hat{S}_z$, $2\hat{I}_y\hat{S}_z$, $2\hat{I}_z\hat{S}_x$, $2\hat{I}_z\hat{S}_y$, $2\hat{I}_z\hat{S}_z$, $2\hat{I}_x\hat{S}_x$, $2\hat{I}_y\hat{S}_y$, $2\hat{I}_x\hat{S}_y$, and $2\hat{I}_y\hat{S}_x$) as remarkably useful tools for analyzing and exploiting a variety of pulse sequences used to interrogate the effects of chemical shift and J coupling. Discussion of MR relaxation process has been left for later chapters. The end of this chapter is both a high and a low point (see Figure 5.28). The high point is that we have now discussed all the basic mathematical tools needed to analyze a wide variety of *in vivo* MR phenomenon. Furthermore, we would develop simple graphical tools to analyze many sophisticated pulse sequences. On the other hand, whatever happened to the bras, kets, wavefunctions, and all that other stuff that was so much fun?!?

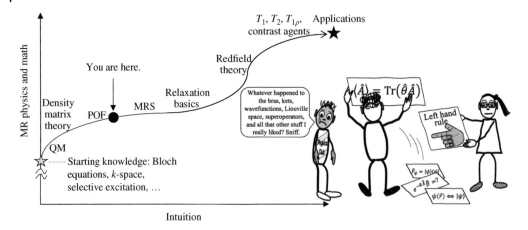

Figure 5.28 The roadmap at the end of Chapter 5.

Exercises

E.5.1 Miscellaneous Spectra

Assuming weak coupling, what is the real part of the observed spectrum following a selective 90°-excite acquisition of the I spin in an I–S spin system if the density operator at the start of acquisition is (a) $\hat{\sigma}_0 = \hat{I}_x$, (b) $\hat{\sigma}_0 = 2\hat{I}_x\hat{S}_z$, (c) $\hat{\sigma}_0 = \hat{I}_x + 2\hat{I}_x\hat{S}_z$, or (d) $\hat{\sigma}_0 = \hat{I}_x + 2\hat{I}_y\hat{S}_z$.

Note, for this problem and the ones that follow assume a finite T_2 decay time and use the sign conventions given in this chapter. Namely,

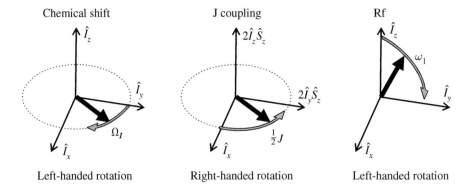

E.5.2 Proton-Observe Carbon-Edit (POCE)

Let $I = {}^1\text{H}$ spins be weakly coupled to ^{13}C nuclei, and $R =$ uncoupled ^1H spins. The initial density operator is

$$\hat{\sigma}(0) = \frac{\hbar B_0}{4kT}(\gamma_H\hat{I}_z + \gamma_C\hat{S}_z + \gamma_H\hat{R}_z).$$

For the pulse sequence shown below in which the carbon Rf pulse is alternated between 0 (off) and 180° (on), fill in the table below for the coherences observed at the start of data acquisition. Combine the data as recon = (Acq. 1 − Acq. 2)/2. Suggest a use for this sequence.

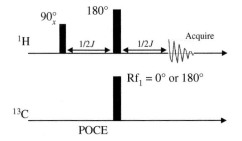

	Rf$_1$ flip angle	Coherences
Acquisition 1	0	?
Acquisition 2	180	?

E.5.3 Proton Observed Carbon Edited Effectively Decoupled (PROCEED)

Let $I = {}^1$H spins weakly coupled to ^{13}C nuclei, and R = uncoupled ^{1}H spins. The initial density operator is

$$\hat{\sigma}(0) = \frac{\hbar B_0}{4kT}\left(\gamma_H \hat{I}_z + \gamma_C \hat{S}_z + \gamma_H \hat{R}_z\right)$$

a) For the pulse sequence shown below in which the carbon Rf pulses are both alternated between 0 (off) and 180° (on), fill in the table below for the coherences observed at the start of data acquisition.

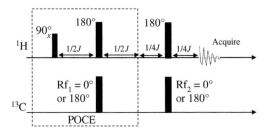

	Rf$_1$	Rf$_2$	Coherences
Acquisition 1	0	0	?
Acquisition 2	180	0	?
Acquisition 3	0	180	?
Acquisition 4	180	180	?

b) Suggest a data reconstruction algorithm using Acquisitions 1–4, and sketch the resulting spectrum.

E.5.4 INEPT with a CH$_2$ Spin System

Use branch diagrams to compute and compare the resulting spectra from these two pulse sequences for a CH$_2$ spin system. Ignore chemical shift evolution over the $1/2J$ and $1/2J$ time intervals.

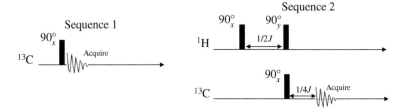

E.5.5 Distortionless Enhanced by Polarization Transfer (DEPT)

For this problem, we are going to examine the inverse DEPT pulse sequence, which is useful if we have ^{13}C signals that we wish to detect via coupling to more sensitive ^{1}H nuclei.

a) Given a system of ^{13}C-^{1}H J-coupled spin pairs, show the full branch diagram and compute the relative sensitivity of direct ^{13}C detection versus the sequence shown below known as inverse DEPT.

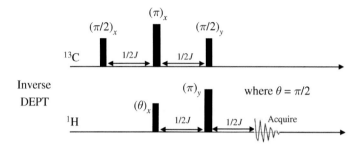

b) In general, the value of θ in an inverse DEPT sequence needs to be optimized for different spin systems. Show that the inverse DEPT coherence transfer efficiency as a function of θ for CH, CH$_2$, and CH$_3$ systems is given by the plot below. What is the relative SNR of inverse DEPT compared to direct ^{13}C detection?

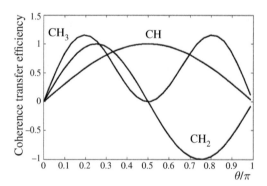

E.5.6   C2 Lactate Imaging

The ^{13}C nucleus in a lactate molecule labeled at the 2nd carbon position, [2-^{13}C]Lac, is coupled to the attached ^{1}H nucleus with a coupling constant $J = 140$ Hz. The effect of J coupling on the ^{13}C (or ^{1}H) spectrum is to split the peak into a doublet.

a) What is the effect of the J coupling on a single-shot imaging sequence like the Echo Planar Imaging (EPI) sequence? Assume readout time is ~10 ms.

b) Propose a method to eliminate the artifact seen in part (a).

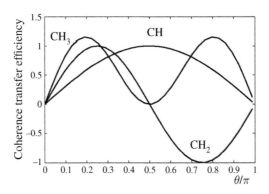

c) Suppose you only excite a narrow RF band (~100 Hz), centered at the center of lactate doublet, what do you expect to see? What would happen if this narrow band excitation is now centered at one of the peaks? Simulate the two cases using full density matrix with J coupling included in the Hamiltonian and show the results.

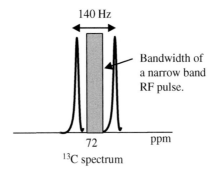

E.5.7 J Editing

This chapter discussed *J* editing for the doublet resonance of lactate. Other *in vivo* peaks (*e.g.*, GABA) are more complicated (triplets, quartets, etc.). In this problem, we will consider *J* editing for IS_2 spin systems. An example spectrum is shown on the right.

The goal is to edit for the *I* spin triplet. One proposal is to use the lactate or GABA schemes and subtract the following two acquisitions.

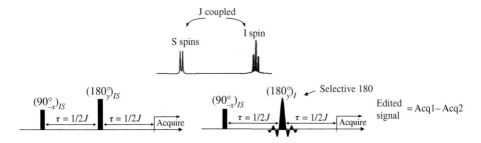

What fraction of the *I* spin signal is observed using this approach? Is there a different value of t that detects more of the desired signal?

E.5.8 Selective Double Quantum Filtering (selDQF) for Lactate

The conventional double quantum filter is of limited utility for measuring lactate *in vivo*. While water is removed by the filter, lipids, like lactate, are coupled with $J \approx 7$ Hz, and hence pass through the filter unattenuated. *In vivo*, large lipid peaks can easily overwhelm the lactate methyl signal (at 1.3 ppm). A proposed solution is the use of a frequency-selective readout pulse for the last 90° pulse of the DQ-filter (hence the name "selective double quantum filter"). Representative (and simplified) lactate and lipid spectra along with the proposed pulse sequence are shown below. Source: Figure adapted with permission from Adalsteinsson et al. (1993).

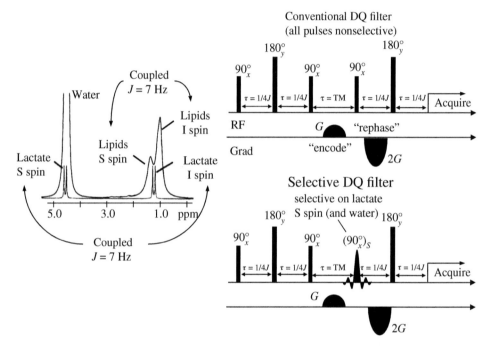

a) Using the POF shows that not only does the selDQF sequence suppress both the unwanted water and lipid resonances but also generates twice the lactate methyl signal as a conventional DQ-filter. As is shown in the diagram, assume both lipids and lactate are weakly coupled IS spin systems.

b) Analyze the performance of the pulse sequence if a positive instead of a negative rephasing gradient is used. How does this compare with the conventional DQ filter?

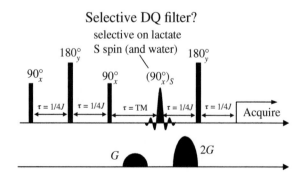

Historical Notes

Kurt Wüthrich, a Swiss scientist, made groundbreaking contributions to nuclear magnetic resonance (NMR) and its application in studying large biological molecules, especially proteins. Born on October 4, 1938, in Aarberg, Switzerland, Wüthrich's pioneering work in the 1980s revolutionized the understanding of protein structures and earned him the Nobel Prize for Chemistry in 2002, which he shared with John B. Fenn and Tanaka Koichi.

After obtaining his PhD in organic chemistry from the University of Basel in 1964, Kurt Wüthrich pursued postdoctoral training in Switzerland and the United States. He later joined the Swiss Federal Institute of Technology in 1969, becoming a professor of biophysics in 1980. In 2001, he moved to the Scripps Research Institute in La Jolla, California, where he served as a visiting professor.

Wüthrich's groundbreaking work involved adapting NMR technology to study large biological molecules, especially proteins. NMR, developed in the late 1940s, provides detailed information about molecular structures by subjecting a sample to a strong magnetic field and radio waves. Certain atomic nuclei in the molecules emit their own radio waves, which can be analyzed to reveal structural details.

However, applying NMR to large molecules like proteins posed significant challenges. The presence of numerous atomic nuclei led to a complex tangle of radio signals, making data interpretation difficult. In the early 1980s, Wüthrich made a significant breakthrough with the development of a solution known as "sequential assignment," an innovative technique systematically that matched each NMR signal with the corresponding hydrogen nucleus in the analyzed protein. Furthermore, Wüthrich demonstrated how this information could determine distances between multiple pairs of hydrogen nuclei, effectively creating a three-dimensional picture of the molecule. This groundbreaking work allowed researchers to visualize the intricate structure of proteins, offering unprecedented insights into their functions and mechanisms.

In 1985, Wüthrich's team achieved the first complete determination of a protein structure using his sequential assignment method. Over time, this pioneering technique became an essential tool in structural biology, enabling the determination of numerous protein structures with NMR. By 2002, approximately 20% of known protein structures had been determined using NMR, highlighting the significant impact of Wüthrich's contributions.

(Source: Photo from https://commons.wikimedia.org/wiki/File:Kurt_Wuthrich_in_2022_02 .jpg, licensed under the Creative Commons Attribution-Share Alike 4.0 International.)

6

In vivo MRS

Chapter 5 saw the introduction of some of the metabolites that can be observed *in vivo*. In this chapter, emphasizing studies in human, we will go into more detail regarding what compounds are accessible by *in vivo* magnetic resonance spectroscopy (MRS), the methods used for both single-voxel and magnetic resonance spectroscopic imaging (MRSI) studies, and some of the clinical applications. Figure 6.1 provides an overview of the information available from *in vivo* human MRI and MRS.

6.1 ^{1}H MRS

6.1.1   Acquisition Methods

In vivo ^{1}H MR signals are dominated by water and, in some tissues, lipid components, with resonances from other metabolites of interest typically on the order of 1000 times smaller. To robustly detect these compounds, some form of water and lipid suppression is typically required (see Figure 6.2). For simplicity, we shall restrict our description of the most common methods for ^{1}H MRS data acquisition to studies of the brain, the organ most often targeted by *in vivo* MRS investigations due to the lack of mobile lipids and minimal physical motion.

Within each cubic centimeter of brain tissue, there are innumerable chemical compounds containing hydrogen, each of which generates a ^{1}H MRS signal. Whether this is a blessing or an important limitation, not all such compounds are effectively measurable in an *in vivo* study. Roughly speaking, the concentration limit for robust ^{1}H MRS detection in humans is around 1 mM. Furthermore, macromolecules are largely undetectable due to their short T_2 relaxation times, considerably simplifying *in vivo* data. Hence, the technical requirements for a successful *in vivo* ^{1}H MRS brain exam include addressing the following issues: (i) the spatial homogeneity of the main magnet field, (ii) some form of spatial localization (largely to avoid subcutaneous fat), (iii) robust water suppression, and (iv) accurate spectral quantification. We will start with achieving spatial localization.

The prototypical ^{1}H MRS brain exam involves exciting a single large voxel of tissue (typically 1–20 cc in volume) located away from any subcutaneous lipids (Öz et al. 2020), and the most common method, known as point-resolved spectroscopy (PRESS) (point resolved spectroscopy), is to use a double spin-echo 90° – 180° – 180° acquisition with the initial 90° excitation pulse and two refocusing 180°s, each selective along different spatial axes (Bottomley 1987). As shown in Figure 6.3, the PRESS acquisition is typically preceded by some form of water suppression, and

Fundamentals of In Vivo Magnetic Resonance: Spin Physics, Relaxation Theory, and Contrast Mechanisms, First Edition.
Daniel M. Spielman and Keshav Datta.
© 2024 John Wiley & Sons, Inc. Published 2024 by John Wiley & Sons, Inc.
Companion website: www.wiley.com/go/Spielman

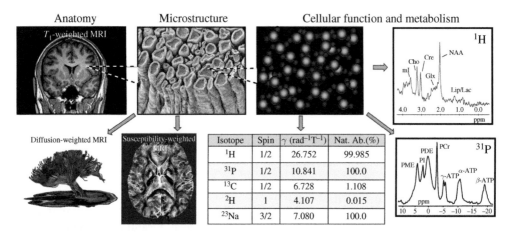

Figure 6.1 An overview of the information available from *in vivo* human MRI and MRS studies, the primary nuclei of interest, and their associated sensitivities and natural abundances. Source: Adapted with permission from Novikov et al. (2019) Figure 1.

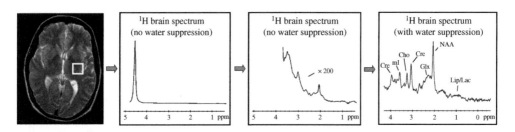

Figure 6.2 Localized ^{1}H brain spectroscopy requires effective water suppression to robustly detect *in vivo* metabolites.

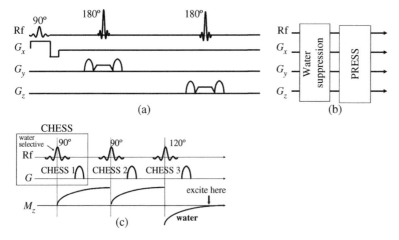

Figure 6.3 Single-voxel spectroscopy using (a) the PRESS technique proceeded by (b) a water suppression module (c) illustrates the CHESS methods whereby the flip angle on the final CHESS pulse is fine tuned to minimize residual water signals.

a common choice is the use of the chemical shift selective imaging sequence (CHESS) technique (Haase et al. 1985). This method uses repeated application of a frequency-selective 90° pulse to selectively excite the water signal followed by a spoiler gradient to dephase the resulting water magnetization.

We should also note that an increasingly popular alternative to PRESS is the sLASER (semi-adiabatic localization by adiabatic selective refocusing) method, which provides improved spatial selectivity and robustness using a slice-selective 90° in combination with two pairs of spatially-selective adiabatic 180°s (Deelchand et al. 2021). A second alternative is to use STEAM (stimulated-echo acquisition method), a method allowing shorter echo times and decreased chemical-shift localization errors at the cost of a 50% signal-to-noise ratio (SNR) loss (Frahm et al. 1989).

In general, single-voxel ^1H MRS methods are widely available on commercial MRI scanners with the ability to graphically prescribe regions-of-interest (ROI) and automated shimming. Data collection times of 2–5 minutes are typically used to acquire localized ^1H-MRS data from 2 to 8 cc voxels (Wilson et al. 2019). Primary clinical applications include the study of focal lesions or diffuse diseases where the precise location of the voxel is relatively unimportant.

The reliability of single-voxel ^1H MRS is generally quite high, but some technical challenges remain, starting with robust and accurate spectral quantification. There are multiple commercial software packages for analyzing *in vivo* ^1H MRS data. Popular choices include LCmodel (Provencher 2001), jMRIU (Naressi et al. 2001), and TRAQUIN (Naressi et al. 2001). To first order, these methods typically approach spectral quantification using a least-squares analysis. A given *in vivo* spectrum is modeled as a vector containing a linear combination of metabolite contributions, and quantification consists of projecting this vector onto a basis set comprised of simulated (or acquired) individual metabolite spectra. Proprietary details in the various commercial software packages largely consist of adjustments needed to account for unmodeled components such as broad macromolecule or lipid signals. Final metabolic concentrations are usually expressed as ratios relative to *in vivo* water or the relatively stable total creatine signal (see Figures 6.3 and 6.4).

Some important issues to keep in mind when interpreting results include choice of echo time, contributions from different tissue types (*e.g.*, gray matter, white matter, and cerebral spinal fluid (CSF)) within the interrogated spectroscopic voxel, variations across the normal brain, and developmental age of the subject (see Figure 6.5 for examples).

There are also additional challenges to acquiring high-quality spectra. Subtle shifts in the voxel location can inadvertently include large unwanted contributions from extracranial lipids, and, perhaps most important of all, as illustrated in Figure 6.6, a highly homogenous B_0 field is critical.

The need to minimize B_0 inhomogeneities brings us to what have been called the three most important factors for a successful *in vivo* MRS exam – homogeneity, homogeneity, and homogeneity (SNR should probably also be somewhere in this list). Hence, shimming, the process of making fine adjustments to the magnetic field to minimize spatial inhomogeneities, is extremely important. The B_0 fields of modern MRI scanners are compensated with passive and superconducting shims to very high orders (*e.g.*, 14th order zonal shims), achieving typical homogeneities of better than 1 ppm over a 30 cm spherical volume. In addition, scanners are equipped with linear gradients for both imaging and shimming as well as higher order resistive shims such as z^2 and xy.

If superconducting shims are already adjusted to maximize field uniformity, why do we need additional adjustable resistive shims? The answer is that any object placed within a magnet field changes the field itself. Known as magnetic susceptibility, and denoted by the dimensionless constant χ_m, all materials are magnetized to some degree such that the observed field is given by:

$$B = \mu_0(1 + \chi_m)H, \tag{6.1}$$

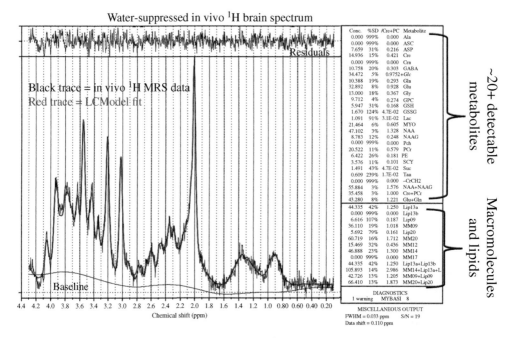

Figure 6.4 A representative 3 T single-voxel *in vivo* ¹H MRS spectrum from a neonatal pig (sLASER, TR/TE = 2000/30 ms, 128 averages, 4 minutes acquisition, $12 \times 12 \times 15$ mm³ voxel) and corresponding spectral fit using LCModel analysis software. A very optimistic set of detectable ¹H-MRS detectable metabolites includes alanine (Ala), ascorbate (Asc), creatine (), choline + phosphocholine (Cho), dehydroascorbic acid (DHA), γ-aminobutyric acid (GABA), glucose (Glc), glutamine (Gln), glutamate (Glu), glycine (Gly), glutathione - reduced (GSH), glutathione-oxidized (GSSG), lactate (Lac), myo-inositol (mI), N-acetyl-aspartate (NAA), N-acetylaspartylglutamate (NAAG), phosphocreatine (PCr), phosphatidylethanolamine (PE), scyllo-inositol (SCY), succinate (Suc), and taurine (Tau).

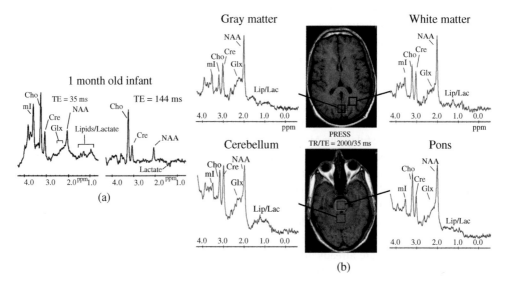

Figure 6.5 Representative ¹H MRS brain spectra acquired from (a) a one-month-old infant, and (b) selected regions across the adult brain.

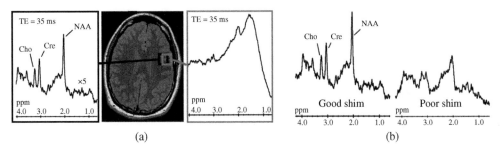

(a)   (b)

Figure 6.6 ^1H MRS challenges include large extracranial lipids and magnetic field inhomogeneities. (a) A slight shift in location for voxels near the skull can significantly change lipid contamination, and (b) poor B_0 homogeneity broadens spectral peaks and hinders accurate quantification.

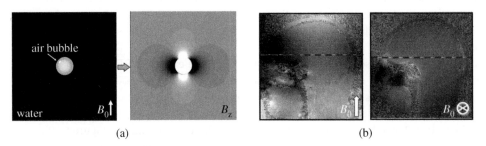

(a)   (b)

Figure 6.7 Magnetic susceptibility and B_0 inhomogeneity (a) the magnetic field surrounding an air bubble in a water phantom placed in a large B_0 magnetic field, (b) representative magnetic field maps from an adult human head for two different orientations of B_0.

where μ_0 is the permeability of free space. The magnetic susceptibility of air $= 0.000004$, while that of water $= -0.000002$. Maximum field shifts seen *in vivo*, largely at air tissue interfaces (and absent artificial materials such as metallic implants), are about ± 10 ppm (see Figure 6.7).

To improve SNR and increase chemical shift separation between peaks, there is significant interest in high magnetic field clinical scanners (≥ 3 T). However, care must be used to ensure excellent magnetic field homogeneity on such systems. Magnetic susceptibility effects scale with B_0, and improved shimming is needed to fully exploit the advantages offered by high-field scanners. If linewidths are dominated by static macroscopic magnetic field inhomogeneities (denoted as T_2^* rather than T_2), SNR only scales as $\sqrt{B_0}$ (see Figure 6.8).

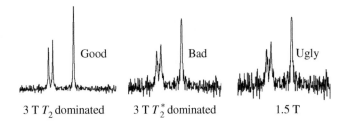

Figure 6.8 High-field MRI systems require better B_0 homogeneity to take full advantage of the higher SNR.

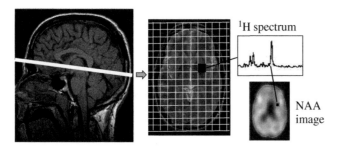

Figure 6.9 ¹H MRSI, whereby spatially resolved spectral information is acquired and typically displayed as metabolite images.

Rather than limiting acquisitions to sequentially acquired single-voxels, MRS sequences can be extended to provide regional information via spectroscopic imaging methods whereby a large volume of tissue is excited, and gradients are then used to provide spatial encoding. As shown in Figure 6.9, images of individual metabolites can be constructed using the spectral information available from each voxel.

A simple approach is to use the PRESS technique. Here, the excited "PRESS box" is enlarged to cover an extended region of interest, and phase encoding is used to provide spatial localization. Spatial saturation pulses can also be added to improve the suppression of extracranial lipids. Typical parameters for a clinical 3D ¹H PRESS MRSI study are TR/TE = 1100/144 ms, 16 × 16 × 8 matrix, 1.5-cm slice, 24-cm FOV, 3.4-cc voxels, and an 18-minute scan time (see Figure 6.10).

The PRESS MRSI methods are generally quite robust, but there are some limitations. Some tissue is invariably missed due to the age-old problem of fitting a square peg (the PRESS box) into a round hole (the skull). In addition, the field-of-view (FOV), spatial resolution, and overall imaging time are not independent. For example, a volumetric PRESS MRSI study covering 16 × 16 × 16 voxels would require over an hour of scan time using a TR = 1.5 s.

More efficient MRSI methods are largely motivated by SNR considerations. Fundamentally, the SNR for a given voxel depends only on the voxel volume and square root of the acquisition time. Namely, SNR∝ $V\sqrt{T_{AD}}$, independent of the number of voxels. Hence, in principle, the achievable spectral SNR should be the same for a single-voxel, single-slice, or volumetric studies given fixed voxel volume and acquisition time.

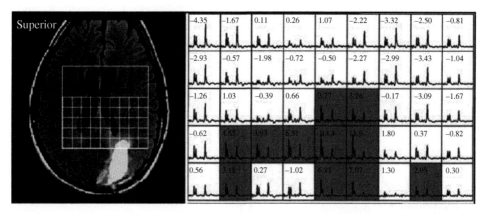

Figure 6.10 Single-slice from a 3D ¹H PRESS MRSI study of a patient with a grade IV glioma, who underwent a subtotal resection along with external beam radiation treatment. Voxels shaded in gray were deemed metabolically abnormal. Source: Adapted with permission from Osorio et al. (2007) Figure 4.

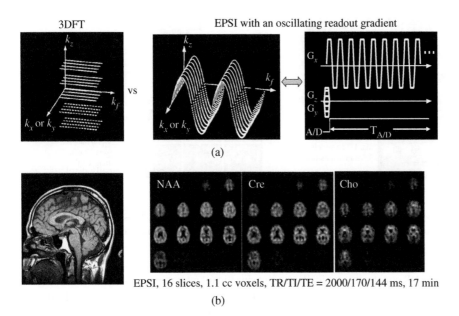

3DFT

EPSI with an oscillating readout gradient

vs

(a)

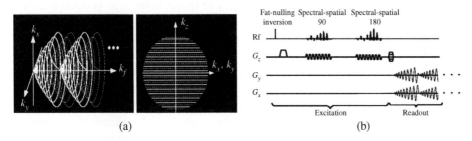

NAA   Cre   Cho

EPSI, 16 slices, 1.1 cc voxels, TR/TI/TE = 2000/170/144 ms, 17 min

(b)

Figure 6.11 ¹H echo-planar spectroscopic imaging (EPSI). (a) Gradient and sampling scheme for EPSI, whereby *k*-space is covered in a series of planar acquisitions. (b) Representative *in vivo* 1.5 T ¹H EPSI study of the healthy human brain.

The k-space view of MRSI is that of extending a 3D (k_x, k_y, k_z) MRI acquisition to a 4D (k_x, k_y, k_z, k_f) one, where k_f = time. An obvious strategy is to use time-varying readout gradients to more efficiently cover k-space, and one approach is to replace one of the phase encoding gradients with an echo-planar imaging (EPI)-like readout. This EPSI (echo-planar spectroscopic imaging) approach can be used to acquire 3D (single-slice) or 4D (volumetric) k-space information in a series of planes (Posse et al. 1997). These acquisitions can either be acquired with a large PRESS box, or, more commonly, inversion recovery for lipid suppression in order to achieve whole-brain coverage. EPSI and additional k-space coverage schemes are shown in Figure 6.11.

Subject to amplitude and slew rate constraints, imaging gradients allow arbitrary movement along k_x, k_y, and k_z. However, physics constrains us to move linearly along $k_f = t$. More efficient than EPSI, spiral MRSI (Adalsteinsson et al. 1998) uses oscillating gradients along two axes, and a highly efficient spiral ¹H acquisition example, using spectral–spatial pulses for water suppression, and inversion recovery for lipid suppression, is shown in Figure 6.12.

(a)   (b)

Figure 6.12 Representative spiral-MRS with spherical coverage in k_x, k_y, and k_z (a) and pulse sequence diagram (b). A typical protocol for a human brain study would include 16 slices of 18 × 18 pixels each, with a field-of-view (FOV) of $24 \times 24 \times 10 \, cm^3$ and TR/TI/TE = 2000/170/144 ms timing parameters. For a spectral bandwidth of 400 Hz and 5-Hz spectral resolution, a minimum of 46 TRs are needed to cover 4D k-space (minute scan time ≈1.5 minute) ss = spectral-spatial Rf pulse.

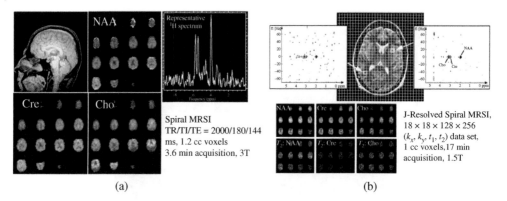

(a)   (b)

Figure 6.13 Spiral MRSI. (a) A volumetric ^{1}H spiral MRSI brain study with TR/TI/TE = 2000/180/144 ms, and 1.2 cc voxels acquired in a 3.6-minute scan time. (b) 2D-J ^{1}H spiral MRSI at 1.5 T. Source: (a) Adapted with permission from Adalsteinsson et al. (1998) Figure 4.

Given that *in vivo* SNR constraints typically require significant averaging, why bother scanning so rapidly? The answer is that more rapid scanning provides increased flexibility to acquire additional information (Adalsteinsson and Spielman 1999). By decoupling the imaging time, voxel size, and FOV requirement, this high efficiency allows implementation of features such as interleaving to increase the FOV and/or spectral bandwidth, Rf phase cycling, spatially-resolved *in vivo* 2D NMR studies. Examples of such fast ^{1}H MRSI studies are shown in Figure 6.13.

Overall, both ^{1}H MRS and ^{1}H MRSI are best viewed as an adjunct to MRI that provide unique metabolic information. Ongoing technical development efforts include improved water/lipid suppression, improved *in vivo* shimming, parallel imaging using phased-array coils, motion-insensitive approaches, automated processing, quantification tools, high-field methods, and multinuclear studies (Maudsley et al. 2021).

6.1.2 Detectable Metabolites and Applications

In this section, we will briefly survey some of the most prominent metabolites detectable *in vivo* using ^{1}H MRS, describe their biochemical roles, and discuss some associated clinical applications. A good overview of this topic can be found in Rae (2014).

Most cells share a common set of metabolic processes critical to supporting life. Among these are converting glucose into energy, maintain ionic and osmotic balances, and ensuring cellular membrane integrity. In addition, depending on the organ or tissue, many cells perform specific functions, an example being neurons sending and receiving action potentials (see Figure 6.14).

The largest peak in the *in vivo* ^{1}H brain spectrum is from the nervous system-specific metabolite N-acetylaspartate (NAA) (Moffett et al. 2007), and there are numerous MRS studies showing changes in NAA levels under multiple neuropathological and psychopathological conditions. However, NAA's precise biochemical role remains unclear. Synthesized in neurons from aspartate, and acetyl-coenzyme A, NAA is a precursor for the synthesis of the neuronal dipeptide N-acetylaspartylglutamate (NAAG). Further, NAA is the second-most-concentrated molecule in the brain after the amino acid glutamate and may also play an important role as a neuronal osmolyte. From a clinical perspective, NAA is primarily found in healthy neurons and oligodendrocytes, and the loss of NAA is often used as a marker of neuronal injury or death (Figure 6.15).

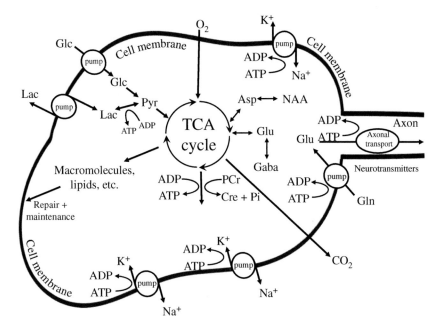

Figure 6.14 Schematic representation of a neuron and some of the major metabolic pathways accessible via *in vivo* MRS. TCA cycle = Tricaboxylic Acid Cycle, also know as the Krebs cycle.

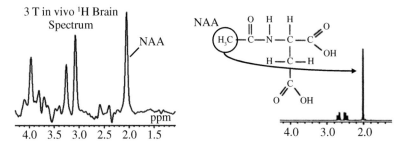

Figure 6.15 NAA is the most prominent peak in an *in vivo* ^1H MRS brain spectrum.

Creatine (Cre) and its phosphorylated partner phosphocreatine (PCr) play fundamental roles in cellular energetics via the creatine kinase reaction, a fast reaction used to buffer cellular ATP levels (see Figure 6.16a). The primary ^1H peaks for Cre resonate at 3.9 and 3.0 ppm, though the 3.9 ppm resonance is often lost *in vivo* due to water suppression. The nominal Cre signal has contributions from both creatine and phosphocreatine and is often labeled as total creatine, tCr; the brain concentration is approximately 5–10 mM. However, as shown in Figure 6.16b, under ideal conditions, it is actually possible to distinguish the individual Cre and PCr resonances, allowing the direct study of the creatine kinase reaction *in vivo*. For the example shown in Figure 6.16b, a study of brain metabolism during cardiopulmonary bypass surgery using deep hypothermia circulatory arrest (DHCA) (Spielman et al. 2022), stopping blood flow to the brain results in the conversion of the remaining brain glucose to lactate (anaerobic metabolism) and a conversion of PCr to Cre as a means of ATP production.

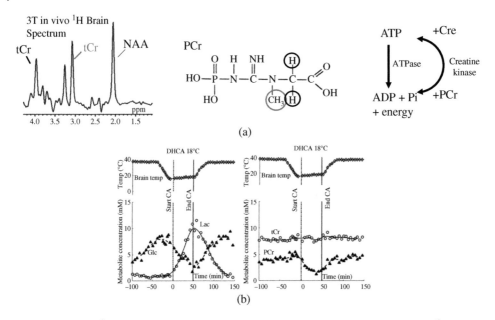

Figure 6.16 The ^{1}H MRS of creatine. (a) Cre chemical structure, and biochemical role. (b) ^{1}H MRS measurement of the temporal changes in brain Lac, Glc, tCr (=Cre + PCr), and PCr for a porcine model undergoing 18 °C deep hypothermia circulatory arrest (DHCA). Source: (b) Adapted with permission from Spielman et al. (2022) Figure 3.

The "Cho" peak detected at 3.2 ppm using *in vivo* ^{1}H MRS actually contains contributions from multiple choline-containing compounds, primarily cell membrane constituents such as phosphocholine and glycerphophocholine. An elevated Cho signal is largely believed to reflect cell membrane repair and synthesis, with brain contributions from chemicals such as the neuro-transmitter acetylcholine very small. Brain choline concentrations are typically in the 1–2.5 mM range, but the presence of nine ^{1}H nuclei improves detection sensitivity.

An elevated Cho signal is seen in multiple tumor types believed to reflect altered mem-brane metabolism and is correlated with cellular density and indices of cellular proliferation (Figure 6.17a). Hence, ^{1}H-MRSI can provide a valuable clinical tool depicting metabolic changes reflective of cellular density, anaplasia, and mitotic index. In the brain, NAA decreases with tumor infiltration and substitution of normal neural and glial cells, hence the Cho/NAA ratio can be an important metric, particularly for primary brain tumors (Shi et al. 2022). ^{1}H-MRSI has also been used to provide guidance for targeted biopsy, surgery, or therapy (Bonm et al. 2020). Outside of the brain, the prostate contains multiple ^{1}H-MRS detectable compounds including citrate (Cit), Cre, Cho, and polyamines, with prostate cancer associated with decreased Cit and increased Cho (Jung et al. 2004) (see Figure 6.17b).

Another prominent peak at 3.6 ppm in the *in vivo* ^{1}H brain spectrum comes from myo-Inositol (mI), which has been suggested as a glial marker and osmolyte (see Figure 6.18). In the brain, the concentration of mI is 4–8 mM. With respect to research and clinical applications, glial cells provide structural support for neurons in addition to involvement in neurotransmitter processes, and elevated mI/Cr ratios have been reported with dementia.

^{1}H MRS is highly sensitive to metabolic changes associated with hypoxic or ischemic injury to the brain, and a key marker of anaerobic metabolism is lactate. The primary ^{1}H MRS lactate

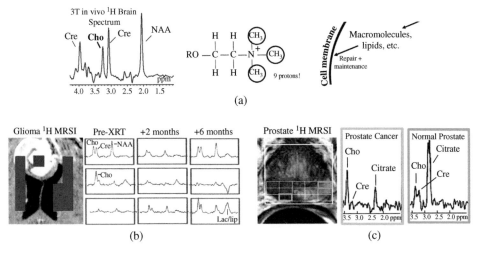

Figure 6.17 ¹H MRS of choline. (a) The ¹H the Cho peak is a biomarker for measuring synthesis or breakdown of cellular membranes. (b) Changes in ¹H MRSI spectra following glioma radiation therapy for three different spatial locations corresponding to normal appearing white matter (red), tumor response (blue), and tumor progression (mauve). (c) ¹H MRSI for the assessment of prostate cancer, whereby an elevated Cho to citrate ratio is indicative of disease. Source: (b) and (c) Adapted with permission from Nelson et al. (2002) Figure 6 and Kurhanewicz et al. (2002) Figure 2, respectively.

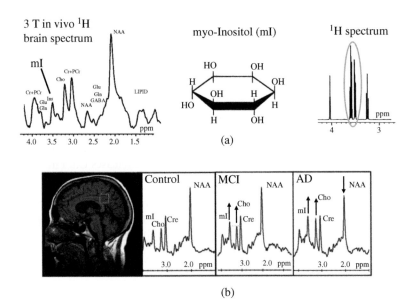

Figure 6.18 ¹H MRS of mI. (a) ¹H MRS brain spectrum, mI chemical structure, and spectral pattern. (b) Elevated mI/Cr ratios are seen with mild cognitive impairment (MCI) and Alzheimer's disease (AD). Source: (b) Adapted with permission from Kantarci (2013) Figure 3.

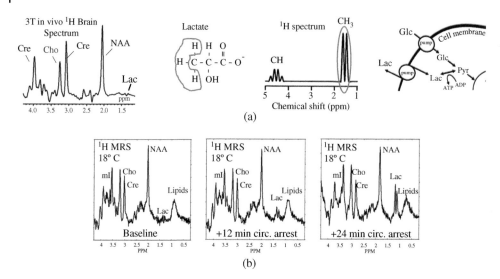

Figure 6.19 ^{1}H MRS of lactate. (a) Brain spectrum, chemical structure, and role in anaerobic metabolism (b) ^{1}H MRS data from a porcine model (acquired at 18 °C) showing the rise of lactate due to ischemia caused by circulatory arrest.

contribution is a doublet centered at 1.3 ppm (see Figure 6.19). Very low in normal brain tissue (~0.4 mM), lactate is elevated during acute hypoxia or ischemia, and may also increase during "secondary" energy failure after reperfusion.

The fundamental function of the brain is to receive and process signals reflecting both external stimuli and internal processes. The electrical signals generated within brain nerve cells, known as action potentials, are passed from one neuron to another via the secretion and later absorption of key chemicals into the synapses located between neuronal dendritic branches. These chemicals are known as neurotransmitters, and multiple neurotransmitters can be found in the brain, some enhancing the transmission of neuronal signals and others retarding them. Glu and GABA are the brain's primary excitatory and inhibitory neurotransmitters, respectively, and both can be detected with *in vivo* ^{1}H MRS.

As shown in Figure 6.20, neurotransmission via Glu involves a complex interplay between neurons and glial cells. Upon receiving an action potential, the primary neuron releases Glu into the

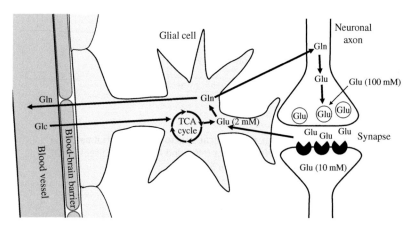

Figure 6.20 Neuronal signals are passed between neurons via the release and uptake of neurotransmitters (*e.g.*, Glu or GABA).

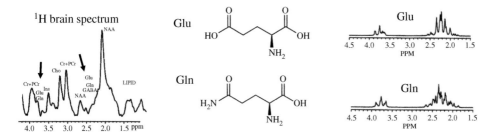

Figure 6.21 ^{1}H MRS of glutamate (Glu) and glutamine (Gln). Gln and Glu have similar chemical structures resulting in overlapping resonances at clinical field strengths, *e.g.*, 3 T.

neuronal cleft. The Glu binds to the secondary neuron, with excess Glu reabsorbed by supporting glial cells. Glu is then converted to Gln, transported back to the primary neuron, and reconverted back to Glu, where it is stored in vesicles awaiting the next action potential. The *in vivo* concentrations of Glu and Gln are 8–10 and 2–3 mM, respectively. The Glu and Gln ^{1}H MRS resonances are heavily overlapped (see Figure 6.21) and often reported as a combined Glx = Glu + Gln.

GABA, as the major inhibitory neurotransmitter in the brain (see Figure 5.26 for the chemical structure and corresponding spectrum), is the subject of considerable research. One of the most active areas is the study of autism spectrum disorder (ASD), where a leading hypothesis for the underlying neuropathological change is an imbalance between excitatory and inhibitory processes (Canitano and Palumbi 2021). The brain concentration of GABA is 1–2 mM, and *J* editing, such as via the MEGA-PRESS method, is the most common ^{1}H MRS measurement technique (Harris et al. 2017).

Oxidative stress, an imbalance between production and removal of reactive oxygen species (ROS) in cells and tissues, and the ability of a biological system to detoxify these products, is thought to drive many diseases of the brain (Kumar et al. 2023). Both the antioxidants glutathione (GSH) and ascorbate (ASC) work in tandem to remove excess ROS species in the brain (see Figure 6.22), and

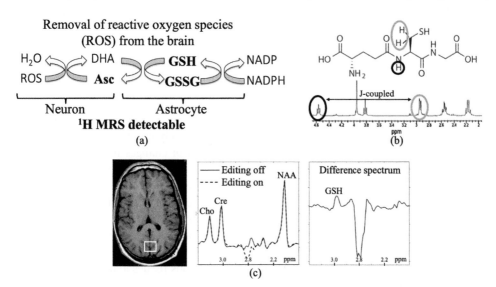

Figure 6.22 Oxidative stress in the brain. (a) The role of the antioxidants glutathione (GSH) and ascorbate (ASC) in eliminating ROS species. Though challenging at clinical field strengths, GSH, GSSG, and ASC are detectable by *in vivo* ^{1}H MRS. (c) ^{1}H MRS spectral editing (MEGA-PRESS) for GSH at 3 T.
GSH/GSSG = reduced/oxidized glutathione. ROS = reactive oxygen species, NADP = nicotinamide adenine dinucleotide phosphate.

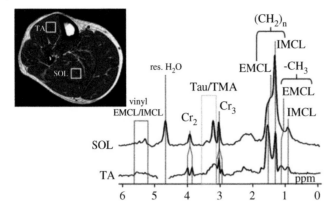

Figure 6.23 Representative ^{1}H MR spectra of skeletal muscle recorded from the tibialis anterior (TA) and soleus muscle (SOL) at 3 T (PRESS, TE = 30 ms). The different fiber orientations in each muscle group give rise to different spectral appearances due to residual dipolar coupling (Cre at 3.05 and 3.9 ppm, Tau and TMA at 3.2 ppm), or susceptibility effects (IMCL/EMCL at 0.9–1.5 ppm). Source: Adapted with permission from Popadic Gacesa et al. (2017) Figure 1.

the robust ^{1}H MRS *in vivo* detection of both of these compounds is an active area of research (Harris et al. 2017).

Outside of the brain, ^{1}H MRS is generally more challenging both due to large lipid signals and motion artifacts (particularly prominent in liver and cardiac studies). One tissue that has seen some notable successes is skeletal muscle (Krššák et al. 2021). To date, ^{1}H-MRS of muscle has a limited clinical role but is used as a research tool to assess intramyocellular lipids, insulin resistance, and type 2 diabetes mellitus. The major peaks seen in a ^{1}H muscle spectrum come from trimethylamines (TMA) and lipids. The lipid signals are particularly interesting in that they exhibit peak splitting due to susceptibility effects (see Figure 6.23). Other resonances show splitting due to residual dipolar coupling (see Exercises E.6.1, E.6.2, and E.6.4).

6.2 ^{31}P-MRS

Given the potential to quantitatively measure real-time metabolic changes *in vivo*, ^{31}P MRS currently has an important role in biomedical research, but, largely due to limited sensitivity, much less so in clinical practice. There are numerous studies of changes in the high energy phosphate constituents (PCr, ATP, and inorganic phosphate) associated with exercise in skeletal muscle (Meyerspeer et al. 2020) (see Figure 6.24). ^{31}P studies of the heart have also made important contributions to our understanding of cardiac energetics, particularly with respect to measurements of the creatine kinase fluxes, ATP/PCr ratios, and intracellular pH in cardiomyocytes (Gupta 2023). With respect to focal lesions, ^{31}P MRS can also be used to measure PCr and phospholipid components of cellular membranes, such as phosphomonoesters (PME) and phosphodiesters (PDE) to help differentiate normal and cancerous tissues and monitor therapy (Sharma and Jagannathan 2022).

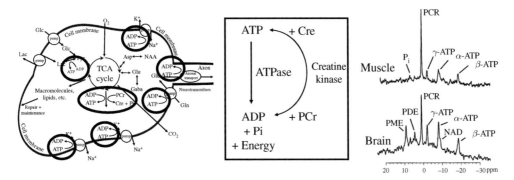

Figure 6.24 ^{31}P MRS provides an important measure of ATP production as well as assessment of cellular membrane constituents, including adenosine triphosphate (ATP), phosphocreatine (PCr), inorganic phosphate (P$_i$), phosphodiesters (PDE), phosphomonoesters (PME), and nicotine adenine dinucleotide (NAD).

6.3 ^{13}C-MRS

Although ^{13}C natural abundance is low, MRS of infused ^{13}C-labeled compounds provide a unique noninvasive tool for measuring critical *in vivo* processes including neurotransmission and energy metabolism. Here, we highlight some of the key methods used and their applications.

6.3.1 Acquisition Methods

The natural abundance of ^{13}C is low, and most *in vivo* studies start with the infusion or injection of key ^{13}C-labeled metabolic precursors such as glucose, acetate, or pyruvate. A major advantage of this approach is that ^{13}C spectra of many biological compounds of interest cover a very broad chemical shift range of almost 200 ppm, greatly facilitating interpretation of complex *in vivo* spectra. The poor sensitivity resulting from the low gyromagnetic ratio of ^{13}C can be addressed by exploiting adjacent *J*-coupled hydrogens using techniques such as INEPT or POCE (discussed in Chapter 5). An additional commonly used method to enhance *in vivo* multinuclear studies is a technique known as **decoupling** (Figure 6.25).

To understand key concepts associated with decoupling, let us start with a homonuclear case that is of some interest to conventional ^1H MRI, namely the behavior of lipid signals in a popular acquisition method known as fast spin echo (FSE). FSE, also called rapid imaging with refocused echoes (RARE) or turbo spin-echo (TSE), depending on the scanner manufacturer, is a scan sequence designed to rapidly obtain T_2-weighted image contrast (Hennig and Friedburg 1988). The method, shown in Figure 6.26, reduces acquisition times by acquiring multiple *k*-space lines in each repetition time (TR) interval. However, investigators soon noted that signals from lipids in FSE images where much brighter than anticipated, having apparent T_2s of 100–150 ms, rather than 30–40 ms when measured using a conventional spin-echo sequence.

Let us analyze this effect using our prior analysis of spin product operators, while gaining some practice with multi-spin systems. Following the development given by Stables et al. (1999), the total x, y, and z coherences for a system with multiple spins can be written as:

$$\hat{F}_x = \sum_j \hat{I}_{xj}, \ \hat{F}_y = \sum_j \hat{I}_{yj}, \ \text{and} \ \hat{F}_z = \sum_j \hat{I}_{zj}. \tag{6.2}$$

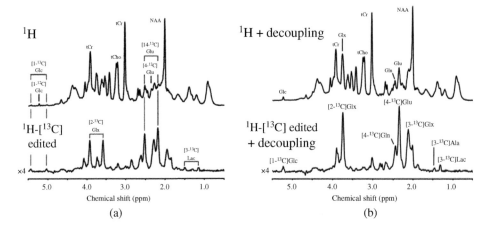

Figure 6.25 Representative ^{13}C-labeled glucose infusion study of the rat brain at 9.4 T (180 µl voxel, TR/TE = 4000/8.5 ms, 9.4 T, 2 hours post [1,6-^{13}C$_2$]glucose infusion) demonstrating the advantages of (a) polarization transfer and (b) decoupling for enhancing and simplifying *in vivo* spectra. Source: Adapted with permission from de Graaf (2019) Figures 8.24 and 8.27. pages 410 and 416, respectively.

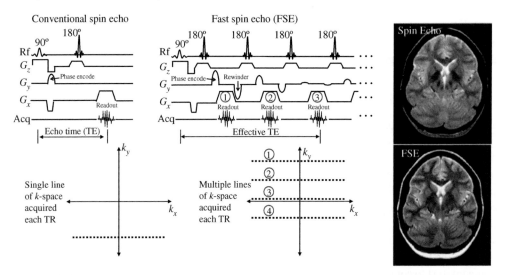

Figure 6.26 FSE imaging. A conventional SE scan acquired a single line of *k*-space each TR, versus the FSE method acquired multiple lines of *k*-space each TR. This reduction in the required TR interval shortens the overall scan time, and the effective FSE image contrast is determined by the T_2-weighting for the central *k*-space lines. While this approach achieves the expected T_2 contrast for brain tissue and CSF, subcutaneous fat is anomalously bright in the FSE images.

Ignoring relaxation, the Hamiltonian during free precession can be written as:

$$\hat{H} = \sum_j \delta_j \hat{I}_{zj} + \sum_{j<k} 2\pi J_{jk} \vec{\hat{I}}_j \cdot \vec{\hat{I}}_k. \tag{6.3}$$

We wish to consider two cases: case 1) Magnetically equivalent spins, *i.e.*, all chemical shifts δ_js are equal, and case 2) Non-equivalent spins with unequal δ_js. For the equivalent spin case,

$$\hat{H} = \sum_j \delta \hat{I}_{zj} + \sum_{j<k} 2\pi J_{jk} \vec{\hat{I}}_j \cdot \vec{\hat{I}}_k = \hat{H}_1 + \hat{H}_2. \tag{6.4}$$

One can readily show that:

$$\left[\sum_j \hat{I}_{pj}, \vec{I}_j \cdot \vec{I}_k\right] = 0 \text{ for } p = x, y, z. \tag{6.5}$$

For example, in 2-spin case, we have:

$$[\hat{I}_z + \hat{S}_z, \hat{I}_x\hat{S}_x + \hat{I}_y\hat{S}_y + \hat{I}_z\hat{S}_z] = \hat{I}_y\hat{S}_x - \hat{I}_x\hat{S}_y + \hat{I}_x\hat{S}_y - \hat{I}_y\hat{S}_x = 0$$

$$[\hat{I}_x + \hat{S}_x, \hat{I}_x\hat{S}_x + \hat{I}_y\hat{S}_y + \hat{I}_z\hat{S}_z] = \hat{I}_z\hat{S}_y - \hat{I}_y\hat{S}_z + \hat{I}_y\hat{S}_z - \hat{I}_z\hat{S}_y = 0$$

$$[\hat{I}_y + \hat{S}_y, \hat{I}_x\hat{S}_x + \hat{I}_y\hat{S}_y + \hat{I}_z\hat{S}_z] = \hat{I}_z\hat{S}_x - \hat{I}_x\hat{S}_z + \hat{I}_x\hat{S}_z - \hat{I}_z\hat{S}_x = 0. \tag{6.6}$$

Consider the theorem: If $\hat{H} = \hat{H}_1 + \hat{H}_2$ and $[\hat{H}_1, \hat{H}_2] = [\hat{F}_x, \hat{H}_2] = [\hat{F}_y, \hat{H}_2] = [\hat{F}_z, \hat{H}_2] = 0$, then the observed signal is independent of $\hat{H}_2$. The proof is as follows:

$$\text{Tr}(\hat{F}_p e^{-i\hat{H}t}\sigma(0)e^{i\hat{H}t}) = \text{Tr}\left(\hat{F}_p e^{-i(\hat{H}_1 t + \hat{H}_2 t)}\sigma(0)e^{i(\hat{H}_1 t + \hat{H}_2 t)}\right)$$

$$= \text{Tr}\left(\hat{F}_p e^{-i\hat{H}_2 t}e^{-i\hat{H}_1 t}\sigma(0)e^{i\hat{H}_1 t}e^{i\hat{H}_2 t}\right)$$

$$= \text{Tr}\left(e^{i\hat{H}_2 t}\hat{F}_p e^{-i\hat{H}_2 t}e^{-i\hat{H}_1 t}\sigma(0)e^{i\hat{H}_1 t}\right)$$

$$= \text{Tr}\left(\hat{F}_p e^{i\hat{H}_2 t}e^{-i\hat{H}_2 t}e^{-i\hat{H}_1 t}\sigma(0)e^{i\hat{H}_1 t}\right)$$

$$= \text{Tr}\left(\hat{F}_p e^{-i\hat{H}_1 t}\sigma(0)e^{i\hat{H}_1 t}\right) \text{ QED.} \tag{6.7}$$

However, lipids, as shown in Figure 6.27, consist of multiple J-coupled resonances (Hamilton et al. 2009), and lipid ^{1}H nuclei are not equivalent, hence the δ_js not all equal. Thus, for Case 2 we have:

$$\left[\sum_j \delta_j \hat{I}_{zj}, \vec{I}_j \cdot \vec{I}_k\right] \neq 0, \tag{6.8}$$

and J coupling can NOT be ignored.

Let us now consider the pulse sequence shown in Figure 6.28a (also known as a Carr–Purcell–Meiboom–Gill (CPMG) acquisition) for the case of a multi-spin system with non-equivalent spins. Assuming non-selective Rf pulses with $\omega_1 \gg \delta_j, J_{jk}$ for all j, our goal is to find the Hamiltonian for each time interval. Again, we can ignore any relaxation effects, but there will be no need in this case for the weak coupling assumption.

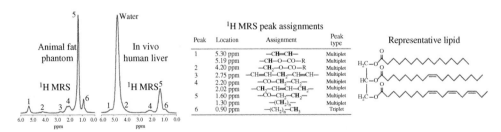

Figure 6.27 *In vivo* ^{1}H spectra from an animal fat phantom and the human liver highlighting the major lipid peaks, their associated J coupling patterns, and representative chemical structures. Source: Adapted with permission from Hamilton et al. (2009) Figure 1 and Table 1.

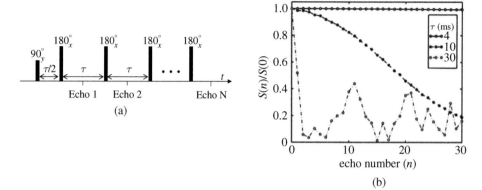

Figure 6.28 (a) A representative Carr–Purcell–Meiboom–Gill (CPMG) Rf pulse train (note similarity to fast spin echo). (b) Effects of J coupling on a A_3B_3 spin system showing signal versus echo number of CPMG sequence where τ is the spacing between echoes. The intrinsic T_2 decay is ignored, and the notation A_nB_ms denotes a strongly coupled system with n A and m B spins. Source: (b) Adapted with permission from Stables et al. (1999) Figure 1.

During the $90°_y$ Rf pulse, $\hat{H}_1 = \omega_1 \sum_j \hat{I}_{yj} = \omega_1 \hat{F}_y$ where $\omega_1 = -\gamma B_1 \gg \delta_j$, J_{jk} for all j, k, whereas during the 180_x pulses $\hat{H}_2 = \omega_1 \hat{F}_x$. Finally, between Rf pulses, the Hamiltonian is $\hat{H}_3 = \sum_j \delta_j \hat{I}_{zj} + \sum_j 2\pi J_{jk} \vec{\hat{I}}_j \cdot \vec{\hat{I}}_k$. Assume the system is thermal equilibrium at the start of the sequence, namely $\hat{\sigma}(0) \propto \hat{F}_z$. After the 90_y, $\hat{\sigma} \propto \hat{F}_z$. Before the first $180°_x$,

$$\hat{\sigma} \propto e^{-\frac{i}{2}\hat{H}_3\tau} \hat{F}_x e^{\frac{i}{2}\hat{H}_3\tau}, \tag{6.9}$$

which after the $180°_x$ becomes:

$$\hat{\sigma} \propto e^{-i\pi\hat{F}_x} e^{-\frac{i}{2}\hat{H}_3\tau} \hat{F}_x e^{\frac{i}{2}\hat{H}_3\tau} e^{i\pi\hat{F}_x}. \tag{6.10}$$

Let us rewrite Eq. (6.10) as $\hat{\sigma} \propto \hat{B}\hat{A}\hat{F}_x\hat{A}^{-1}\hat{B}^{-1}$ where $\hat{A} = e^{-\frac{i}{2}\hat{H}_3\tau}$ and $\hat{B} = e^{-i\pi\hat{F}_x}$. At the first echo, $\hat{\sigma}_1 \propto (\hat{A}\hat{B}\hat{A})\hat{F}_x(\hat{A}\hat{B}\hat{A})^{-1}$. Continuing, the spin density at the nth echo will be $\hat{\sigma}_n \propto (\hat{A}\hat{B}\hat{A})\hat{F}_x(\hat{A}\hat{B}\hat{A})^{-n}$. Now examine $\hat{\sigma}_n \propto (\hat{A}\hat{B}\hat{A})\hat{F}_x(\hat{A}\hat{B}\hat{A})^{-n}$ more closely for the case where τ is short, namely $|J_{jk}\tau|$, $|\delta_j\tau| \ll 1$ for all spin groups j, k. Expanding $\hat{A}$ to first order in a Taylor series yield:

$$\hat{A} = e^{-\frac{i}{2}\hat{H}_3\tau} \cong 1 - \frac{i}{2}\hat{H}_3\tau = 1 - \frac{i}{2}\sum_j \delta_j \hat{I}_{zj}\tau - \frac{i}{2}\sum_{j<k} 2\pi J_{jk} \vec{\hat{I}}_j \cdot \vec{\hat{I}}_k\tau. \tag{6.11}$$

Substituting, we get:

$$\hat{A}\hat{B}\hat{A} = e^{-\frac{i}{2}\hat{H}_3\tau} e^{-i\pi\hat{F}_x} e^{-\frac{i}{2}\hat{H}_3\tau} \xrightarrow{\text{short }\tau} \left(1 + i\sum_{j<k} 2\pi J_{jk} \vec{\hat{I}}_j \cdot \vec{\hat{I}}_k\tau\right) e^{-i\pi\hat{F}_x}. \tag{6.12}$$

However, $\hat{F}_x$ and $e^{-i\pi\hat{F}_x}$ commute. Hence, in a CPMG sequence with $|J_{jk}\tau|$, $|\delta_j\tau| \ll 1$, $\hat{\sigma}_n \approx \hat{F}_x$, independent of n, δ_j, and J_{jk}! Therefore, for this rapidly refocused CPMG sequence, lipids will decay with their true T_2s, free from the additional dephasing due to J coupling. The effects of J coupling have been removed, i.e., the system is effectively decoupled. The phantom study in Figure 6.28 shows the effects of changing the refocusing interval τ.

The suppression of J coupling during an FSE sequence is considered a nuisance in clinical MRI, often requiring fat suppression to be added to the acquisition sequence. However, this effect is often exploited in other NMR studies. Consider the more general technique known as ***spin locking***.

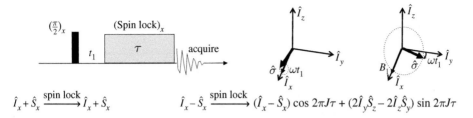

$$\hat{I}_x + \hat{S}_x \xrightarrow{\text{spin lock}} \hat{I}_x + \hat{S}_x \qquad \hat{I}_x - \hat{S}_x \xrightarrow{\text{spin lock}} (\hat{I}_x - \hat{S}_x)\cos 2\pi J\tau + (2\hat{I}_y\hat{S}_z - 2\hat{I}_z\hat{S}_y)\sin 2\pi J\tau$$

Figure 6.29 A spin locking pulse sequence and associated spin operators. Spin locking is the basis for several NMR pulse sequences such as TOCY, HOHAHA, and ROESY. Source: Adapted with permission from van de Ven (1996), Figures 4.42 and 4.43, pp. 242–243.

As illustrated in the rotating frame picture of Figure 6.29, consider the application of a long, strong, continuous Rf pulse along a specified axis, in this case x. Chemical shift is suppressed, and spins are rendered effectively equivalent. Coherences along x are retained, while those along y are dephased due to Rf inhomogeneities. In general, the observable magnetization for truly equivalent spins does not evolve under J coupling, however, spins rendered temporarily equivalent can show much more complex behavior, as they can enter the spin lock period in a variety of initial states.

In the heteronuclear case, for example, with *in vivo* ^{1}H-^{13}C studies, line splitting reduces the ability to distinguish and quantify individual metabolites, and decoupling is often required.

Decoupling involves the use of a long strong Rf pulse on the coupled partner of the spin being observed. However, the use of hard 180°s, separated by a time interval τ, as illustrated in Figure 6.30 requires significant Rf peak power. If the refocusing rate is on a time scale $\ll 1/J$, the timings of the 180°s become unimportant, in which case continuous lower power Rf decoupling methods can be used.

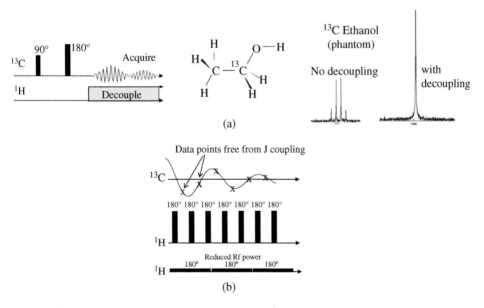

Figure 6.30 Heteronuclear decoupling. (a) Example of ^{1}H decoupling collapsing peak splitting from to ^{1}H-^{13}C J couplings. (b) Decoupling can be achieved via the use of hard 180 s or even with an effectively continuous Rf to reduce the peak Rf power requirements.

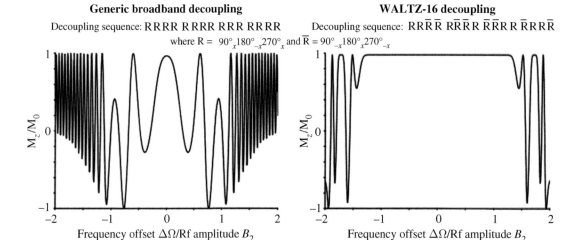

Figure 6.31 Broadband decoupling via the use of supercycles of composite Rf pulses (a). The improved performance of a WALTZ-16 sequence is also shown in (b). Source: Adapted with permission from (de Graaf 2019) Figure 8.29, p. 412.

The problem with long continuous Rf decoupling pulses is that the corresponding very narrow bandwidth is unlikely to cover the required chemical shift range for all spins of interest. Phase cycling the 180° pulses, such as use of the pulse train $R = 180°_x 180°_{-x} 180°_x \cdots$ for decoupling, considerably improves this off-resonance behavior. Composite 180° pulses provide even better broadband decoupling, *e.g.*, $R = 90°_x 180°_{-x} 270°_x = 1\bar{2}3$. Such composite pulses are usually used in in supercycles such as $1\bar{2}3\ 1\bar{2}3\ \overline{1}2\overline{3}\ \overline{1}2\overline{3}$, leading to the most common broadband decoupling scheme known as Wideband Alternating-phase Low-power Technique for Zero-residue Splitting (WALTZ) (see Figure 6.31). For *in vivo* studies, reducing Rf power deposition, as measured by the specific absorption rate (SAR), to minimize tissue heating, is usually the limiting constraint (Freeman and Hurd 1997).

6.3.2 ^{13}C Infusion Studies

Via the infusion of istopically-labeled substrates, *in vivo* ^{13}C NMR is now well established as an extremely powerful tool for the assessment of neuroenergetics, neurotransmission, and associated metabolic fluxes in both preclinical and human studies. Despite several technical challenges, including overcoming the low intrinsic SNR while minimizing Rf power deposition, *in vivo* studies have established the relationship between neuroenergetics and neural function, measured TCA cycling and neurotransmitter synthesis rates, and assessed how these critical processes are altered in multiple neurological and psychiatric diseases (de Graaf et al. 2011). Figure 6.32 shows some representative data from a review paper by de Graaf et al. (2003).

6.3.3 Hyperpolarized ^{13}C

A fundamental limitation for *in vivo* ^{13}C MRS studies is the fact that ^{13}C nuclei are low-natural-abundance weak magnets, resulting in long acquisition times from relatively large volumes of

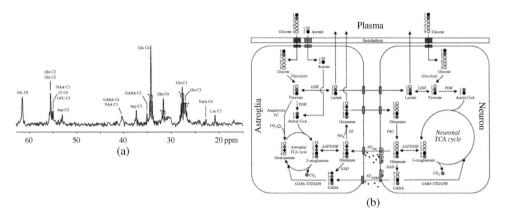

Figure 6.32 Proton-decoupled ^{13}C infusion study of brain metabolism. (a) 7T ^{13}C spectrum from rat brain spectrum acquired 60–90-minute post intravenous [1,6-^{13}C$_2$]-glucose infusion. (b) A model of cerebral metabolism in neuronal and astroglial compartments and metabolite ^{13}C labeling patterns following the intravenous infusion of [1-^{13}C]-glucose (black circle), [2-^{13}C]-glucose (gray circle) or [2-^{13}C]-acetate (black circle). Source: Adapted with permission from de Graaf et al. (2003) Figure 1.

tissue. The introduction of hyperpolarized metabolic MRI with ^{13}C-labeled substrates (HP^{13}C) has emerged as a powerful technique for *in vivo* assessments of real-time metabolism. The basic idea is to magnetically prepare an injectable biological substrate, using a process known as dynamic nuclear polarization (DNP), such that the available MR signal is enhanced >10^4 fold, enabling *in vivo* detection of both the substrate and downstream metabolic products (Ardenkjaer-Larsen et al. 2003).

As shown in Figure 6.33, one of the fundamental metabolic alternations that allows tumor cells to maintain high rates of cellular proliferation is a shift of the catabolism of glucose from the most energetically favorable oxidative phosphorylation (OXPHOS) pathway to aerobic glycolysis (GLY) as a means of generating energy while preserving biomass (a process known as the **Warburg effect**, or more generally **metabolic reprogramming**) (Vander Heiden et al. 2009).

In addition to altered metabolism, tumors generally exhibit significantly increased glucose uptake, which can be measured using ^{18}F-fluorodeoxyglucose PET (FDG-PET), the most clinically successful metabolic imaging technique to date. However, following cellular uptake, FDG is trapped during the first step of glucose metabolism and is thus, in contrast to hyperpolarized ^{13}C-labeled pyruvate (Pyr), unable to image the Warburg effect in tumors, nor detect the hallmarks of ischemia or viability in other organs such as the heart.

In conventional MRI, effectively only a few spins per million are preferentially aligned with the main magnetic field, and there are several methods that have been proposed to dramatically increase polarization. These include optical pumping (limited to Noble gases (Perron and Ouriadov 2023)) and parahydrogen-induced polarization (PHIP (Buntkowsky et al. 2022)). However, the most successful technology to date is hyperpolarization via DNP. So, let us explore how this method achieves the large SNR increase required for dynamic *in vivo* HP^{13}C studies.

The thermal equilibrium magnetization can be written as the product of (spins/unit volume)×(magnetic moment)×(spin polarization). Namely,

$$M_o = \rho \frac{\gamma^2 \hbar^2 B_0}{4kY} = \rho \left(\frac{\gamma \hbar}{2} \right) \left(\frac{\gamma \hbar B_0}{2kT} \right). \tag{6.13}$$

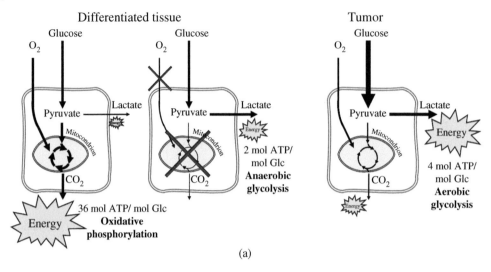

(a)

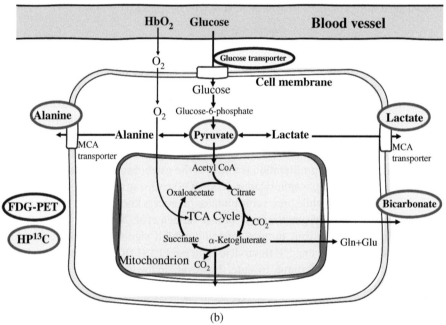

(b)

Figure 6.33 The Warburg Effect. (a) Tumors show Glc uptake and a metabolic shift from OXPHOS to GLY. (b) Glc uptake can be imaged using ^{18}F-FDG-PET, whereas HP^{13}C-Pyr and its products (^{13}C-Lac, ^{13}C-Ala, ^{13}C-Bic) allow assessing the balance between OXPHOS and GLY.

Outside of buying a higher field MRI scanner (of limited utility as M_o only scales linearly with B_0 whereas the financial costs of MRI scanners generally scale at a much faster rate), methods to improve sensitivity include (1) increasing ρ (*e.g.*, injecting isotopically enriched ^{13}C substrates), (2) operating at a lower temperature, and (3) increasing polarization via changing the "effective" γ. *In vivo* HP^{13}C exploits all three of these methods.

Beyond the serious biological implications, lowering the temperature of the subject is not very effective as temperatures need to reach the milli-Kelvin range to appreciably increase ^{1}H or ^{13}C

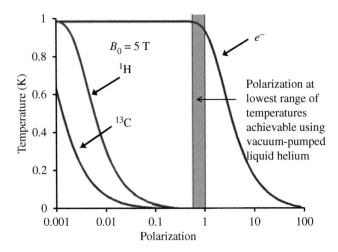

Figure 6.34 Brute-force polarization. Electrons are almost fully polarized at 5 T and 1 K, however, even at this field strength and very low temperature, ^{1}H and ^{13}C nuclei are not appreciably polarized.

polarizations. Fortunately, Figure 6.34 contains the clues needed for an effective solution. Preparing the sample in a separate magnet before injecting it into the patient holds some promise. Namely, at temperatures achievable using vacuum-pumped liquid helium systems (slightly lower than 1 K), unpaired electrons are almost fully polarized (due to their larger γ), although ^{13}C nuclei remain poorly polarized. To analyze how this situation may be exploited, we first need to introduce the concept of the temperature of a spin system.

The probability, P_n, of finding a system in a specific state is dependent on the energy $\mathscr{E}_n$ of that state as given by the Boltzmann distribution $P_n = \frac{1}{Z}e^{-\mathscr{E}_n/kT}$, where k is the Boltzmann constant, T is the absolute temperature, and Z normalization factor. If the energy levels of a spin system are populated in accordance with the Boltzmann distribution, we can associate a well-defined temperature, T_s, with the system (see Figure 6.35). As an interesting note, for systems with a finite number of energies levels, *e.g.*, spin systems, the temperature, as defined, can be negative.

Now consider a spin system in a magnetic field, B_0, consisting of unpaired electrons and nearby ^{13}C nuclei. If the density of unpaired electrons is high enough, the electrons are coupled to each other (via dipolar coupling) as well as to nearby nuclei, e^--e^- dipolar coupling splits the energy levels, resulting in continuous bands for the electron energies (see Figure 6.36).

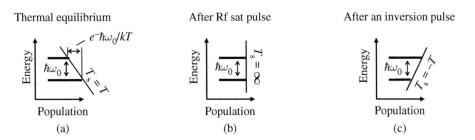

Figure 6.35 Spin temperature. The temperature of a system of spin-½ particles at (a) thermal equilibrium ($M_z = M_0$), (b) after an Rf saturation pulse ($M_z = 0$), and (c) after an inversion pulse ($M_z = -M_0$).

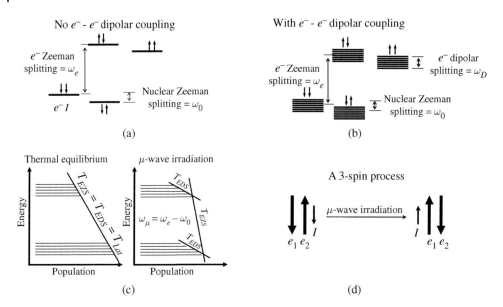

Figure 6.36 Dynamic nuclear polarization by thermal mixing. (a) Energy levels for a dipolar coupled electron-nuclear ($e^- - I$) spin system. (b) With $e^- - e^-$ dipolar coupling, the energy levels for the electrons broaden. (c) At thermal equilibrium, the temperatures of the lattice (T_{Lat}), electron Zeeman system (T_{EZS}), and the electron dipolar-coupled system (T_{EDS}) are equal. Heating the electron Zeeman system, via μ-wave irradiation, cools the electron dipolar-coupled spin system. (d) Overall, cooling the nuclear Zeeman system is a three-spin process whereby two electron spin flips drive one nuclear spin flip.

If the e^- dipolar splitting, ω_e, is greater than the nuclear Zeeman splitting, ω_D, then microwave (μ-wave) irradiation can induce what is known as the ***thermal mixing effect*** (Abragam 1983; Abragam and Goldman 1978). For completeness, there are other types of physical interactions that can also be induced, including the ***Overhauser effect*** and the ***solid effect*** (Abragam and Goldman 1978; Abragam 1983), but thermal mixing is the dominate process exploited in HP^{13}C. Assuming a static magnetic field, such as $B_0 = 5$ T, and thermal equilibrium, all spin systems are at the temperature of the lattice (T_{Lat}). Then μ-waves can be used to heat the e^- Zeeman system (temperature = T_{EZS}), which, with rapid e^- spin T_1 relaxation, cools the e^- dipolar system (temperature = T_{EDS}). Assuming the e^- dipolar system is in "thermal" contact (here thermal contact is essentially defined as being able to exchange energy) with nuclear Zeeman system (temperature = T_{NZS}), this results in a cooling of the nuclear Zeeman system and an enhanced nuclear polarization. Overall, this is a three-spin process whereby two electron spin flips drive one nuclear spin flip (see Figure 6.36). In effect, with the wave irradiation, the high electron polarization is transferred to the coupled nuclei.

The overall process of dissolution DNP as used in HP^{13}C is depicted in Figure 6.37. The substrate of choice, for example [1-^{13}C]pyruvate (Pyr), is first mixed with a free radical (source of unpaired electrons), and then placed in the polarizer consisting of a high field magnet (*e.g.*, 5 T) equipped with a μ-wave irradiator and vacuum-pumped liquid helium cooling system. Operating at a temperature of 0.8 K, materials such as [1-^{13}C]Pyr, take 1–2 hours. to reach ~50% ^{13}C polarization, after which the sample is removed via the injection of a solvent (typically water) at high temperature and pressure. The goal is to rapidly remove the sample from the polarizer, after which the polarization will decay with its intrinsic T_1 relaxation time. For use in clinical imaging, quality controls must be performed on the sample to ensure purity and sterility before it is injected into the

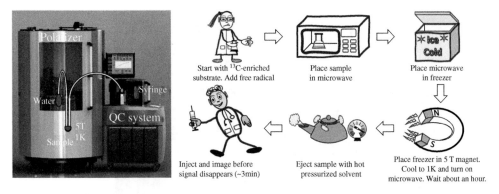

Figure 6.37 Dissolution DNP. Schematic of the polarizer and polarization/dissolution steps needed to perform clinical HP^{13}C.

patient via an intravenous (iv) line. Note, the final set of steps, including sample removal, quality control checks, IV injection, and MRI data acquisition must be performed rapidly, as typically T_1s are about 60 s *in vitro* and closer to 30 s *in vivo* (Hurd et al. 2012).

Not every substrate is suitable for DNP studies. *In vivo* imaging requirements include low toxicity (substrates are injected at mM concentrations), long T_1 relaxation times to allow the substrate to both reach the organ(s) of interest, be rapidly taken into cells, and then metabolized. There is also need for sufficient chemical shift separation between the substrate and metabolic products to allow robust MRS quantification. To date, the focus has been on low molecular weight endogenous compounds having long T_1s, with the first substrate to reach human studies being [1-^{13}C]Pyr (*in vivo* $T_1 = 30$ s), for which a 25% polarization level achieves a ~30,000-fold signal gain! A depiction of representative *in vivo* hyperpolarized [1-^{13}C] Pyr data is shown in Figure 6.38. Quantitative interpretation of these studies includes accounting for blood-brain barrier transport (for brain studies (Hurd et al. 2013)), robust metabolic modeling (Larson and Gordon 2021), and differentiating true metabolic fluxes from isotopic exchange processes (Witney et al. 2011).

An increasing number of clinical studies have demonstrated that clinical HP^{13}C is not only viable but may provide important metabolic information previously inaccessible using other techniques. Multiple rapid spectroscopic imaging methods have been proposed for *in vivo* imaging, including dynamic echo-planar spectroscopic imaging (EPSI) (Yen et al. 2009), spiral MRSI (Mayer

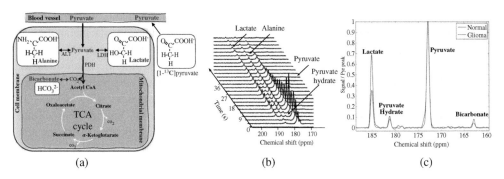

Figure 6.38 *In vivo* HP^{13}C. (a) ^{13}C labeling of lactate, alanine, and bicarbonate with the intravenous injection of [1-^{13}C]pyruvate. (b) Stack plot of dynamic ^{13}C-MRS spectra from a rat liver acquired every 3 s following the bolus injection of HP [1-^{13}C]pyruvate. (c) HP ^{13}C spectra from a rat C6 glioma model showing increased lactate and decreased bicarbonate labeling in the tumor, reflecting the Warburg effect.

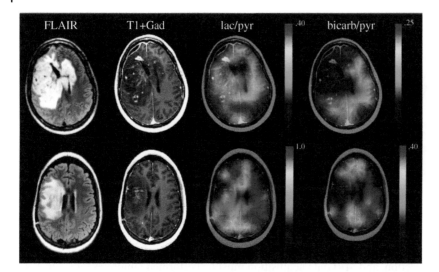

Figure 6.39 *In vivo* HP^{13}C. Representative HP^{13}C metabolite ratio image overlays from two patients with glioma. Lesion Lac/Pyr ratios differed between subjects, but, in both cases, the Bic/Pyr was higher in regions of normal appearing brain. Source: Adapted with permission from Park et al. (2018) Figure 7.

et al. 2010), and metabolite-interleaved EPI/spiral imaging (Lau et al. 2010). Some of the primary clinical targets to date include oncology (particularly studies of brain tumors (Wang et al. 2019) and prostate cancer (Kurhanewicz et al. 2019)), heart disease (Schroeder et al. 2013; Cunningham et al. 2016), and diabetes as well as metabolic diseases of the liver and kidney (von Morze et al. 2021). Figure 6.39 shows representative clinical Warburg imaging results of glioblastoma. Other HP^{13}C probes being investigated in preclinical studies (the exception being [2-^{13}C]Pyr, which has been used in humans), include [1-^{13}C]bicarabonate, [2-^{13}C]Pyr, [1-^{13}C]lactate, ^{13}C-urea, [1-4-^{13}C$_2$]fumarate, [1-^{13}C]dehydroascorbate, and [1-^{13}C]alpha-ketoglutarate (Wang et al. 2019).

6.4 Deuterium Metabolic Imaging

While HP^{13}C has generated much interest in the medical imaging community, obstacles associated with hyperpolarization, substrate administration, high costs, and supraphysiological substrate dosages remain significant challenges. The introduction of deuterium metabolic imaging (DMI) (De Feyter et al. 2018) (see Figure 6.40) offers an attractive alternative by allowing the metabolic fate of deuterated substrates, such as [6,6'-^{2}H$_2$]Glc, [^{2}H$_3$]-acetate, and [2,3-^{2}H$_2$]fumarate (Hesse et al. 2022), to be monitored *in vivo*. Furthermore, the low natural abundance of deuterium avoids the ^{1}H MRS need for water and lipid suppression, and the sparsity of the DMI ^{2}H spectrum minimizes complications arising from B_0 inhomogeneities (De Feyter and de Graaf 2021). Although the ^{2}H magnetic moment is approximately 6.5× lower than that of ^{1}H, a short T_1 relaxation time (due to quadrupolar coupling of the spin-1 ^{2}H nucleus) improves sensitivity by allowing short imaging repetition times, partially offsetting the lack of the enhanced polarization as exploited in HP^{13}C (Polvoy et al. 2021).

With respect to clinical cancer imaging, following oral consumption of deuterated glucose, DMI can map the balance of glycolysis and mitochondrial metabolism; analogous to the HP^{13}C Lac/Bic ratio, the ratio of ^{2}H-Lac/^{2}H-Glx (glutamate + glutamine) provides a direct measure of the Warburg effect (see Figure 6.41). In preclinical models, deuterated fumarate, administered by both

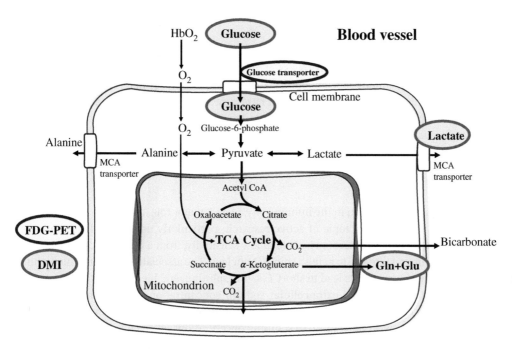

Figure 6.40 Deuterium metabolic imaging. Glucose metabolic products detectable using DMI of $[6,6'-^2H_2]$-Glc, as compared to FDG-PET (Compare with Figure 6.33b).

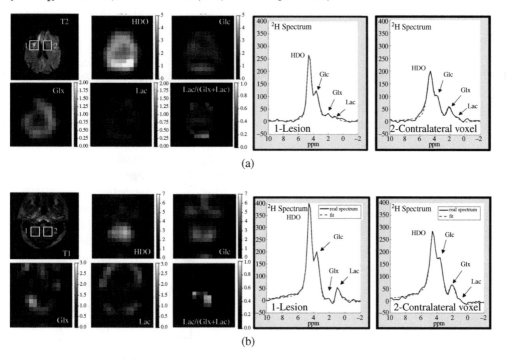

Figure 6.41 Representative DMI studies at 3 T of patients with brain tumors. (a) MRI and ^2H metabolic maps of (a) patient diagnosis with an anaplastic oligodendroglioma, Grade III, but currently without evidence of active disease (No elevated lactate is observed), and (b) patient with cerebellar pilocytic astrocytoma with anaplastic features. Decreased Glx and elevated Lac are observed in the lesion. Scan parameters: 42 minutes DMI acquisition, starting 45 minutes after oral ingestion of 60 g of $[6,6'-^2H_2]$Glc, 3D phase-encoded CSI with spherical k-space encoding, TR = 350 ms, 90°-degree flip angle.

intravenous injection and oral gavage, has also been demonstrated to measure tumor cell death via the production of malate (Hesse et al. 2022).

Alternatively, deuterated substrates can also be detected by observing changes in the ^{1}H spectrum using either indirect methods (Niess et al. 2023) or decreases in the ^{1}H peaks arising from ^{2}H-to-^{1}H exchange among labeled and unlabeled compounds. ^{1}H detection has the major advantage of requiring no additional multinuclear hardware added to a conventional MRI scanner, although SNR is limited, and precise spectral fitting algorithms in combination with excellent B_0 homogeneity are required for robust *in vivo* measurements (Cember et al. 2022).

6.5 ^{23}Na-MRI

Sodium, the most abundant cation in the human body, plays a major role in multiple cellular functions, and *in vivo* ^{23}Na MRI is a topic of active research, particularly using high-field scanners ($\geq$3 T). Though challenging to detect *in vivo* due to low sensitivity, from a MR physics perspective, ^{23}Na is particularly interesting due to being a spin-3/2 nucleus. The associated static and fluctuating quadrupolar coupling components lead to short T_1s, line splitting, and biexponential T_2 decays.

The balance between intra- and extra-cellular sodium concentrations (approximately 10 and 150 mM, respectively) is critical for maintaining cellular homeostasis, ionic balances, action potentials, and cell volumes. Disruptions in these sodium concentrations are seen in multiple diseases such as stroke and tumors (Madelin and Regatte 2013). In cartilage, positive sodium ions balance the negatively charged proteoglycans responsible for proper hydration, a process disrupted in osteoarthritis (Choi and Gold 2011).

6.6   Summary

This chapter discussed a wide range of compounds that can be targeted with *in vivo* ^{1}H, ^{31}P, and ^{13}C MR spectroscopy to provide novel insights into *in vivo* metabolic processes present under both normal and pathological conditions. We have now reached the midpoint of this textbook (see Figure 6.42). The remaining chapters build upon the MR physics developed in these early chapters

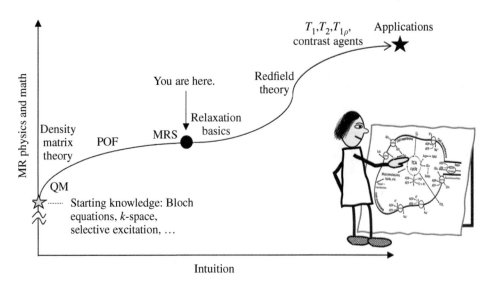

Figure 6.42   The roadmap at the end of Chapter 6.

to gain a deeper understanding of relaxation processes and how these are exploited to generate unique *in vivo* information and image contrast.

Exercises

E.6.1 A Strongly Coupled Spin System

At 3 T, citrate is an example of a strongly coupled two-spin system with $\Omega_I - \Omega_S = 0.15$ ppm, $J = 15.4$ Hz, and $T_2 = 150$ ms. Shown below are representative *in vitro* and *in vivo* ^{1}H spectra, along with the pulse sequence most often used to measure citrate in the human prostrate. (Source: Spectra adapted with permission from Kobus et al. (2015), Figures 3b and 3e.)

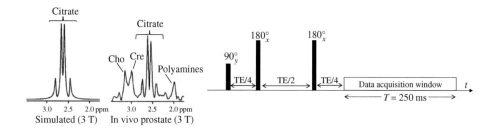

a) Using the full density matrix, simulate (*e.g.*, via Matlab) the measured citrate spectrum for a series of TEs ranging from 50 to 300 ms. Use a first order approximation for the effect of T_2 by simply assuming the received signal is weighted by a factor of e^{-t/T_2} (*i.e.*, consider relaxation only during the data acquisition window).

b) What echo time, TE, would you choose to maximize the detection of citrate?

E.6.2 Creatine in Skeletal Muscle

The $-CH_3$ and $-CH_2$ groups of creatine, a metabolite involved in cellular energetics, gives rise to a singlet peaks at 3.0 and 4.0 ppm in an *in vivo* ^{1}H-MRS brain spectrum. However, the same compound, when measured in skeletal muscle, gives rise to split resonances (note: TMA = trimethylamines). Source for figure: Adapted with permission from Boesch and Kreis (2001) Figure 2.

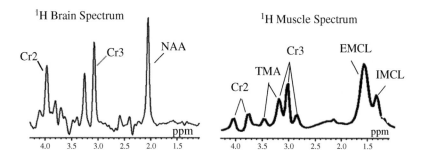

a) Suggest an explanation for this effect (Hint: The effect is *not* due to *J* coupling).
b) How might you test your hypothesis?

E.6.3 Magnetic Susceptibility

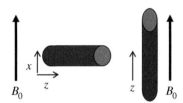

a) You are asked to scan a cylindrical water-filled phantom. Calculate the frequency shift of the water peak if the long axis of cylinder is aligned parallel to the main magnetic field (*i.e.*, along the z axis) as compared to perpendicular to B_0 (*e.g.*, along the x axis). For the purposes of this problem, you may ignore end effects by assuming an infinitely long cylinder.

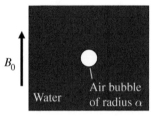

b) You are asked to scan a large water-filled phantom that has a spherical air bubble in the center. Calculate the magnetic field surrounding the air bubble. For the purposes of this problem, you may ignore end effects by assuming an infinitely large water phantom.

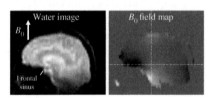

c) A sagittal field map (*i.e.*, image in which pixel intensity is proportional to the strength of the local magnetic field) of a human brain is shown below. Use the results from (a) and (b) to explain the observed B_0 inhomogeneity. Will the homogeneity in the frontal lobes change if the patient's head is positioned at a different angle with respect to the main magnetic field?

E.6.4 Lipids and Skeletal Muscle

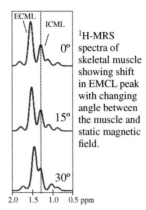

¹H-MRS spectra of skeletal muscle showing shift in EMCL peak with changing angle between the muscle and static magnetic field.

Skeletel muscle contains lipids both within the muscle cells (intramyocellular lipids [IMCL]) and in the extracellular space (extramyocellular lipids. [ECML]). As shown in the figure on the right, the IMCL peaks have a different chemical shift from those arising from EMCL. However, this chemical shift difference is a function of the angle between the muscle fibers and the applied magnetic field B_0. In the case shown below, an angle of $\alpha = 0°$ represents muscle fibers parallel to B_0, and $\alpha = 90°$ corresponds to fibers perpendicular to the B_0. The best separation between the IMCL and EMCL lipid peaks is achieved when the muscle fibers are approximately parallel to the static magnetic field. How do you explain this effect? Source for figure: Adapted from Khuu et al. (2009), Figure 3.

E.6.5 Chemical Shift Localization Error

Conventional slice selective excitation assumes that all spins have the same Larmor frequency. For spins with different resonant frequencies, the selected slice is spatially shifted. Cho and NAA are separated by 1.2 ppm (154 Hz at 3 T and 360 Hz at 7 T) and the relative shift in space between the selected slices for these two metabolites is given by:

$$\delta_x = \frac{\delta \gamma B_0 S_{th}}{BW},$$

where δ_x is the spatial shift caused by chemical shift, δ is the relative resonance frequency difference between spins (in ppm), B_0 is the main magnetic field strength, S_{th} is the slice thickness, γ is the gyromagnetic ratio in Hz/T, and BW is the spatial bandwidth in Hz.

a) Given that there is an Rf peak amplitude limit on the scanner, determine whether the PRESS or the STEAM sequence is more susceptible to chemical shift localization error.

b) What are some ways to eliminate or reduce the chemical shift localization error?

E.6.6 Lactate, Stroke, and ¹³C-glucose

Investigators studying the ¹H-MRS spectra from a stroke victim obtains the following data. Source: Figure adapted with permission from Rothman et al. (1991) Figure 3.

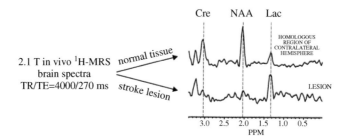

2.1 T in vivo ¹H-MRS   normal tissue
brain spectra
TR/TE=4000/270 ms   stroke lesion

The investigators want to know if the observed lactate signal in the lesion is being actively metabolically produced (hence possibly coming from viable but poorly-oxygenated tissue) or from a pool of metabolically inactive lactate. To answer this question, they performs a second experiment in which a ¹H-MRS spectrum is acquired after the infusion of ¹³C-labeled glucose into the patient's bloodstream. They observe the data shown below and identify the two new resonances at ±63.5 Hz around the original lactate signal as coming from lactate associated with the glucose infusion.

a) What is the basis for this claim?
b) What new information can be obtained from the second experiment that was unavailable without the glucose infusion?
c) Design an editing scheme that **only** detects the satellite peaks and suppresses all other resonances.

7

Relaxation Fundamentals

This chapter introduces the dominant physical processes that drive *in vivo* relaxation, which, in the case of MR, means the gain or loss of phase coherence among the spins. By introducing the underlying mechanisms, presenting useful analysis tools, and examining some of the consequences for representative tissues, the goal is to develop an intuitive understanding of this topic. The more rigorous mathematical derivations of the associated relaxation rates will be left for Chapter 8.

7.1 Basic Principles

7.1.1 Molecular Motion

In vivo, molecular motion is the key source of spatially and temporally varying magnetic fields, and the time scales for these motions determine the corresponding physical effects. Molecular orientation is time-dependent, and the Hamiltonian can be written as $\hat{H}(\Theta(t))$. In Chapter 4, we used temporal averaging to simplify the nuclear spin Hamiltonian. Specifically, we replaced time-dependent terms with their spatial averages as calculated by

$$\overline{\hat{H}} = \frac{1}{\tau}\int_0^\tau \hat{H}(\Theta(t))dt = \int_\Theta \hat{H}(\Theta)P(\Theta)d\Theta, \tag{7.1}$$

where $\Theta(t)$ is the generalized angle (here used in place of the more cumbersome notation $(\theta(t), \phi(t))$) and $P(\Theta)$ is the probability for a molecule having orientation Θ. For isotropic materials, this expression reduces to

$$\overline{\hat{H}}_{iso} = \frac{1}{4\pi}\int \hat{H}(\Theta)d\Theta. \tag{7.2}$$

The right-hand side of Eq. (7.1) comes from an ***ergodicity*** assumption. For a collection of nuclear spins, the basic assumption is that the observed ensemble average (*i.e.*, the average across spins) equals the time average for any one spin, in which case dynamical descriptions can be replaced with much simpler probabilistic ones. We no longer want to make this approximation. Instead, temporal variations in the Hamiltonian will be analyzed as perturbations, and each of the terms in the full spin Hamiltonian, $\hat{H} = \hat{H}_Z + \hat{H}_D + \hat{H}_J + \hat{H}_Q$, have associated characteristic temporal variations.

With respect to chemical shift, electron shielding is in general anisotropic, *i.e.*, the degree of shielding depends on the molecular orientation. The full shielding tensor can be written as the

Fundamentals of In Vivo Magnetic Resonance: Spin Physics, Relaxation Theory, and Contrast Mechanisms, First Edition. Daniel M. Spielman and Keshav Datta.
© 2024 John Wiley & Sons, Inc. Published 2024 by John Wiley & Sons, Inc.
Companion website: www.wiley.com/go/Spielman

sum of three terms:

$$\underline{\sigma} = \begin{bmatrix} \sigma_{xx} & \sigma_{xy} & \sigma_{xz} \\ \sigma_{yx} & \sigma_{yy} & \sigma_{yz} \\ \sigma_{zx} & \sigma_{zy} & \sigma_{zz} \end{bmatrix} = \sigma_{iso} \begin{bmatrix} 1 & 0 & 0 \\ 0 & 1 & 0 \\ 0 & 0 & 1 \end{bmatrix} + \underline{\sigma}^{(1)} + \underline{\sigma}^{(2)}, \tag{7.3}$$

where $\sigma_{iso} = \frac{1}{3}(\sigma_{xx} + \sigma_{yy} + \sigma_{zz})$ is the primary isotropic component that we have previously discussed. Of the new terms, $\underline{\sigma}^{(1)}$ is antisymmetric and $\underline{\sigma}^{(2)}$ is symmetric and traceless, and both $\underline{\sigma}^{(1)}$ and $\underline{\sigma}^{(2)}$ are time-varying due to molecular tumbling. $\underline{\sigma}^{(1)}$ causes only 2nd order effects and is typically ignored; however, $\underline{\sigma}^{(2)}$, sometimes written as $\Delta\underline{\sigma}(t)$, gives rise to a relaxation mechanism called chemical shift anisotropy (CSA) (Kowalewski and Mäler 2006). A topic to be discussed in more detail in Chapter 8.

Because J is unchanged with molecular tumbling, J coupling typically does not contribute to relaxation. That is, $J(\Theta(t)) = J$. However, there are a few cases where J can become "effectively" time-varying. The two primary cases are (1) when the S spin is engaged in chemical exchange, and (2) when the T_1 of the S spin is $\ll 1/J$. These cases are called scalar relaxation of the first and second kind, respectively, and both are important for the study of MRI contrast agents.

As previously mentioned, a nucleus with spin $I > \frac{1}{2}$ has an electrical quadrupolar moment due to its nonuniform charge distribution, and this electrical quadrupole moment interacts with local electric field gradients. Static E-field gradients result in shifts of the resonance frequencies of the observed peaks, whereas dynamic (time-varying) E-field gradients result in relaxation. The quadrupolar coupling Hamiltonian (secular approximation) is given by

$$\hat{H}_Q = \frac{3eQ}{4I(2I-1)\hbar} V_0 \left(3\hat{I}_z^2 - \hat{\vec{I}} \cdot \hat{\vec{I}} \right). \tag{7.4}$$

where e is the electron charge, Q is known as the quadrupolar coupling constant, and V_0 is the electric field gradient that is dependent on molecular orientation. In essence, this term looks like an interaction of a spin with itself and will come up in the discussion of paramagnetic contrast agents (see Chapter 9).

However, the dominant source of random magnetic field variations *in vivo* is due to dipolar coupling in combination with molecular tumbling. Let us look at this process in more detail. Consider an *in vivo* water molecule in a large magnetic field, as shown in Figure 7.1. The motion of the water molecule is called "tumbling" rather than "rotating" since molecules are constantly bumping into each other. With tumbling, spins remain aligned with B_0, but see a small perturbation from their nearby neighbors. Although the temporal average of this perturbation is zero,

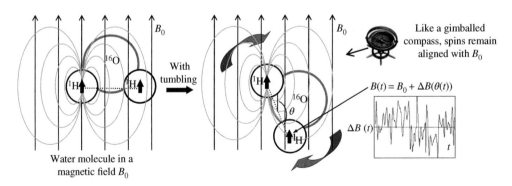

Figure 7.1 Molecular tumbling and associated magnetic fields. In addition to the large B_0 field, each dipolar-coupled nucleus is exposed to a small perturbing magnetic field $\Delta B(t)$.

i.e., $\langle \Delta B(t) \rangle = 0$, the instantaneous effect is *not* negligible, and we will need to examine $\Delta B(t)$ in some detail.

Let us first lay out our general strategy. Our goal is to write the Hamiltonian as the sum of a large static component plus a small time-varying perturbation.

$$\hat{H} = \hat{H}_0 + \hat{H}_1(t). \tag{7.5}$$

We will then look for an equation of the form

$$\frac{\partial}{\partial t}\hat{\sigma} = -i\hat{\hat{H}}_0\hat{\sigma} - \hat{\hat{\Gamma}}(\hat{\sigma} - \hat{\sigma}_B), \tag{7.6}$$

where the first term on the right-hand side generates the rotations we have been studying. The second term represents relaxation processes where $\hat{\hat{\Gamma}}$ is known as the **relaxation superoperator** and $\hat{\sigma}_B$ is the thermal equilibrium density operator as given by the Boltzmann equation. Equation (7.6) is known as the **Master Equation of NMR**, and we note its similarity to the classical Bloch equations

$$\frac{d\vec{M}}{dt} = \gamma \vec{M} \times B_0 \vec{z} - \frac{M_x \vec{x} + M_y \vec{y}}{T_2} - \frac{(M_z - M_0)\vec{z}}{T_1}, \tag{7.7}$$

which also contains terms for both rotations and relaxation.

The key concept for understanding MR relaxation is that temporal and spatial magnetic (also electric for spins $> \frac{1}{2}$) field variations cause spin systems to lose or gain phase coherences. Relaxation is the process by which phase coherences among spins return to their equilibrium values. Restoration of longitudinal magnetization, $M_z = \gamma\hbar\langle I_z\rangle$, is characterized by a time constant, T_1, and disappearance of transverse magnetization, $M_{xy} = \gamma\hbar(\langle I_x\rangle + i\langle I_y\rangle)$, is characterized by a time constant, T_2. We will first look at some simple relaxation models to help build intuition.

QUESTION: Do the other coherences, such as $\langle 2\hat{I}_z\hat{S}_z\rangle$, $\langle 2\hat{I}_x\hat{S}_x\rangle$, $\langle 2\hat{I}_y\hat{S}_y\rangle$... also relax?

7.1.2 Stochastic Processes

The perturbing magnetic field, $\Delta B(t)$, is modeled as a **stochastic process** and represents a family of time functions. Consider a collection of nuclear spins, I_i, for $i = 1, ..., N$. Let $\Delta B_i(t)$ be the time-varying perturbation field seen by the ith spin. At any time, t_0, $\Delta B(t_0)$ is a random variable with zero mean and variance $= \langle B^2 \rangle$. $\Delta B(t)$ is called **stationary** if its statistics are independent of time. $\Delta B_i(t)$ is a random function of time and assumed ergodic, meaning averages over time t equal averages over i. That is $\langle \Delta B_i^2(t_0) \rangle_i = \langle \Delta B_i^2(t) \rangle_t = \langle B^2 \rangle$.

However, the quantity we really care about is the **statistical correlation** between $\Delta B_i(t)$ and $\Delta B_i(t + \tau)$, namely, $G_i(t, \tau) = \langle \Delta B_i(t)\Delta B_i(t + \tau) \rangle$. Averaging over all spins i yields $G(t, \tau)$. For a stationary process (*i.e.*, independent of t): $G(t, \tau) = G(\tau)$, a second highly useful function is the Fourier transform of $G(\tau)$,

$$S(\omega) = \int_{-\infty}^{\infty} G(\tau)e^{-i\omega\tau}\,d\tau. \tag{7.8}$$

$S(\omega)$ is called the **power spectrum** and represents the energy available at each frequency. This function plays a fundamental role in MR relaxation theory.

Now, consider a water molecule (viewed as a two-spin system with equivalent spins) undergoing isotropic tumbling via Brownian motion (see Figure 7.2a). For simplicity, assume the inter-nuclear distance is fixed, and we will arbitrarily define the I spin to be at the origin of our coordinate system. Let $\Theta = 0$ at $t = t_0$. Over time, the S spin will traverse a random path on the surface of a sphere. Now consider a large ensemble of water molecules each initially aligned so that $\Theta = 0$ at $t = t_0$. As illustrated in Figure 7.2b, after a short time interval τ, the water molecules will be distributed randomly across the sphere centrally clustered around $\Theta = 0$. As τ increases, the spread of the molecules increases, approaching a uniform distribution as $\tau \to \infty$. From this physical picture, we can define the **rotational correlation time**, τ_c, as the average time for a given molecule to rotate over one radian. Hence, τ_c is a measure of rotational coherence.

Driven primarily by molecular tumbling, almost all MR relaxation processes are well described by an exponential correlation function,

$$G(\tau) = \langle \Delta B(t) \Delta B(t + \tau) \rangle = G(0)e^{-|\tau|/\tau_c}, \tag{7.9}$$

implying the correlation between the position of a molecule at two points in time falls off exponentially. To provide some support for this claim, let's find $G(\tau)$ for a simple case.

Consider a (nearly) spherical molecule undergoing isotropic tumbling and let $\Delta B(t) = F(t)$. Then,

$$G(\tau) = \langle F(t)F^*(t + \tau) \rangle = \frac{1}{4\pi} \int\limits_{\Theta} \int\limits_{\Theta_0} F(\Theta_0)F^*(\Theta)P(\Theta_0 \mid \Theta, \tau)d\Theta_0 d\Theta, \tag{7.10}$$

where $P(\Theta_0 \mid \Theta, \tau)$ is the probability at time τ of finding the molecule at an angle Θ given it started at angle Θ_0. To derive an explicit expression for $G(\tau)$, start with Fick's 2nd law of diffusion, given by

$$\frac{\partial f}{\partial t} = D\left(\frac{\partial^2 f}{\partial x^2} + \frac{\partial^2 f}{\partial y^2} + \frac{\partial^2 f}{\partial z^2}\right) = D\nabla^2 f, \tag{7.11}$$

where D is the diffusion constant and ∇^2 is the Laplacian. In spherical coordinates, the Laplacian operator is:

$$\nabla^2 = \frac{1}{r^2}\frac{\partial}{\partial r}\left(r^2\frac{\partial}{\partial r}\right) + \frac{1}{r^2 \sin\theta}\frac{\partial}{\partial \theta}\left(\sin\theta\frac{\partial}{\partial \theta}\right) + \frac{1}{r^2\sin^2\theta}\frac{\partial^2}{\partial \phi^2}. \tag{7.12}$$

In our case, fixing the radius r and just considering the angular components yields

$$\frac{\partial P(\Theta_0 \mid \Theta, \tau)}{\partial \tau} = D_r\nabla_2^2 P(\Theta_0 \mid \Theta, \tau), \tag{7.13}$$

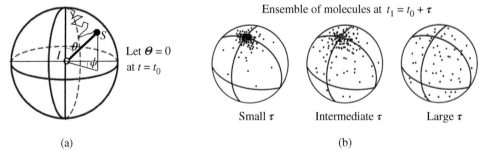

Let $\Theta = 0$
at $t = t_0$

Ensemble of molecules at $t_1 = t_0 + \tau$

Small τ Intermediate τ Large τ

(a)   (b)

Figure 7.2 Rotational correlation times. (a) Temporal evolution of a tumbling molecule containing dipolar coupled spins I and S, where the origin is defined at the I spin. Θ is assumed to be 0 at $t = t_0$. (b) The distribution of an ensemble of tumbling molecules at time $t_1 = t_0 + \tau$.

where D_r is the rotational diffusion constant, and ∇_2^2 is the Legendre operator (Laplacian with fixed r). This equation is known as Fick's law of rotational diffusion.

To solve this, we note that the spherical harmonics, Y_l^m, are eigenfunctions of ∇_2^2, and form a complete orthonormal basis set. Hence, the solution for Fick's law of rotational diffusion can be written as:

$$P(\Theta_0 \mid \Theta, \tau) = \sum_l \sum_m Y_l^{m*}(\Theta_0) Y_l^m(\Theta) e^{-l(l+1)D_r\tau}. \tag{7.14}$$

The spherical harmonics functions, $Y_l^m(\theta, \phi)$, are orthonormal over the surface of a sphere,

$$\langle Y_l^m | Y_{l'}^{m'} \rangle = \begin{cases} 1 & \text{for } l = l' \text{ and } m = m' \\ 0 & \text{for } l \neq l' \text{ or } m \neq m' \end{cases}, \tag{7.15}$$

and arise in multiple physical applications, including atomic orbitals, and (as we will see) MR relaxation theory. The first few spherical harmonics are shown in Figure 7.3.

Substituting back into Eq. (7.10),

$$G(\tau) = \frac{1}{4\pi} \sum_l \sum_m e^{-l(l+1)D_r\tau} \int_{\Theta_0} Y_l^{m*}(\Theta_0) F(\Theta_0) d\Theta_0 \int_\Theta F^*(\Theta) Y_l^m(\Theta) d\Theta. \tag{7.16}$$

In general, solving Eq. (7.16) can be difficult, except when the functions, $F(\Theta)$, can also be expressed as sums of spherical harmonics. An important example is the secular approximation of the dipolar coupling Hamiltonian

$$F(\Theta) = \frac{\mu_0 \gamma^2 \hbar}{4\pi r^3} \sqrt{\frac{3}{2}} (3\cos^2\theta - 1) = \frac{\mu_0 \gamma^2 \hbar}{4\pi r^3} \sqrt{\frac{24\pi}{5}} \, Y_2^0. \tag{7.17}$$

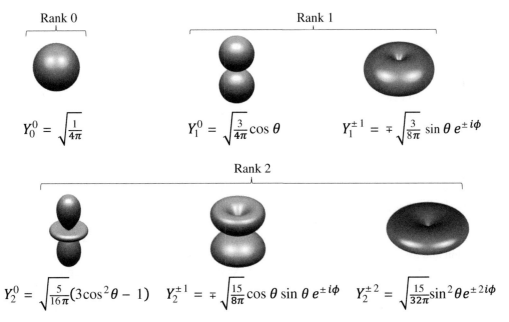

Figure 7.3   The first six spherical harmonic functions.

Tissue or compound	Approximate rotational correlation time (s)
Water: cerebral spinal fluid (CSF)	10^{-11}
Water: muscle	10^{-9}
Water: bone	10^{-7}
Albumin (representative protein)	3×10^{-8}
Gd-DTPA	6×10^{-11}
Water: ice at $-2\,°C$	10^{-6}

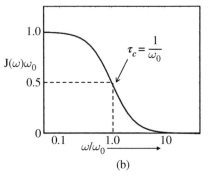

(a)   (b)

Figure 7.4 Rotational correlation times. (a) Table of rotation correlation times for some materials of interest, and (b) plots of spectral density functions as a function of τ_c and ω_0.

Because of the orthogonality of Y_l^ms, all the terms in Eq. (7.16) are zero except one. Hence,

$$G(\tau) = \frac{3\mu_0^2\gamma^4\hbar^2}{40\pi^2 r^6}e^{-6D_r\tau}\left|\left\langle Y_2^{0*}|Y_2^0\right\rangle\right|^2 = \frac{3\mu_0^2\gamma^4\hbar^2}{40\pi^2 r^6}e^{-\tau/\tau_c}. \tag{7.18}$$

A simple decaying exponential, with

$$\tau_c = \frac{1}{6D_r} = \frac{4\pi\eta a^3}{3kT}. \tag{7.19}$$

This is known as the Stokes–Einstein–Debye equation expressed for nearly spherical molecules, where a is the molecular radius, and η is the viscosity of the material.

Given a correlation function of the form $G(\tau) = G(0)e^{-|\tau|/\tau_c}$, let us define a spectral density function as

$$J(\omega) = \frac{1}{2}\int_{-\infty}^{\infty}e^{-|\tau|/\tau_c}e^{-i\omega\tau}d\tau = \frac{\tau_c}{1+\omega^2\tau_c^2}. \tag{7.20}$$

In general, small molecules have shorter rotational correlation times as compared to larger molecules, which tumble more slowly. For the systems we will typically encounter in MR, pure water molecules have $\tau_c \cong 10^{-12} - 10^{-10}$ s, as compared to water in more solid tissues (where the water interacts with macromolecules and other materials) with $\tau_c \cong 10^{-8} - 10^{-6}$ s. Figure 7.4 lists the rotational correlation times for some representative materials and plots $J(\omega)$.

7.1.3 A Simple Model of Relaxation

While dipolar coupling is the most important source of *in vivo* relaxation, it is not the simplest to analyze. Looking carefully at Figure 7.1, the perturbational magnetic field due to molecular tumbling, $\Delta B(t)$, seen by a given nuclei is actually not independent from that seen by its dipolar-coupled partner. Rather these perturbations are *correlated*. For now, we will ignore this complication and assume each nuclei sees an independent, time-varying random field $\Delta B(t)$ with corresponding spectral density given by Eq. (7.20). As previously discussed, $J(\omega)$ represents the amount of energy from the lattice available at frequency ω.

Assume an isotropic randomly fluctuating magnetic field perturbation given by

$$\Delta \vec{B}(t) = B_x(t)\vec{x} + B_y(t)\vec{y} + B_z(t)\vec{z}, \tag{7.21}$$

where $\langle B_x^2 \rangle = \langle B_y^2 \rangle = \langle B_z^2 \rangle = \langle B^2 \rangle$. The relaxation of M_z can then be shown to be (a derivation is given in Section 7.2)

$$\frac{1}{T_1} = \gamma^2 \left(\langle B_x^2 \rangle + \langle B_y^2 \rangle \right) \mathrm{J}(\omega_0) = 2\gamma^2 \langle B^2 \rangle \frac{\tau_c}{1 + \omega_0^2 \tau_c^2}. \tag{7.22}$$

The first term, $2\gamma^2 \langle B^2 \rangle$, represents the energy of the interaction, while the second term, $\tau_c / \left(1 + \omega_0^2 \tau_c^2 \right)$, corresponds to the fraction of lattice energy at frequency $\omega = \omega_0$. With respect to T_1 being known as ***spin-lattice relaxation***, the word "lattice" is a solid-state term, short for "crystal lattice." We still call T_1 the spin-lattice relaxation time even though *in vivo* there is typically no actual crystal.

There are several key features to note regarding T_1 relaxation. Given the main magnetic field B_0 is in the z-direction, changes in M_z involve spin transitions between energy states. As a resonant system, energy exchange occurs at $\hbar\omega_0$. Viewed from this perspective, T_1 relaxation and Rf excitation are much the same processes as transverse magnetic fields rotating at ω_0 are needed to induce these transitions. For Rf excitation, we provide a coherent rotating B_1 via an excitation coil. For T_1 relaxation, the lattice provides the needed fluctuating B field(s).

Let us now consider T_2 relaxation, also known as ***spin-spin relaxation***. However, we should note that this term is also somewhat misleading as relaxation can occur without any spin–spin interactions. T_2 relaxation concerns the loss of transverse coherences $\overline{\langle \hat{I}_x \rangle}$ and $\overline{\langle \hat{I}_y \rangle}$, and changes in $\overline{\langle \hat{I}_x \rangle}$ and $\overline{\langle \hat{I}_y \rangle}$ do not require energy transfer. In the case of transverse magnetization, magnetic fields in the z direction cause dephasing, and the slower the fluctuations the more efficient this relaxation mechanism. This dephasing drives a relaxation time constant known as T_2' and is given by

$$\frac{1}{T_2'} = \gamma^2 \left(\langle B_z^2 \rangle \right) \mathrm{J}(0). \tag{7.23}$$

Spin transitions also cause loss of transverse phase coherence and hence are a second factor in T_2 relaxation.

$$\frac{1}{T_2} = \frac{1}{2T_1} + \frac{1}{T_2'}. \tag{7.24}$$

Note, technically $2T_1 > T_2$, rather than the usually quoted $T_1 > T_2$, although is typically only a concern when dealing with highly anisotropic materials. Thus,

$$\frac{1}{T_2} = \gamma^2 \langle B_z^2 \rangle \mathrm{J}(0) + \frac{\gamma^2}{2} \left(\langle B_x^2 \rangle + \langle B_y^2 \rangle \right) \mathrm{J}(\omega_o), \tag{7.25}$$

and the full expression for the case of interest is

$$\frac{1}{T_2} = \gamma^2 \langle B^2 \rangle (\mathrm{J}(0) + \mathrm{J}(\omega_0)) = \gamma^2 \langle B^2 \rangle \left(\tau_c + \frac{\tau_c}{1 + \omega_0^2 \tau_c^2} \right), \tag{7.26}$$

probing the spectral density function at both 0 and ω_0.

Due to differential local environments, water T_1s and T_2s vary with tissue type. In general, water in bone exhibits the longest rotational correlation times, with water in fluids, such as CSF, the shortest. In accordance with Eqs. (7.22) and (7.26), T_1 is minimized when the scanner resonant frequency matches the reciprocal of the rotational correlation time, $\omega_0 \tau_c = 1$, while T_2 has a much weaker field-strength dependence.

Given the field from a dipole falls off as $1/r^3$, the dipolar coupling contribution to both $1/T_1$ and $1/T_2$ decrease as $1/r^6$. This very rapid decrease with distance results in dipolar coupling being largely an intramolecular rather than intermolecular effect. Finally, the condition known as ***extreme narrowing*** occurs when $\omega_0\tau_C \ll 1$, in which case

$$\frac{1}{T_1} = \frac{1}{T_2}. \tag{7.27}$$

Relationships between $J(\omega)$, τ_c, T_1, T_2, and ω_0 are shown in Figure 7.5.

In summary, thermal processes in the lattice provide random time-varying magnetic fields, for which the x and y components induce the spin energy transitions that drive T_1 (and also T_2) relaxation. The z component causes dephasing, driving T_2 relaxation. From our simple model of an uncorrelated, random perturbation $\Delta B(t)$, $1/T_1 = 2\gamma^2\langle B^2\rangle J(\omega_0)$ and $1/T_2 = \gamma^2\langle B^2\rangle(J(0) + J(\omega_0))$. Hence, T_1 relaxation depends on transverse fields having energy at the Larmor frequency, while T_2 relaxation depends on both $J(0)$ and $J(\omega_0)$. These simplified expressions provide an intuitive understanding for the observed *in vivo* relaxation times in most tissues.

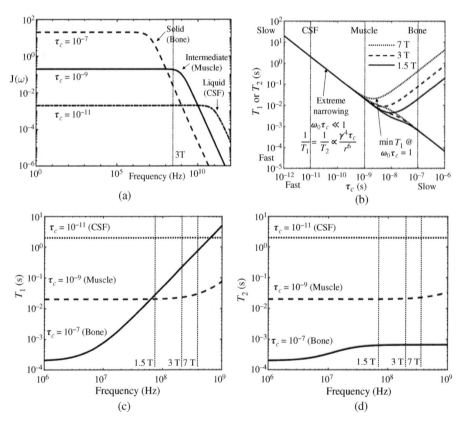

Figure 7.5 Relationships between the spectral density function $J(\omega)$, the rotational correlation time τ_c, water ^1H relaxation times T_1 and T_1, and B_0 for three representatives *in vivo* tissues. The astute reader will note that these plots differ slightly from Eqs. (7.22) and (7.26). These small variations can be explained using the full dipolar coupling Hamiltonian rather than the isotropic random perturbation model used in this section.

7.2 Dipolar Coupling

Equations (7.22) and (7.26), derived from a random fields model, are only approximations for the relaxation rates $1/T_1$ and $1/T_2$. From a historical perspective, the seminal work of Bloembergen and colleagues in 1948 (Bloembergen et al. 1948) first recognized the importance of dipolar coupling to NMR relaxation. Named after the three authors, Bloembergen, Purcell, and Pound, this paper is often referred to as BPP in the NMR literature.

A more complete derivation including the correlated perturbations among coupled spins was later provided by Solomon (1955). For identical spins, as seen with hydrogen in water, the more complete equations for dipolar coupling are

$$\frac{1}{T_1} = \frac{3}{20}\left(\frac{\mu_0}{4\pi}\right)^2 \frac{\gamma^4 \hbar^2}{r^6}(J(\omega_0) + 4J(2\omega_0))$$

$$= \frac{3}{20}\left(\frac{\mu_0}{4\pi}\right)^2 \frac{\gamma^4 \hbar^2}{r^6}\left(\frac{\tau_c}{1 + \omega_0^2 \tau_c^2} + \frac{4\tau_c}{1 + 4\omega_0^2 \tau_c^2}\right), \tag{7.28}$$

and

$$\frac{1}{T_2} = \frac{3}{40}\left(\frac{\mu_0}{4\pi}\right)^2 \frac{\gamma^4 \hbar^2}{r^6}(3J(0) + 5J(\omega_0) + 2J(2\omega_0))$$

$$= \frac{3}{40}\left(\frac{\mu_0}{4\pi}\right)^2 \frac{\gamma^4 \hbar^2}{r^6}\left(3\tau_c + \frac{5\tau_c}{1 + \omega_0^2 \tau_c^2} + \frac{2\tau_c}{1 + 4\omega_0^2 \tau_c^2}\right). \tag{7.29}$$

The new term, $J(2\omega_0)$, corresponds to both I and S spins flipping together, that is, a double quantum coherence. Let's look at a derivation of Eqs. (7.28) and (7.29) (details can be found in Solomon (1955)).

7.2.1 The Solomon Equations

Consider a general dipolar-coupled two-spin system. Note, nuclei I and S are close enough in space that dipolar coupling is significant, but they do not share a chemical bond (*i.e.*, in this case $J = 0$). As shown in Figure 7.6a, at a given point in time, if the energy of the spins were measured, the energy levels would be occupied by the number of spins given by N_{++}, N_{+-}, N_{-+}, and N_{--}. The W_is correspond to the transition rates, where W_I and W_S are probability per unit time spin I or S change energy levels, and W_0 and W_2 are probabilities per unit time of zero and double quantum transitions.

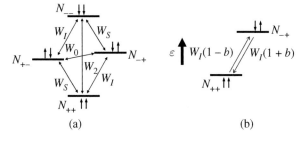

(a) (b)

Figure 7.6 A dipolar-coupled spin system. (a) Energy diagram, and (b) Boltzmann correction for spin transition rates.

Given the transition rates and the populations, let us compute the dynamics. Namely,

$$\frac{dN_{++}}{dt} = -(W_S + W_I + W_2)N_{++} + W_S N_{+-} + W_I N_{-+} + W_2 N_{--}, \tag{7.30}$$

$$\frac{dN_{+-}}{dt} = -(W_0 + W_S + W_I)N_{+-} + W_0 N_{-+} + W_I N_{--} + W_S N_{++}, \tag{7.31}$$

$$\frac{dN_{-+}}{dt} = -(W_0 + W_S + W_I)N_{-+} + W_0 N_{+-} + W_I N_{++} + W_S N_{--}, \tag{7.32}$$

and

$$\frac{dN_{--}}{dt} = -(W_S + W_I + W_2)N_{--} + W_S N_{-+} + W_I N_{+-} + W_2 N_{++}. \tag{7.33}$$

Before proceeding further, we need to make a small addition, known as the finite temperature correction. The energy diagram and associated Eqs. (7.30)—(7.33) assumed an equal transition probability for a spin flipping from the up-to-down state or down-to-up state. Under this assumption, the system will evolve until the energy states are equally populated, which, using the Boltzmann distribution, corresponds to an infinite temperature! To achieve a finite temperature, we can make an ad hoc correction reflecting the slightly increased probability of a transition that decreases the energy of the system (see Figure 7.6b). This Boltzmann factor, $b = \frac{\gamma \hbar B_0}{kT}$, can be derived explicitly if we treat both the spin system and the lattice as quantum mechanical systems (see Abragam 1983).

Let us now look at T_1 relaxation. Namely,

$$\overline{\langle \hat{I}_z \rangle} \propto (N_{++} - N_{-+}) + (N_{+-} - N_{--}) \tag{7.34}$$

and

$$\overline{\langle \hat{S}_z \rangle} \propto (N_{++} - N_{+-}) + (N_{-+} - N_{--}). \tag{7.35}$$

Substituting yields a set of coupled differential equations indicating longitudinal magnetization recovers via a combination of two exponential terms,

$$\frac{d\overline{\langle \hat{I}_z \rangle}}{dt} = -(W_0 + 2W_I + W_2)\left(\overline{\langle \hat{I}_z \rangle} - I_z^{eq}\right) - (W_2 - W_0)\left(\overline{\langle \hat{S}_z \rangle} - S_z^{eq}\right) \tag{7.36}$$

and

$$\frac{d\overline{\langle \hat{S}_z \rangle}}{dt} = -(W_0 + 2W_S + W_2)\left(\overline{\langle \hat{S}_z \rangle} - S_z^{eq}\right) - (W_2 - W_0)\left(\overline{\langle \hat{I}_z \rangle} - I_z^{eq}\right). \tag{7.37}$$

The first term, $(W_0 + 2W_I + W_2)$, is referred to as **direct relaxation**, while the second term is known as **cross-relaxation**.

These equations can also be written in a more compact form as

$$\frac{d}{dt}\begin{bmatrix} \overline{\langle \hat{I}_z \rangle} \\ \overline{\langle \hat{S}_z \rangle} \end{bmatrix} = -\begin{bmatrix} \rho_I & \sigma_{IS} \\ \sigma_{IS} & \rho_S \end{bmatrix}\begin{bmatrix} \overline{\langle \hat{I}_z \rangle} - I_z^{eq} \\ \overline{\langle \hat{S}_z \rangle} - S_z^{eq} \end{bmatrix}, \tag{7.38}$$

with $I_z^{eq} = \hbar \gamma_I B_0 / 2kT$ and $S_z^{eq} = \hbar \gamma_S B_0 / 2kT$. Written even more compactly,

$$\frac{d}{dt}\vec{V} = -R(\vec{V} - \vec{V}^{eq}), \tag{7.39}$$

where R is called the **relaxation matrix** with elements given by $\rho_I = W_0 + 2W_I + W_2$, $\rho_S = W_0 + 2W_S + W_2$, and $\sigma_{IS} = W_2 - W_0$. The general solution is of the form,

$$\overline{\langle \hat{I}_z \rangle}(t) = \alpha_{11} e^{-\lambda_1 t} + \alpha_{12} e^{-\lambda_2 t} \text{ and } \overline{\langle \hat{S}_z \rangle}(t) = \alpha_{21} e^{-\lambda_1 t} + \alpha_{22} e^{-\lambda_2 t}, \tag{7.40}$$

which are clearly not single exponentials.

Consider the special case where S and I are equivalent, i.e., $\omega_I = \omega_s$ and $W_I = W_S = W_1$. Then,

$$\frac{d(\overline{\langle \hat{I}_z \rangle} + \overline{\langle \hat{S}_z \rangle})}{dt} = -2(W_1 + W_2)\left(\overline{\langle \hat{I}_z \rangle} + \overline{\langle \hat{S}_z \rangle} - I_z^{eq} - S_z^{eq}\right) \tag{7.41}$$

has the solution of a pure exponential given by

$$\frac{1}{T_1} = 2(W_1 + W_2). \tag{7.42}$$

We now only need explicit expressions for the transition probabilities W_1, and W_2.

7.2.2 Calculating Transition Rates

To compute the transition probabilities, we will use a branch of QM known as **time-dependent perturbation theory**. Consider the case where the Hamiltonian can be written

$$\hat{H}(t) = \hat{H}_0 + \hat{H}_1 \text{ with } |\hat{H}_1| \ll |\hat{H}_0|. \tag{7.43}$$

In this case, $\hat{H}_1(t)$ can be treated as a perturbation. Namely, let $|m_j\rangle, j = 1...N$, be the eigenkets of the unperturbed Hamiltonian with energies ε_j,

$$\hat{H}_0 |m_j\rangle = \frac{1}{\hbar} \varepsilon_j |m_j\rangle \text{ with } j = 1 \ldots N. \tag{7.44}$$

Assuming the system starts in state $|m_j\rangle$, then, to first order, the probability of being in state $|m_k\rangle$ at time t is given by

$$P_{kj} = \frac{1}{\hbar^2}\left| \int_0^t \langle m_k(0)|\hat{H}_1(t')|m_j(0)\rangle e^{-i\omega_{kj}t'} dt' \right|^2, \tag{7.45}$$

where $\omega_{kj} = (\varepsilon_k - \varepsilon_j)/\hbar$ (see Exercise E.7.4 for the proof).

Now consider a sinusoidal perturbation,

$$\hat{H}_1(t) = \hat{V}_1 \cos \omega t = \frac{\hat{V}_1}{2}(e^{i\omega t} + e^{-i\omega t}), \tag{7.46}$$

which is quite general as we can always analyze any perturbation via a Fourier decomposition. The **transition rate** is then given by,

$$W_{kj} = \lim_{t\to\infty} \frac{P_{kj}}{t} = \lim_{t\to\infty} \frac{1}{t\hbar^2}\left| \int_0^t \langle m_k(0)|\hat{H}_1(t')|m_j(0)\rangle e^{-i\omega_{kj}t'} dt' \right|^2. \tag{7.47}$$

Note, we can take the limit of $t \to \infty$ if our measurements are much longer than $1/\omega_{kj}$. This yields

$$W_{kj} = \frac{|\langle m_k|\hat{V}_1|m_j\rangle|^2}{4}(\delta_{\varepsilon_k - \varepsilon_j, -\hbar\omega} + \delta_{\varepsilon_k - \varepsilon_j, +\hbar\omega}). \tag{7.48}$$

Equation (7.48) is a famous result known as **Fermi's Golden Rule**. The first part of this solution, $\frac{1}{4}|\langle m_k|\hat{V}_1|m_j\rangle|^2$, is known as the interaction term, and the second, $(\delta_{\varepsilon_k - \varepsilon_j, -\hbar\omega} + \delta_{\varepsilon_k - \varepsilon_j, +\hbar\omega})$, enforces the conservation of energy (see Figure 7.7).

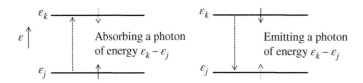

Figure 7.7 Fermi's Golden Rule for the emission or absorption of a photon driving spin transitions.

To gain insight, let us use this result to analyze an Rf excitation with amplitude B_1 as a perturbation to the much larger B_0 field. For a two-spin system, the Hamiltonian during Rf excitation is $\hat{H} = \hat{H}_0 + \hat{H}_1$, where

$$\hat{H} = \gamma_I B_0 \hat{I}_z + \gamma_s B_0 \hat{S}_z + 2\pi J (\hat{I}_x \hat{S}_x + \hat{I}_y \hat{S}_y + \hat{I}_z \hat{S}_z) + \omega_1^I \hat{I}_x \omega_1^S \hat{S}_x. \tag{7.49}$$

Assuming the simple case of $\gamma_I = \gamma_S$, the interaction term expressed in matrix form using the eigenkets of $\hat{H}_0$ as the basis is given by

$$\hat{\underline{H}}_1 \propto \begin{array}{c} \\ |++\rangle \\ |+-\rangle \\ |-+\rangle \\ |--\rangle \end{array} \overset{\begin{array}{cccc} |++\rangle & |+-\rangle & |-+\rangle & |--\rangle \end{array}}{\begin{bmatrix} 0 & 1-i & 1-i & 0 \\ 1+i & 0 & 0 & 1-i \\ 1+i & 0 & 0 & 1-i \\ 0 & 1+i & 1+i & 0 \end{bmatrix}}. \tag{7.50}$$

There are a couple of things to note about Eq. (7.50). As expected from the conservation of energy term in Eq. (7.48), there is no excitation if the Rf is off-resonance. Second, zeros along the anti-diagonal show that Rf excitation cannot directly excite double or zero quantum coherences (this follows from the interaction term of Eq. (7.48) being zero in this case).

Now, let us return our attention to dipolar coupling. The complete dipolar coupling Hamiltonian is

$$\hat{H}_D = \frac{\mu_0}{4\pi} \frac{\gamma_I \gamma_S \hbar}{r^3} \left(\hat{\vec{I}} \cdot \hat{\vec{S}} - \frac{3}{r^2} (\hat{\vec{I}} \cdot \vec{r})(\hat{\vec{S}} \cdot \vec{r}) \right), \tag{7.51}$$

where $\vec{r}$ is the vector from spin I to spin S. Defining the raising and lowering operators, $\hat{I}_+ = \hat{I}_x + i\hat{I}_y$ and $\hat{I}_- = \hat{I}_x - i\hat{I}_y$, respectively, the Hamiltonian can be rewritten in polar coordinates as:

$$\hat{H}_D = \frac{\mu_0}{4\pi} \frac{\gamma_I \gamma_S \hbar}{r^3} (A + B + C + D + E + F), \tag{7.52}$$

where

$$\begin{aligned} A &= \hat{I}_z \hat{S}_z F_0 \\ B &= -\frac{1}{4} (\hat{I}_+ \hat{S}_- + \hat{I}_- \hat{S}_+) F_0 \\ C &= (\hat{I}_+ \hat{S}_z + \hat{I}_z \hat{S}_+) F_1 \\ D &= (\hat{I}_- \hat{S}_z + \hat{I}_z \hat{S}_-) F_1 \\ E &= \hat{I}_+ \hat{S}_+ F_2 \\ F &= \hat{I}_- \hat{S}_- F_2, \end{aligned} \tag{7.53}$$

and

$$F_0(t) = 1 - 3\cos^2\theta$$

$$F_1(t) = -\frac{3}{2}\sin\theta\cos\theta e^{-i\phi},$$
(7.54)

$$F_2(t) = -\frac{3}{4}\sin^2\theta e^{-2i\phi}$$

where A and B correspond to zero quantum coherences, C and D correspond to single quantum coherences, and E and F correspond to double quantum terms. We also note that, with molecular tumbling, both θ and ϕ in Eq. (7.54) are stochastic functions of time (actually rank 2 spherical harmonics as illustrated in Figure 7.3).

To calculate the transition rates, for example W_1, assume the correlation function for $F(t)$ is of the form

$$\langle F(t)F^*(t+\tau)\rangle = \langle |F(0)|^2\rangle e^{-|\tau|/\tau_c},$$
(7.55)

where $\langle |F(0)|^2\rangle$ is a time average. For a given pair of like spins,

$$W_1 = \lim_{t\to\infty}\frac{1}{t}\left|\int_0^t \left(\frac{\mu_0}{4\pi}\right)^2 \frac{\gamma^2\hbar}{2r^3}F_1(t')e^{-i\omega_0 t'}\,dt'\right|^2 = \left(\frac{\mu_0}{4\pi}\right)^2\frac{\gamma^4\hbar^2}{4r^6}\int_0^\infty \langle |F_1(0)|^2\rangle\, e^{-|\tau|/\tau_c}e^{-i\omega_0\tau}\,d\tau,$$
(7.56)

and, for an ensemble of spins, we have

$$W_1 = \left(\frac{\mu_0}{4\pi}\right)^2\frac{\gamma^4\hbar^2}{4r^6}\overline{\langle |F_1(0)|^2\rangle}\frac{\tau_c}{1+\omega_0^2\tau_c^2},$$
(7.57)

where

$$\overline{\langle |F_1(0)|^2\rangle} = \frac{1}{4\pi}\int_0^{2\pi}\int_0^\pi \frac{9}{4}\sin^2\theta\cos^2\theta\sin\theta\, d\theta d\phi.$$
(7.58)

Integration yields a final transition rate of

$$W_1 = \left(\frac{\mu_0}{4\pi}\right)^2\frac{\gamma^4\hbar^2}{4r^6}\frac{3}{20}\frac{\tau_c}{1+\omega_0^2\tau_c^2}.$$
(7.59)

Using similar equations, the full set of transition rates are

$$W_I = \frac{3}{2}q\mathrm{J}(\omega_I),$$
(7.60)

$$W_2 = 6q\mathrm{J}(\omega_I+\omega_S),$$
(7.61)

$$W_0 = q\mathrm{J}(\omega_I-\omega_S),$$
(7.62)

where

$$\mathrm{J}(\omega) = \frac{2\tau_c}{1+\omega^2\tau_c^2},$$
(7.63)

and

$$q = \frac{1}{10}\left(\frac{\mu_0}{4\pi}\right)^2\frac{\gamma_I^2\gamma_S^2\hbar^2}{r^6}.$$
(7.64)

$$\tau_c = 5.0 \times 10^{-12} \text{ s}$$

$$K = \left(\frac{\mu_0}{4\pi}\right)^2 \frac{3}{10} \frac{\gamma^4 \hbar^2}{r^6} = 1.02 \times 10^{10}$$

$$\frac{1}{T_1} = \frac{K}{2}\left(\frac{\tau_c}{1 + \omega_0^2 \tau_c^2} + \frac{4\tau_c}{1 + 4\omega_0^2 \tau_c^2}\right)$$

$$\frac{1}{T_2} = \frac{K}{4}\left(3\tau_c + \frac{5\tau_c}{1 + \omega_0^2 \tau_c^2} + \frac{2\tau_c}{1 + 4\omega_0^2 \tau_c^2}\right)$$

Water

@3 T: $T_1 = T_2 \cong 3.02$ s
(extreme narrowing
condition)

Figure 7.8 Theoretical values for the T_1 and T_2 of pure water (^{1}H nuclei).

Note the rapid fall off with distance for the constant q, and that W_2 and W_0 probe J(ω) at the sum and difference frequencies of the chemical shifts, respectively. For the case of S and I identical (*i.e.*, $\omega_I = \omega_S$ and $W_I = W_S = W_1$),

$$\frac{1}{T_1} = 2(W_I + W_2) = \frac{3}{20}\left(\frac{\mu_0}{4\pi}\right)^2 \frac{\gamma^4 \hbar^2}{r^6}(J(\omega_0) + 4J(2\omega_0))$$

$$= \frac{3}{20}\left(\frac{\mu_0}{4\pi}\right)^2 \frac{\gamma^4 \hbar^2}{r^6}\left(\frac{\tau_c}{\omega_0^2 \tau_c^2} + \frac{4\tau_c}{1 + 4\omega_0^2 \tau_c^2}\right). \tag{7.65}$$

which, under the extreme narrowing condition $\omega_0^2 \tau_c^2 \ll 1$, reduces to

$$\frac{1}{T_1} = \left(\frac{\mu_0}{4\pi}\right)^2 \frac{3\gamma^4 \hbar^2 \tau_c}{4r^6}. \tag{7.66}$$

If we crunch through the numbers…

$$\frac{1}{T_2} = \frac{3}{40}\left(\frac{\mu_0}{4\pi}\right)^2 \frac{\gamma^4 \hbar^2}{r^6}(3J(0) + 5J(\omega_0) + 2J(2\omega_0))$$

$$= \frac{3}{40}\left(\frac{\mu_0}{4\pi}\right)^2 \frac{\gamma^4 \hbar^2}{r^6}\left(3\tau_c + \frac{5\tau_c}{1 + \omega_0^2 \tau_c^2} + \frac{2\tau_c}{1 + 4\omega_0^2 \tau_c^2}\right). \tag{7.67}$$

Under the extreme narrowing condition reduces to

$$\frac{1}{T_2} = \left(\frac{\mu_0}{4\pi}\right)^2 \frac{3\gamma^4 \hbar^2 \tau_c}{4r^6}. \tag{7.68}$$

In Chapter 8, we will develop Redfield theory and not have to "crunch the numbers" every time. Figure 7.8 gives the theoretical numbers for the T_1 and T_2 of water, which is quite close to the values measured *in vivo* for cerebral spinal fluid at 3 T.

7.2.3 Nuclear Overhauser Effect

Let us now try to gain some more physical insight and intuition into the phenomenon of cross-relaxation. Consider the Solomon equations as given by Eq. (7.38). We will start by examining the results of a series of theoretical **saturation recovery** experiments in which the initial z magnetization of the I or S spins (or both) are saturated (the process of driving the magnetization to zero using Rf pulses), and we then observe the recovery of the longitudinal magnetization over time.

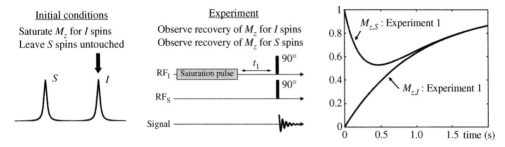

Figure 7.9 Theoretical experiment for observing the evolution of M_z for a dipolar coupled spin system after initial saturation of $M_{z,I}$. Data is collected for a series of t_1 values and results plotted on the right.

Experiment 1 (Figure 7.9). Start by saturating M_z for the I spins and observe the recovery of M_z for both the I and S spins for a series of time intervals t_1. In this case, the recovery of $M_{z,I}$ is not a simple exponential, and $M_{z,S}$ is also affected due to the coupling, with the precise shape of curves depending on the coupling strength.

Experiment 2 (Figure 7.10). Start by saturating M_z for both the I and S spins and observe the recovery of I-spin M_z coherence for a series of time intervals t_1. However, in contrast to Experiment 1, magnetization for the S spin continues to be saturated throughout the experiment. For this experiment, M_z for the I spins **does** recover exponential with the true T_1 of spin I. However, it **does not** recover to the Boltzmann thermal equilibrium value of M_z^{eq}!

The results shown in Figure 7.10 demonstrate what is known as the Nuclear Overhauser Effect (NOE), first proposed for electron–nuclear interactions by Albert Overhauser in 1953 (Overhauser 1953). Specifically, the NOE is the change in the equilibrium magnetization of one nucleus with the Rf irradiation of a nearby nucleus (nearby defined in terms of dipole coupling). This change in magnetization can be positive (generally with small rapidly tumbling molecules) or negative (as with slower tumbling molecules). We will now describe the NOE mathematically, graphically, and with an *in vivo* example.

To calculate the NOE, start with Eq. (7.36) and saturate S_z, i.e., set $\overline{\langle \widehat{S}_z \rangle} = 0$. At steady state $d\langle \widehat{I}_z \rangle / dt = 0$, which implies

$$\frac{\overline{\langle \widehat{I}_z \rangle}}{I_z^{eq}} = 1 + \frac{S_z^{eq}}{I_z^{eq}} \left(\frac{W_2 - W_0}{W_0 + 2W_I + W_2} \right). \tag{7.69}$$

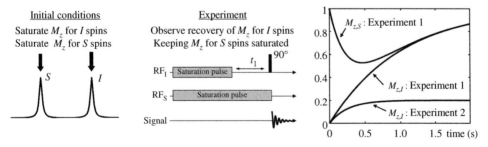

Figure 7.10 Theoretical experiment for observing the evolution of M_z for a dipolar coupled spin system after initial saturation of both $M_{z,I}$ and $M_{z,S}$. Data is collected for a series of t_1 values while the S spins continue to be saturated. Results are plotted on the right.

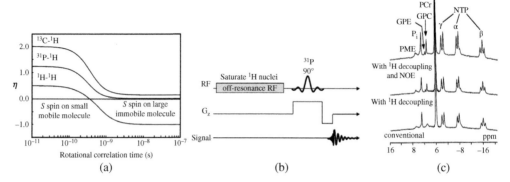

Figure 7.11 The Nuclear Overhauser effect (NOE). (a) The NOE, η, varies with the types of coupled nuclei as well as the corresponding molecular rotational correlation time. (b) A representative ^{31}P–^{1}H-pulse sequence utilizing NOE, and (c) representative *in vivo* ^{31}P spectra acquired with combinations of ^{1}H decoupling and NOE. *Source:* (a) and (c) Adapted with permission from van de Ven (1996), Figure 2.8, p. 78, and Brown et al. (1995) Figure 1, respectively.

Rewriting Eq. (7.69) in a more convenient form and letting I_e be the steady state magnetization. Then,

$$I_e = (1 + \eta)I_z^{eq} \quad \text{where } \eta = \frac{\gamma_S}{\gamma_I}\left(\frac{W_2 - W_0}{W_0 + 2W_I + W_2}\right), \tag{7.70}$$

which is often just expressed as

$$NOE = 1 + \frac{\gamma_S}{\gamma_I}\left(\frac{W_2 - W_0}{W_0 + 2W_I + W_2}\right) = 1 + \eta. \tag{7.71}$$

As shown in Figure 7.11, η, the NOE enhancement factor, is positive for the heteronuclear case and can be positive or negative in homonuclear situation. Typical *in vivo* NOE enhancement factors for ^{31}P–^{1}H and ^{13}C–^{1}H interactions are 1.4–1.8 and 1.3–2.9, respectively, and NOE is an important tool for achieving improved signal-to noise ratio (SNR) in such studies.

7.3 Chemical Exchange

7.3.1 Introduction

Cross-relaxation can lead to exchange of magnetization between coupled spins I and S. However, uncoupled spins can also manifest themselves as an apparent coupled spin system if the spins are engaged in chemical exchange. Let us look at this more closely. Consider spins A and B on two molecules undergoing chemical exchange with rate constants k_A and k_B, respectively.

$$A \underset{k_B}{\overset{k_A}{\rightleftharpoons}} B, \quad \text{where } \frac{1}{\tau_{ex}} = k_{ex} = \frac{k_A}{k_B}, \tag{7.72}$$

where τ_{ex} is the exchange time and k_{ex} is the exchange rate. For simplicity, let us assume each transition from molecule A to molecule B is instantaneous but happens at an average rate of $k_{ex} = 1/\tau_{ex}$. Figure 7.12 shows several important *in vivo* examples of chemical exchange processes.

Chemical exchange generates stochastic magnetic field modulations that, in turn, drive relaxation processes. However, the exchange times are very important, with exchange rates, typically

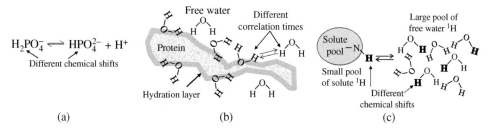

Figure 7.12 Examples of *in vivo* chemical exchange. (a) Shows an example of exchange processes involving inorganic phosphate (measurable using ^{31}P MRS) and (b) illustrates water interactions with a hydration layer surrounding a protein, respectively. (c) Illustrates an example of intermolecular exchange between water and smaller solute pool.

on μs to ms time scales, being much slower than molecular tumbling. Such processes are typically too slow to affect anisotropic interactions such as CSA or dipole coupling but can affect isotropic interactions such as those due to chemical shift or J coupling.

For example, let spins I and S be J-coupled, but the chemical bond between them regularly broken via a chemical exchange process. This modulates the J coupling. If $1/\tau_{ex} \gg J$, then the correlation function $G(\tau) = \langle J(t)J(t+\tau)\rangle = J^2 e^{-|\tau|/\tau_{ex}}$, and

$$\frac{1}{T_{1,sc}} = \frac{(2\pi J)^2}{2} \frac{\tau_{ex}}{1 + (\omega_I - \omega_S)^2 \tau_{ex}^2}. \tag{7.73}$$

Hence, the exchange time can look just like a rotational correlation time! The name for this particular effect is ***scalar relaxation of the 1st kind***, and we will see why $J(\omega)$ is probed at the difference frequency, $\omega_I - \omega_S$, in Chapter 8.

7.3.2 Effects on Longitudinal Magnetization

We can analyze the chemical-exchange-driven flow of longitudinal magnetization between sites using Eq. (7.72) and what are known as the Bloch–McConnell equations (McConnell 1958). Namely,

$$\frac{dM_z^A}{dt} = -\frac{M_z^A - M_{z,0}^A}{T_1^A} - \frac{M_z^A}{\tau_A} + \frac{M_z^B}{\tau_B}, \tag{7.74}$$

and

$$\frac{dM_z^B}{dt} = -\frac{M_z^B - M_{z,0}^B}{T_1^B} - \frac{M_z^B}{\tau_B} + \frac{M_z^A}{\tau_A}. \tag{7.75}$$

Written more compactly using a relaxation matrix,

$$\frac{d}{dt}\begin{bmatrix} M_z^A \\ M_z^B \end{bmatrix} = -\begin{bmatrix} \alpha_A & -1/\tau_B \\ -1/\tau_A & \alpha_B \end{bmatrix}\begin{bmatrix} M_z^A - M_{z,0}^A \\ M_z^B - M_{z,0}^B \end{bmatrix} + \text{constant}, \tag{7.76}$$

with $\alpha_A = 1/T_1^A + 1/\tau_A$ and $\alpha_B = 1/T_1^B + 1/\tau_B$. The diagonal terms in the relaxation matrix, α_A and α_B, are referred to as direct relaxation, while the anti-diagonal terms, $-1/\tau_A$ and $-1/\tau_B$, characterize cross-relaxation rates between molecules A and B. We note that these equations have the same basic form as the Solomon equations for dipolar coupling (Eqs. (7.36) and (7.37)).

There are several interesting limiting cases to examine more closely. Case 1: slow exchange.

$$\frac{1}{\tau_A} + \frac{1}{\tau_B} \ll \frac{1}{T_1^A} + \frac{1}{T_1^B}, \tag{7.77}$$

leading to $\alpha_A \approx 1/T_1^A$ and $\alpha_B \approx 1/T_1^B$, with relaxation occurring at either the A or B spin independent of the exchange rates. Case 2:

$$\frac{1}{T_1^B} \gg \frac{1}{\tau_A}, \frac{1}{\tau_B} \gg \frac{1}{T_1^A}, \tag{7.78}$$

for which $\alpha_A \approx 1/\tau_A$ and $\alpha_B \approx 1/T_1^B$. In this case, almost all the relaxation occurs at site B. Case 3, for which site B has very rapid relaxation and a very small population, *e.g.*, free water rapidly exchanging with a small pool of bound water. Then,

$$\alpha_A = \frac{1}{T_{1A}} + \frac{p_B}{p_A T_1^B + p_B \tau_A}, \tag{7.79}$$

where p_A and p_B are the fractional pool sizes. This is an important case for water in tissue as well as contrast agents.

7.3.3 Effects on Transverse Magnetization

As shown in Figure 7.13, chemical reactions can also have profound effects on NMR linewidths, with the effects strongly dependent on the exchange rate. We can modify the Bloch equations to add chemical exchange. Namely,

$$\frac{dM_x^A}{dt} = -\frac{1}{T_2^A}M_x^A - \cos(\Omega_A t)M_y^A - k_A M_x^A + k_B M_x^B, \tag{7.80}$$

and

$$\frac{dM_y^A}{dt} = -\frac{1}{T_2^A}M_y^A - \sin(\Omega_A t)M_x^A - k_A M_y^A + k_B M_y^B, \tag{7.81}$$

where these are like the Bloch–McConnell equations, but now for transverse magnetization. Note, there are also analogous equations for the B spin.

Letting $M_{xy} = M_x + iM_y$, a more compact notation for both A and B spins is

$$\frac{dM_{xy}^A}{dt} = -\left(i\Omega_A + \frac{1}{T_2^A}\right)M_{xy}^A - k_A M_{xy}^A + k_B M_{xy}^B, \tag{7.82}$$

and

$$\frac{dM_{xy}^B}{dt} = -\left(i\Omega_B + \frac{1}{T_2^B}\right)M_{xy}^B - k_B M_{xy}^B + k_A M_{xy}^A. \tag{7.83}$$

Rewriting in vector notation, $\vec{M}_{xy} = \left[M_{xy}^A, M_{xy}^B\right]'$

$$\frac{d\vec{M}_{xy}}{dt} = \underline{L}\vec{M}_{xy} \text{ where } \underline{L} = -\underline{\Omega} + \underline{k}, \tag{7.84}$$

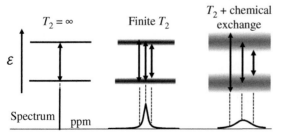

Figure 7.13 Broadening of spectral peaks associated with T_2 decay and chemical exchange.

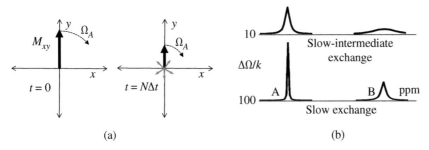

Figure 7.14 M_{xy} for a spin system involved in slow-to-intermediate exchange. (a) At each time interval, Δt, a packet of A spins exchanges to B, but the slow exchange rate results in magnetization packet exchanged back to A from B being out of phase. (b) As the exchange rate increases, these peaks get broader.

for

$$\underline{\Omega} = \begin{bmatrix} i\Omega_A + 1/T_2^A & 0 \\ 0 & i\Omega_B + 1/T_2^B \end{bmatrix} \text{ and } \underline{k} = \begin{bmatrix} -k_A & k_B \\ k_A & -k_B \end{bmatrix}. \tag{7.85}$$

The solution is then

$$\overrightarrow{M}_{xy}(t) = e^{\underline{L}t}\overrightarrow{M}_{xy}(0). \tag{7.86}$$

But before giving the expanded equation, let's look at two special cases.

Line-broadening due to slow-intermediate exchange, where $|\Omega_A - \Omega_B| \gg k_A, k_B$, is shown in Figure 7.14. Here,

$$\frac{dM_{xy}^A}{dt} = -\left(i\Omega_A + \frac{1}{T_2^A} \right) M_{xy}^A, \tag{7.87}$$

and

$$\frac{1}{T_{2_{app}}^A} = \frac{1}{T_2^A} + k_A. \tag{7.88}$$

As the exchange rate k_A increases lines get **broader**; similarly, for the B spin.

In contrast, fast exchange, where $|\Omega_A - \Omega_B| \ll k_A, k_B$, is, in many ways, the opposite of slow exchange (see Figure 7.15). Spins hop back and forth so fast that we observe a single resonance at the weighted average chemical shift. Namely, $\overline{\Omega} = f_A\Omega_A + f_B\Omega_B$ where f_A and f_B are the molar

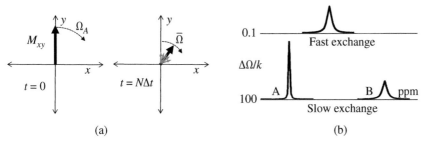

Figure 7.15 M_{xy} for a spin system involved in fast exchange. (a) At each time interval, Δt, a packet of A spins exchanges to B, where the fast exchange rate results in magnetization packet exchanged back to A from B being only slightly out of phase. (b) This results in a single peak at the weighted average between the two exchanging components. As the exchange rate increases, this single peak gets sharper.

fractions of A and B. Here, as k_A and k_B increase, the spectral lines get sharper.

$$\frac{1}{T_{2_{app}}} = \frac{1}{T_2} + \Delta v \text{ with } \Delta v \propto \frac{\Delta\Omega^2}{(k_A + k_B)} \tag{7.89}$$

Let us now look at the more complete equations. Starting with $\vec{M}_{xy}(t) = e^{\underline{L}t}\vec{M}_{xy}(0)$, taking the Fourier Transform, and assuming $k_A \gg 1/T_2^A$ and $k_B \gg 1/T_2^B$ (*i.e.*, chemical exchange is much faster than T_2 relaxation), the real part of spectrum is (after considerable algebra):

$$S(\omega) = \frac{f_A f_B (\Omega_A - \Omega_B)^2 \tau_{ex}^{-1}}{(\Omega_A - \omega)^2 (\Omega_B - \omega)^2 + (\overline{\Omega} - \omega)^2 \tau_{ex}^{-2}} M_0 \tag{7.90}$$

where

$$\overline{\Omega} = f_A \Omega_A + f_B \Omega_B \text{ and } f_A + f_B = 1. \tag{7.91}$$

Here f_A and f_B are the molar fractions of A and B, respectively, and τ_{ex} is a measure of the interconversion time between A and B given by,

$$\tau_{ex} = \frac{1}{k_A + k_B}. \tag{7.92}$$

In terms of the lifetimes of A and B, $\tau_A = 1/k_A$ and $\tau_B = 1/k_B$,

$$\frac{1}{\tau_{ex}} = \frac{1}{\tau_A} + \frac{1}{\tau_B}. \tag{7.93}$$

There are three values of ω which correspond to spectral peaks. The case,

$$S(\omega \cong \Omega_A) = \frac{f_A \tau_A}{(\Omega_A - \omega)^2 + \tau_A^{-2}} M_0 \tag{7.94}$$

generates a Lorentzian line centered at Ω_A with width $1/\tau_A = k_A$. If τ_A is very short, this peak becomes very broad. Using analysis same as for A, a second peak at $S(\omega \cong \Omega_B)$ is also visible under slow exchange. Finally, a third peak forms at

$$S(\omega \cong \overline{\Omega}) = \frac{f_A f_B (\Omega_A - \Omega_B)^2 \tau_{ex}}{f_A^2 f_B^2 \tau_{ex}^2 (\Omega_A - \Omega_B)^4 + (\overline{\Omega} - \omega)^2} M_0, \tag{7.95}$$

which represents a Lorentzian at $\overline{\Omega} = f_A \Omega_A + f_B \Omega_B$ with a linewidth that decreases with k_{ex} as given by $\pi\Delta v = f_A f_B (\Omega_A - \Omega_B)^2 \tau_{ex}$. This peak is visible under fast exchange (see Figure 7.16).

7.3.4 Examples

Example 7.1 *Fast Exchange*
The chemical exchange involving inorganic phosphate can serve as a measure of *in vivo* pH. Under fast exchange, the inorganic phosphate ^{31}P peak will be located at $\overline{\Omega} = f_A \Omega_A + f_B \Omega_B$, where $\Omega_A = 3.2$ ppm and $\Omega_B = 5.7$ ppm are the chemical shifts of $H_2PO_4^-$ and HPO_4^{2-}, respectively. The Henderson–Hasselbalch relationship, expressed in terms of chemical shift (Ackerman et al. 1996),

$$pH = pK_A + \log_{10}\left(\frac{f_A}{1 - f_A}\right) = pK_A + \log_{10}\left(\frac{\omega - \Omega_A}{\Omega_B - \omega}\right), \tag{7.96}$$

allows intracellular pH values to be determined by measuring the chemical shift difference between the phosphocreatine (PCr) and inorganic phosphate (Pi) peaks in the ^{31}P spectrum (see Figure 7.17).

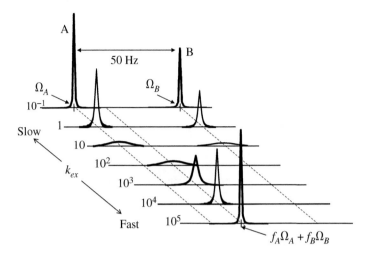

Figure 7.16   2-Spin system with chemical exchange. MR spectral changes as a function of the chemical exchange rate k_{ex}.

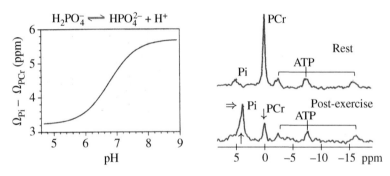

Figure 7.17   *In vivo* ^{31}P spectra from human forearm muscle before and after exercise show both the conversion of PCr to Pi (↑) and decrease in intracellular pH via the Pi chemical shift (⇒). Source: Adapted with permission from Arnold et al. (1984) Figure 1.

Example 7.2   *Fast Exchange*

MRI can be used to map *in vivo* temperatures by measuring changes to the chemical shift of water. In particular, water ^{1}H resonance frequencies are known to shift with temperature, and this effect can be explained via a two-site exchange process. Here, we follow the first calculations published by J.C. Hindman in 1966 (Hindman 1966). Hydrogen bonds decrease the electron density at the involved proton site and hence lead to reduced shielding and a positive MR frequency shift. Furthermore, liquid water can be modeled as a mixture of two components: a hydrogen-bonded "ice-like" fraction and a non-hydrogen-bonded monomeric fraction.

The shielding constants for monomeric water and hydrogen-bonded water are $\sigma_m \cong -0.43 \times 10^{-6}$ and $\sigma_p \cong -5.5 \times 10^{-6}$, respectively, and these two components are in fast exchange. The ^{1}H MR water peak is thus observed at $\overline{\Omega} = f_A \Omega_A + f_B \Omega_B$, where f_A and f_B are the respective molar fractions of the monomeric and hydrogen-bound water molecules. Using water chemical shielding data acquired as a function of temperature, yields a water proton frequency shift of $\Delta \cong 0.01$ ppm/°C (Hindman 1966). An example of MRI-based temperature mapping using the frequency shift of water is given in Figure 7.18.

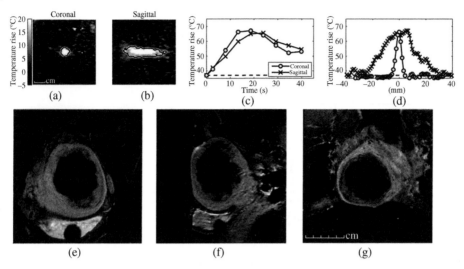

Figure 7.18 Temperature mapping of water using MRI. The temperature-dependent chemical shift of water can be used for MR-guided focused ultrasound (MRgFUS). (a, b) MR thermometry of the focal heating acquired at peak temperature rise of a uterine fibroid. Contours indicate regions that should be ablated based on the thermal dose prediction. (c) Temperature versus time at the focal point for these two sonications. A body temperature of 37 °C was assumed. (d) Spatial temperature distribution for the two sonications in (a,b). (e–g) Contrast-enhanced T1-weighted images acquired immediately after treatment ((e): Coronal, (f): Sagittal, (g): Axial). Almost the entire fibroid appeared nonperfused after MRgFUS, presumably indicating necrosis. Source: Adapted with permission from Jolesz and McDannold (2008) Figure 1.

 QUESTION: What would you expect to see if the hydrogen-bonded and monomeric water were in slow exchange?

Example 7.3 *Fast Exchange*

We note that Eq. (7.90) was derived under the assumptions that $k_A \gg 1/T_2^A$ and $k_B \gg 1/T_2^B$, i.e., chemical exchange being much faster than T_2. However, this will not necessarily be true when we examine contrast agents (see Chapter 9). Some key parameters to consider will be (1) the chemical shift difference between water when free and when coordinated with the agent/metal, (2) the T_2 of water bound to the agent (typically dominated by unpaired electron), and (3) the lifetime, τ_B, of the water in the coordination sphere of the contrast agent.

Example 7.4 *Slow-intermediate Exchange*

This is a particularly important case that is often exploited *in vivo* for measuring an otherwise hard-to-detect pool of spins. The basic idea, known as Chemical Exchange Saturation Transfer (CEST) (to be discussed in detail in Section 7.4.5), is to saturate spins in one environment and then observe the effect on a larger more easily measured exchanging pool (see Figure 7.19).

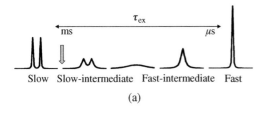

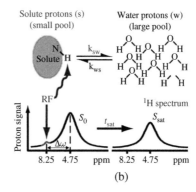

Figure 7.19 CEST. (a) Most CEST applications involve slow-intermediate exchange whereby (b) selective saturation of the smaller solute nuclei is observed as a decrease in the large free water pool. Source: (b) Adapted with permission from van Zijl and Yadav (2011) Figure 1.

7.4 *In Vivo* Water

To recap, MR relaxation is due to interactions between nuclear spins and local fluctuating fields arising from thermal motions of the lattice, molecular tumbling, chemical exchange processes, and, in the case of MR contrast agents, the presence of strong paramagnetic centers (Chapter 9). The effects of these interactions depend on the time scale and nature of the motion, with T_1 most sensitive to fluctuations at the Larmor frequency $\omega = \gamma B_0$, T_2 highly sensitive to fluctuations at very low frequencies such as $\omega = 0$, and $T_{1\rho}$ (a third MR relaxation time that we will discuss in detail in Chapter 8) probes the spectral density function at a user-selected frequency $\omega = \gamma B_1$. Let us now see how these processes can affect *in vivo* water.

7.4.1 Hydration Layers

Relaxation times of *in vivo* water protons found in tissue are typically much shorter than those for pure water. A dominant effect is that a significant fraction of *in vivo* water is associated with macromolecules in the form of hydration layers and other interactions, in which hydrogen bonding to hydrophilic surfaces results in restricted motion, cross-relaxation, and chemical exchange effects (Mathur-De Vré 1979) (as shown in Figure 7.20).

Consider a simple two-compartment model for biological water, consisting of a free water pool with MR relaxation rates $1/T_{1f}$ and $1/T_{2f}$ and a restricted hydration layer with rates $1/T_{1r}$ and $1/T_{2r}$, respectively. Let the exchange rate be $1/\tau_{ex}$, with a free water fraction of f_f and a restricted water fraction of f_r. These two water pools are in fast exchange leading to relaxation rates being averages of the rates for the two pools. Namely,

$$\frac{1}{T_1} = \frac{f_f}{T_{1f}} + \frac{f_r}{T_{1r}} \text{ and } \frac{1}{T_2} = \frac{f_f}{T_{2f}} + \frac{f_r}{T_{2r}}. \tag{7.97}$$

Note, more complex three-compartment models, which add a tightly bound water "ice-like" pool, have also been proposed. The result is that tissue T_1s are dominated by total water content and fraction in the hydration layer, while tissue T_2s depend on the thickness of hydration layer as well as the size of the tightly bound pool (see Fullerton et al. (1982) for more details).

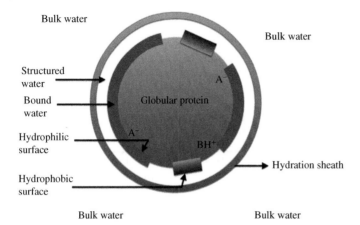

Figure 7.20 A schematic representation of bound, structured, and bulk water compartments on a globular protein in a dilute solution. Basic (BH+) and acidic (A–) protein side chains interact with hydration water to form charged hydrophilic sites. Hydrophobic sites are also shown. Source: Adapted with permission from Gilani and Sepponen (2016) Figure 2.

7.4.2 Tissue Relaxation Times

Let us now examine some typical MR-measured biological water relaxation times. Proton-density-weighted MRI images (referring to water concentrations not the spin density operator) are typically acquired using a short-TE/long-TR acquisition sequence, while T_1- and T_2-weighted images can be obtained using short-TE/short-TR and long-TE/long-TR acquisitions, respectively (see Figure 7.21a). A table of common T_1 and T_2 relaxation times observed for representative tissues as a function of main magnetic field strength are given in Figure 7.21b. In contrast to water, most ^{1}H metabolites discussed in Chapter 6 have T_1s on the order of ~1500 ms at 3 T and T_2s in the range of (200–300 ms) with only minor field dependencies.

QUESTION: What do these images tell us about the tissue in this tumor versus normal brain?

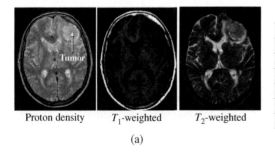

Tissue	1.5 T		3.0 T		7.0 T	
	T_1 (ms)	T_2 (ms)	T_1	T_2	T_1	T_2
WM	750	90	910	90	1290	90
GM	1100	100	1610	100	2060	100
CSF	2500	1500	300	1500	3000	1500
Muscle	900	50	1200	50	1700	50
Fat	250	70	400	70	430	70
Blood	1200	250	193	275	2300	275

Proton density T_1-weighted T_2-weighted

(a)   (b)

Figure 7.21 *In vivo* MRI. (a) Proton density, T_1-weighted and T_2-weighted MRI of a patient with a glioma. (b) Approximate biological water T_1 and T_2 relaxation times. GM, gray matter; WM, white matter. The blood relaxation times are for arterial blood.

7.4.3 Magic Angle Effects

An important observation in clinical MRI is that on moderate to short echo-time sequences, the signal intensities of tendons, ligaments, and cartilage depend on the tissue orientation with respect to the B_0 field. Unlike more fluid materials, these highly ordered tissues contain collagen fibers with bound water that is not free to tumble isotropically, resulting in an angle-dependent residual dipole interaction (see Figure 7.22).

Specifically, the dipolar coupling Hamiltonian (secular approximation) is

$$\hat{H}_D = d(3\hat{I}_z\hat{S}_z - \hat{\vec{I}} \cdot \hat{\vec{S}}), \tag{7.98}$$

where

$$d = -\frac{\mu_0\gamma_I\gamma_S}{4\pi r^3}\hbar(3\cos^2\Theta_{IS} - 1) \tag{7.99}$$

and Θ_{IS} is the time-dependent generalized angle between the main magnetic field and the vector connecting spins I and S. In contrast to isotropic tumbling, the time average for the non-isotropic tumbling case $\hat{H}^D$ depends on Θ_{IS}.

A common technique used in solid-state NMR is to artificially spin the sample in order to average out these angular-dependent dipolar coupling effects (Krushelnitsky and Reichert 2005). Noting that $3\cos^2\theta - 1 = 0$ for $\theta = 54.7°$, residual dipolar coupling effects will disappear if the sample is rapidly spun at this so-called ***magic angle***. Magic angle spinning is also routinely used to analyze *ex-vivo* tissue biopsy samples (DeFeo and Cheng 2010), as illustrated in Figure 7.23.

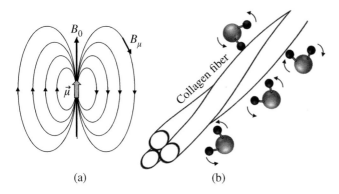

(a) (b)

Figure 7.22 Residual dipolar coupling (a) Magnetic field from a dipole and (b) details of water molecules undergoing restricted tumbling associated with a collagen fiber.

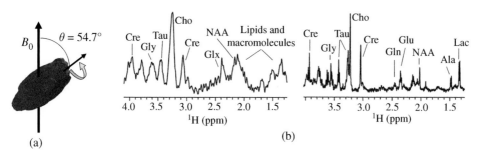

(a) (b)

Figure 7.23 Magic angle spinning (MAS) to eliminate peak broadening from residual dipolar coupling. (a) MAS geometry, (b) Rat brain spectral comparison between *in vivo* ¹H MRS and *ex vivo* ¹H MAS MRS of a tumor biopsy specimen. Source: (b) Adapted with permission from Lucas-Torres et al. (2021) Figure 1.

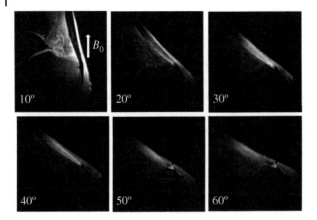

Figure 7.24 T_2-weighted MRI of a ruptured Achilles tendon with the tendon oriented at an increasing angle with respect to the B_0 field. Source: Adapted with permission from Bydder et al. (2007) Figure 2.

Although restricted tumbling is seen *in vivo*, no such spinning is possible for *in vivo* studies. Figure 7.24 shows representative MRI images for a ruptured Achilles tendon, showing a strong T_2-dependence on the angular orientation of the tendon with respect to B_0. In principle, tendon T_2s, and hence signal intensity, should be maximized for tendons angled 54.7° from the main magnetic field due to the vanishing dipolar interaction at this angle.

QUESTION: What about T_1?

7.4.4 Magnetization Transfer Contrast (MTC)

For the general pulse sequence shown in Figure 7.25, if there were no interactions between saturated and observed components, we would get familiar results, such as achieving fat or water

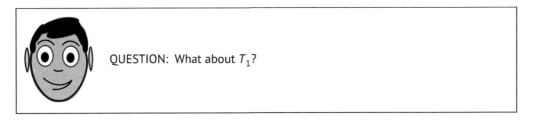

(a)   (b)

Figure 7.25 *In vivo* water. (a) A simple Rf pulse sequence consisting of a long saturation pulse followed by a pulse and acquire acquisition. (b) Schematic of a free water pool interacting with ^{1}H nuclei on macromolecules.

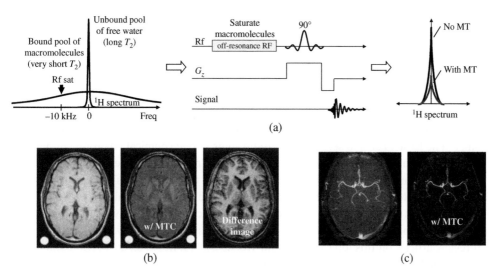

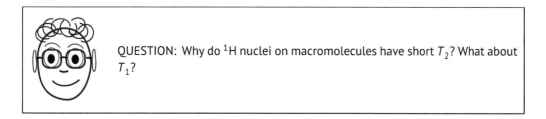

Figure 7.26 Magnetization transfer Contrast (MTC). (a) Saturation of the macromolecule pool, via a highly off-resonance Rf pulse (the macromolecule is symmetric, so saturation can be either up- or down-field of the free water peak), results in an observed decrease in the detected signal from free water. Note, the T_2 of the macromolecules is too short to be imaged directly. Two representative clinical applications of magnetization transfer contrast (MTC) imaging are shown here: (b) myelin mapping of the brain and (c) MRI angiography with and without MTC showing the suppression of non-blood tissues and improved vessel conspicuity. Source: (b) and (c) Adapted with permission from Pike et al. (1993) Figure 5 and Pike et al. (1992) Figure 3, respectively.

suppression depending on the selectivity of the Rf saturation pulse. But what happens if the saturated and observed components interact?

Consider two pools of water protons, as commonly seen in tissues with appreciable macromolecular components: a bound pool of macromolecules (very short T_2) and consequently a very broad ^{1}H spectrum, and an unbound pool of free water (long T_2) exhibiting a relatively narrow spectral peak. Selectively saturate the short-T_2 bound pool. Magnetization is then allowed to exchange between saturated bound protons and unsaturated mobile protons. When we now observe the signal from the unbound water (Figure 7.26), dipolar coupling leads to an NOE effect (see Eq. (7.71)). In this case, we are dealing with slowly tumbling macromolecules, so the NOE is negative.

QUESTION: Why do ^{1}H nuclei on macromolecules have short T_2? What about T_1?

This effect is known as ***magnetization transfer***, and it is used as a tissue-contrast mechanism in MRI (Knutsson et al. 2018). Consider the human brain, which consists of macromolecule-rich (primarily myelin) white matter, gray matter with fewer macromolecules, and blood with even fewer. Two examples of clinical MTC applications are: (1) improved assessment of multiple sclerosis (MS) lesions characterized by loss of myelin, and (2) suppression of brain tissue to improve visibility of blood vessels for vascular imaging applications (see Figure 7.26). As an aside, it is interesting to note that the equations are very similar if MTC is based on dipole–dipole interactions (cross relaxation) or a chemical exchange effect (see Exercise E.7.6).

7.4.5 Chemical Exchange Saturation Transfer (CEST)

Unique image contrast can be generated for spin systems in slow or slow-intermediate chemical exchange. The basic idea is to selectively saturate spins in one chemical environment, which are then exchanged into a second environment that can be more readily measured. For this to be effective, the exchange rate must be slow enough to have two distinct peaks in the MR spectrum, but fast enough to allow magnetization to transfer before T_1 recovery. This is a very powerful *in vivo* method, and, as shown in Figure 7.27, the exchange processes may involve proton exchange, molecular exchange, or even compartmental exchange processes. Fortunately, there are multiple excellent review articles on this topic (Sherry and Woods 2008; Jones et al. 2018; Knutsson et al. 2018), and we shall just give a brief overview here.

The equations for longitudinal magnetization, assuming slow to slow-intermediate exchange of spins in pools A and B, are

$$\frac{dM_z^A(t)}{dt} = \frac{M_z^{A,0} - M_z^A(t)}{T_1^A} - k_{AB}M_z^A(t) + k_{BA}M_z^B(t)$$

$$\frac{dM_z^B(t)}{dt} = \frac{M_z^{B,0} - M_z^B(t)}{T_1^B} - k_{BA}M_z^B(t) + k_{AB}M_z^A(t), \tag{7.100}$$

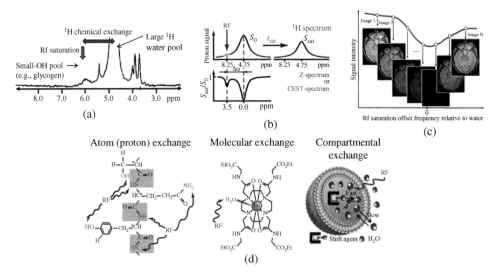

Figure 7.27 Chemical Exchange Saturation Transfer (CEST). (a) Solute protons are in constant exchange with water protons. Rf is used to saturate spins at a particular chemical shift, and the saturation magnetization is, in turn, transferred to the large water pool at 4.75 ppm. (b) By repeating this process at incremented Rf saturation frequencies, a "Z-spectrum" for S_{sat}/S_0 is obtained. (c) A representative CEST imaging acquisition. (d) Classification of CEST contrast based on exchange type. Source: (a), (b), and (d) Adapted with permission from van Zijl and Yadav (2011) Figures 2, 1, and 3, respectively.

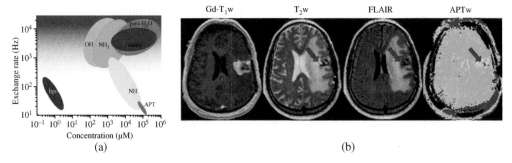

Figure 7.28 *In vivo* CEST applications. (a) Log-plot of concentrations needed to achieve a 5% CEST effect for the different groups of agents. (b) Conventional and APT-weighted (APTw) MRI from a patient with brain tumor. Source: (a) and (b) adapted with permission from van Zijl and Yadav (2011) Figure 4c and Zhou et al. (2019) Figure 5, respectively.

where k_{AB} and k_{BA} are the respective exchange rates. Selectively saturate component B with sufficient Rf irradiation such that $M_z^B \approx 0$. Then the new equilibrium for the A component becomes

$$\frac{M_z^A(\infty)}{M_z^{A,0}} = \frac{1}{1 + k_{AB}T_1^A}. \tag{7.101}$$

The magnitude of this CEST effect depends on the proton exchange rate, the number of exchangeable protons, the pH of the local environment, T_1 and T_2 relaxation times, the saturation efficiency, and the amplitude and duration of saturation pulse. The basic idea is to collect a series of measurements (or images) while stepping the resonant frequency of the Rf saturation pulse. By repeating this process at incremented Rf saturation frequencies, a "Z-spectrum" computed from the ratio of the saturated and unsaturated values, S_{sat}/S_0, at each frequency is obtained. Peaks in the Z-spectrum correspond to the chemical-exchanging components of interest, and the following is a non-exhaustive list of *in vivo* CEST targets, opportunities, and associated challenges (see Figure 7.28).

7.4.5.1 Amide Proton (–NH) Transfer (APT)

The chemical shift of amides is typically ~3.5 ppm below water, and slow exchange rates ($\sim 30\,\mathrm{s}^{-1}$) in combination with relatively high concentrations of the associated compounds makes APT a very attractive method. The peak is easy to saturate, leading to lower Rf power deposition and hence suitability for *in vivo* implementation at fields of 3 T and higher. Importantly, there is a strong pH dependence on exchange rate, providing a valuable biomarker for multiple pathologies, especially tumors (Zhou et al. 2022).

7.4.5.2 Hydroxyl (–OH) CEST

The chemical shifts of hydroxyl groups are ~1 ppm below water, and common *in vivo* targets are imaging of glucose, glycogen, myo-inositol, and glycosaminoglycans (GAG). Moderate exchange rates of $500–1500\,\mathrm{s}^{-1}$ require relatively high-Rf power for effective saturation. Combined with the relatively small chemical shift with respect to water, CEST experiments targeting hydroxyl groups often require the use of high magnetic fields, such as 7 T. Applications include imaging glucose metabolism (glucoCEST) and GAG in cartilage (gagCEST).

7.4.5.3 Amine (–NH₂) CEST

Amines are present in free amino acids, proteins, and peptides. The targeted chemical shifts are ~3 ppm below water, with two example targets being glutamate (gluCEST) and creatine (CrCEST).

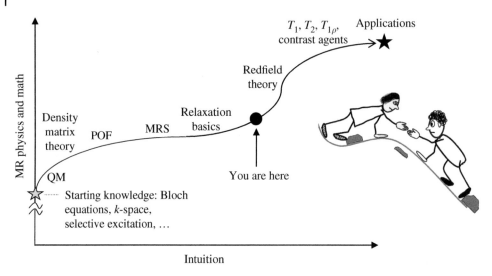

Figure 7.29 The roadmap at the end of Chapter 7.

The faster exchange rate ($\sim$2000–6000 s^{-1}) leads to a high transfer efficiency but requires significant Rf power deposition. Important applications include imaging of protease activity in tumors and measuring *in vivo* tissue pH.

7.5 Summary

This chapter introduces the key idea that local time-varying magnetic fields underly MR relaxation processes. Due primarily to molecular tumbling, fluctuations at the Larmor frequency, in combination with dipolar spin–spin coupling, are the primary source of T_1 (the relaxation time for longitudinal magnetization), while slow fluctuations determine T_2 (the relaxation of transverse magnetization). Chemical exchange processes, being much slower that molecular tumbling, also contribute to T_2. Finally, *in vivo* water molecules are found in multiple local environments, and interactions both with each other and other compounds (*e.g.,* macromolecules) result in widely varying tissue relaxation rates that can be exploited for image contrast by carefully chosen pulse sequences. The next chapter will expand our understanding of MR relaxation processes by developing a more mathematically advanced approach, known as Redfield Theory, which will prove highly useful for multiple applications (see Figure 7.29).

Exercises

E.7.1 Dipolar Coupling Revisited

Molecules in a glass of water undergo ***random isotropic tumbling*** due to Brownian motion. Magnetically, hydrogen nuclei behave as simple dipoles. Hence, if water is placed in a uniform magnetic field, the *B* field at one hydrogen nucleus due to the dipole field of the other hydrogen nucleus is given by $\Delta B(t)$, where $\Delta B(t) = b(3\cos^2\theta(t)-1)$ and b is a constant. What is the average value of $\Delta B(t)$ as defined by

$$\overline{\Delta B(t)} = \frac{1}{\tau}\int_0^\tau \Delta B(t)dt = ?$$

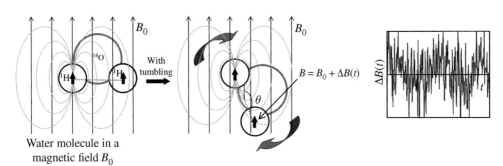

Water molecule in a
magnetic field B_0

E.7.2 Classical Relaxation

For this problem, let us use a simplified model in which each spin, in addition to the main magnetic field B_0, sees a small field $\Delta B\,(\Delta B \ll B_0)$ whose amplitude and orientation change suddenly and randomly at random time intervals of average duration τ_c (τ_c is known as the correlation time).

a) Calculating T_1: consider the longitudinal magnetization M_z and the component of ΔB perpendicular to B_0, $\Delta B_\perp$. During the first-time interval of duration τ_c, the magnetization will precess around the effective field $B_0 + \Delta B_{\perp,1}$, resulting in forming an angle $\Delta\phi_1$ with respect to B_0. This process continues such that, after the n-th time interval, the magnetization precesses around the field $B_0 + \Delta B_{\perp,n}$ making an angle $\Delta\phi_n$ with respect to its prior direction. Assuming $\Delta\phi_1, \Delta\phi_2, ..., \Delta\phi_n$, are independent and identically distributed, show

$$\frac{dM_z(t)}{dt} \cong -\frac{\overline{\Delta\phi^2}}{2\tau_c}M_z(t).$$

Hence,

$$\frac{1}{T_1} = -\frac{\overline{\Delta\phi^2}}{2\tau_c}.$$

Hint:

$$\frac{1}{n}\left(\sum_{i=1}^n \Delta\phi_i\right)^2 \cong \overline{\Delta\phi^2}.$$

b) Use the diagram below and the fact that $\phi \cong \Delta B_\perp/B_0 \ll 1$ to show

$$\frac{1}{T_1} = -\frac{\overline{\Delta B_\perp^2}(1-\cos\gamma B_0\tau_c)}{B_0^2\tau_c}.$$

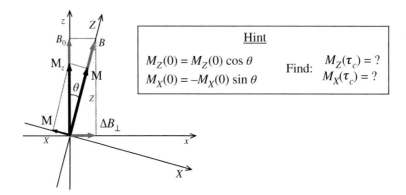

c) What is T_1 for the limiting cases of $\omega_0 t_c \ll 1$ and $\omega_0 t_c \gg 1$? What value of $\omega_0 t_c$ corresponds to the minimum value of T_1?

d) What is missing from this classical derivation of T_1?

e) Calculating T_2: Now consider the transverse magnetization and the component of ΔB parallel to B_0, $\Delta B_\parallel$. The Larmor frequency during the i-th time interval is now given by $\omega_i = -\gamma(B_0 + \Delta B_{\parallel, i}) = \omega_0 + \Delta \omega_i$. Defining the transverse relaxation time T_2 as the time at which the root-mean-square dephasing is equal to 1 rad, show

$$\frac{1}{T_2} = \overline{\Delta\omega^2}\, \tau_c.$$

E.7.3 The Solomon Equations

Using the Solomon equations, derive the relaxation rate of longitudinal two-spin order $2\hat{I}_z\hat{S}_z$. Hint, the energy diagram for this coherence is

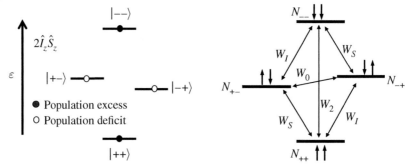

E.7.4 Time-dependent Perturbation Theory

Given a Hamiltonian of the form $\hat{H}(t) = \hat{H}_0 + \hat{H}_1(t)$ where $\hat{H}_1(t)$ is a perturbation small compared to $\hat{H}_0$ and $|m_n\rangle$ for $n = 1, \ldots, N$ are the eigenkets of the unperturbed Hamiltonian $\hat{H}_0$ with eigenvalues $E_n/\hbar$, the goal is to show that if the system starts at time $t = 0$ in the state $|m_j\rangle$, then the probability of finding the system in state $|m_k\rangle$ at time t is given by

$$P_{kj} = \left| \int_0^t \langle m_k(0)|\hat{H}_1(t')|m_j(0)\rangle e^{-i(E_j - E_k)t'/\hbar}\, dt' \right|^2.$$

a) Consider an arbitrary wavefunction $|\phi\rangle = \sum_{n=1}^{N} c_n(t) |m_n\rangle$. Using Schrödinger's equation $d|\psi\rangle/dt = -i\hat{H}|\psi\rangle$, show

$$\frac{dc_n(t)}{dt} = -i\sum_{n=1}^{N} c_n(t)\langle m_k|\hat{H}_1(t)|m_n\rangle.$$

b) Given the state of the system at $t = 0$ is specified by $c_{j=1}$, $c_{n \neq j} = 0$, the perturbation assumption is that $\hat{H}_1(t)$ will only have a small effect on the dynamics, i.e., $c_n(t) \ll c_j(t)$ for $n \neq j$ and $c_j(t) \cong 1$.

Using these assumptions and the results from (a), show

$$P_{kj} = \left| \int_0^t \langle m_k(0)|\hat{H}_1(t')|m_j(0)\rangle e^{-i(E_j - E_k)t'/\hbar} dt' \right|^2.$$

E.7.5 **_In vivo_ Correlation Times**

Design an experiment to estimate the rotational correlation time, τ_c, of water in _in vivo_ human brain gray matter. Any human subjects involved in your experiment should survive the study unscathed. Use the approximate equations for T_1 and T_2. Namely,

$$\frac{1}{T_1} \approx \gamma^2 \langle B^2 \rangle \frac{2\tau_c}{1 + \omega_0^2 \tau_c^2} \quad \text{and} \quad \frac{1}{T_2} \approx \gamma^2 \langle B^2 \rangle \left(\tau_c + \frac{2\tau_c}{1 + \omega_0^2 \tau_c^2} \right).$$

E.7.6 **MTC Imaging**

Design an experiment to help determine if the contrast observed in Magnetization Transfer Contrast (MTC) MRI (see, for example, Henkelman et al. (2001)) is due to chemical exchange or cross-relaxation via dipole–dipole interactions. This may be an animal/tissue study, and the subject(s) of the experiment need not necessarily survive the study. Explain how your experiment works and hypothesize a result.

E.7.7 **Magic Angle Spinning**

Assume the nuclear spins j and k are fixed relative to each other within a molecule. In a liquid, the molecule would freely tumble. However, this is not the case for a solid. Let the molecule rotate about an axis making an angle θ' with respect to the static field B_0, where the frequency of rotation is high compared with the frequencies of interest in an NMR experiment.

a) Consider the simplified case where the radius vector from j to k is perpendicular to the axis of rotation. As the molecule rotates, the angle θ_{jk} (between B_0 and the internuclear vector) varies with time. Show that the dipolar coupling term in the Hamiltonian is proportional

$$\langle 3\cos^2\theta_{jk} - 1 \rangle_\phi = \frac{1}{2}(1 - 3\cos^2\theta'),$$

where $\langle \ \rangle_\phi$ denotes averaging over the angle ϕ ranging from 0 to 2π.

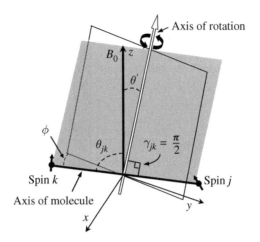

b) What happens to the NMR lineshape if θ' is chosen such that $(1 - 3\cos^2\theta') = 0$?

c) (Extra Credit) Consider the more general case in which the radius vector from j to k makes an angle γ_{jk} with respect to the axis of rotation. Show that for this case, the dipolar coupling term in the Hamiltonian is proportional to

$$\langle 3\cos^2\theta_{jk} - 1\rangle_\phi = \frac{1}{2}(3\cos^2\theta' - 1)(3\cos^2\gamma_{jk} - 1).$$

E.7.8 MRI of Tendons

a) Explain why tendons and ligaments have an unexpectedly long T_2 when obliquely oriented to the main magnetic field. Source: Figure adapted with permission from Oatridge et al. (2001) Figure 1.

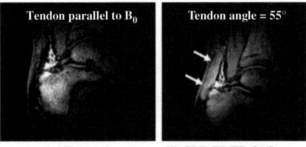

Acute Achilles tendon rupture: 3D GRE, TR/TE=21/7 ms

b) Plot T_2 as a function of θ. For what value of θ is T_2 maximized?

E.7.9 Temperature Mapping

In vivo ^{1}H MRS can be used to map temperatures in the brain by measuring the chemical shift between water and NAA via the formula (Corbett et al. 1995),

$$T = -97.26\, \delta_{water-NAA} + 293.28.$$

where δ is measured in ppm and T is measured in °C.

a) Why is the chemical shift of water temperature dependent, while that for NAA is not?

b) 3 T *in vivo* ^{1}H brain data acquired during hypothermia in a preclinical model recorded a temperature change of 19 °C. Calculate the observed frequency shift of the water?

c) The same study also noted a linear correlation between the intensity of the water peak and temperature. Can you explain this observation?

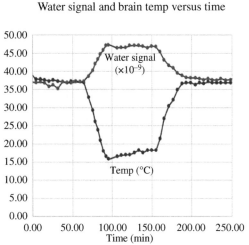

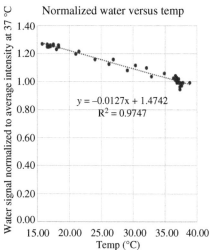

$y = -0.0127x + 1.4742$
$R^2 = 0.9747$

Historical Notes

Nicolaas Bloembergen, a Dutch-born American physicist, left an indelible mark on the fields of spectroscopy and nonlinear optics, earning him the prestigious Nobel Prize in Physics. Born on March 11, 1920, in Dordrecht, Netherlands, he embarked on his academic journey at the University of Utrecht, obtaining his undergraduate and graduate degrees in 1941 and 1943, respectively. His early fascination with nuclear magnetic resonance led him to Harvard University in 1946, where he collaborated with Edward Purcell and Robert Pound, conducting groundbreaking research that laid the foundation for his future achievements.

Having obtained his PhD from the University of Leiden in 1948, Bloembergen returned to Harvard as a professor of applied physics in 1951. Throughout his illustrious career, he rose through the academic ranks, attaining the distinguished position of Gerhard Gade University Professor in 1980, and later retiring as a professor emeritus in 1990. In 2001, he continued to contribute to academia by joining the faculty at the University of Arizona.

Bloembergen's contributions were acknowledged by the United States when he became a naturalized citizen in 1958. His early research in nuclear magnetic resonance sparked his interest in masers, and he went on to design a revolutionary three-stage crystal maser that surpassed the power of its gaseous counterparts, becoming the most widely used microwave amplifier.

However, his most significant breakthroughs came from pioneering the use of lasers in spectroscopic studies. Through his research in laser spectroscopy, Bloembergen achieved high-precision observations of atomic structure, providing unprecedented insights into the interaction of electromagnetic radiation with matter. This groundbreaking work earned him international acclaim and a share of the 1981 Nobel Prize in Physics.

His contributions extended beyond experimental work as he delved into the theoretical aspects of electromagnetic radiation's interaction with matter. Bloembergen's profound research culminated in the formulation of nonlinear optics, a new theoretical approach that transformed the analysis of light-matter interactions.

Albert Warner Overhauser, a renowned American physicist born on December 19, 1925, made pioneering contributions to magnetic resonance, leaving a profound impact on science and technology. His groundbreaking work laid the foundation for significant advancements in nuclear magnetic resonance (NMR) techniques, especially in condensed-matter physics.

Overhauser's academic journey began in 1948 when he earned a Bachelor of Science degree in physics and mathematics. Initially planning to pursue a PhD in nuclear physics at Berkeley, fate intervened when his advisor, Gian-Carlo Wick, left due to the loyalty-oath controversy. This led him to work under Charles Kittel, a leading physicist at the time, who was moving to Berkeley from Bell Labs. Recognizing Overhauser's exceptional talent, Kittel assigned him the thesis topic of studying spin-relaxation mechanisms in metals. Remarkably, Overhauser completed the task in a few months and obtained his PhD in 1951.

In 1953, while at the University of Illinois, Overhauser made one of his most groundbreaking discoveries. During his research on spin systems out of equilibrium, he conceived the idea that formed the basis of the dynamical Overhauser effect (DOE). His ingenious approach involved indirectly polarizing nuclear spins through pumping and saturating the conduction electrons' spin resonance. This technique produced nuclear polarizations thousands of times larger than expected, surpassing the strength of the Fermi nuclear coupling constant. Overhauser's theory was notable for its simplicity and deep physical insight, a characteristic that would pervade his future work.

Although initially met with skepticism, the DOE was soon experimentally confirmed by Charles Slichter and his student Richard Norberg, also at the University of Illinois. Additionally, Ionel Solomon at CNRS in Paris demonstrated that the DOE mechanism also operated between the spins of two nuclear species in ordinary liquids, leading to the nuclear Overhauser effect (NOE). These discoveries proved instrumental in the development of powerful NMR techniques for imaging and determining the structure of proteins and other biological macromolecules in solution.

Over the course of his career, Overhauser displayed versatility, relevance, and creativity in his work, earning numerous prestigious awards. He was honored with the Oliver E. Buckley Prize of the American Physical Society in 1975, followed by the Russell Varian Prize at the EUROMAR Magnetic Resonance Conference in 2009, and the National Medal of Science in 1994. His research papers, known for their clarity and simplicity, have become classics in the field of magnetic resonance.

8

Redfield Theory of Relaxation

The Redfield theory of relaxation, also known as Wangsness, Bloch, and Redfield (WBR) theory (Redfield 1955; Wangsness and Bloch 1953), is more general than that derived by Solomon. Both Solomon and Redfield theory rely on 2nd-order perturbation theory (this can be a limitation, but usually not for liquids and *in vivo* applications), and like Solomon's approach, Redfield theory is semiclassical, using the same Boltzmann correction for thermal equilibrium values. However, rather than directly dealing with spin energy levels and populations, the theory is derived in terms of the density operator. This allows for a more general description of relaxation, permitting the derivation of relaxation rates from multiple mechanisms including dipolar coupling, chemical shift anisotropy (CSA), chemical exchange, and scalar relaxation of the 1st and 2nd kind.

8.1 Perturbation Theory and the Interaction Frame of Reference

When the Hamiltonian can be written as the sum of a large static component plus a small time-varying perturbation, $\hat{H} = \hat{H}_0 + \hat{H}_1(t)$, our goal will be to find an equation of the form

$$\frac{d}{dt}\hat{\sigma} = -i\hat{\hat{H}}_0\hat{\sigma} - \hat{\hat{\Gamma}}(\hat{\sigma} - \hat{\sigma}_B), \tag{8.1}$$

where the first term represents the rotations we have previously studied and $\hat{\hat{\Gamma}}$ is known as the relaxation superoperator. Once we find $\hat{\hat{\Gamma}}$, which differs for different relaxation mechanisms, we will be able to calculate all direct and cross-relaxation terms of interest. The basic form of Eq. (8.1) should be somewhat familiar, as the Bloch equations (see Eq. (7.7)) can similarly be divided into rotation and relaxation terms.

Assume that the spin Hamiltonian can indeed be written as $\hat{H} = \hat{H}_0 + \hat{H}_1(t)$, where $\hat{H}_1(t)$ is a perturbation. Then, to solve the Liouville-von Neumann equation, it's helpful to switch to a rotating frame of reference where the perturbation term is isolated from the static part of $\hat{H}_0$. Namely, let

$$\hat{\sigma}'(t) = e^{i\hat{\hat{H}}_0 t}\hat{\sigma}(t), \tag{8.2}$$

$$\hat{H}'(t) = e^{i\hat{\hat{H}}_0 t}\hat{H}(t), \tag{8.3}$$

and

$$\hat{H}'_1(t) = e^{i\hat{\hat{H}}_0 t}\hat{H}_1(t). \tag{8.4}$$

Fundamentals of In Vivo Magnetic Resonance: Spin Physics, Relaxation Theory, and Contrast Mechanisms, First Edition. Daniel M. Spielman and Keshav Datta. © 2024 John Wiley & Sons, Inc. Published 2024 by John Wiley & Sons, Inc. Companion website: www.wiley.com/go/Spielman

Taking the derivative of $\hat{\sigma}'$ yields

$$\frac{d}{dt}\hat{\sigma}' = \frac{d}{dt}\left(e^{i\hat{\hat{H}}_0 t}\hat{\sigma}(t)\right) = i\hat{\hat{H}}_0 e^{i\hat{\hat{H}}_0 t}\hat{\sigma}(t) + e^{i\hat{\hat{H}}_0 t}\frac{d}{dt}\hat{\sigma}(t), \tag{8.5}$$

which simplifies to (see Exercise E.8.1)

$$\frac{d}{dt}\hat{\sigma}' = -i\hat{\hat{H}}_1'\hat{\sigma}'. \tag{8.6}$$

Hence, in this frame of reference, known as the ***interaction frame***, the time dependence of the density operator depends only on $\hat{H}_1'$.

8.2 The Master Equation of NMR

We are now going to derive Eq. (8.1). Let us start by formally integrating Eq. (8.6),

$$\hat{\sigma}'(t) = \hat{\sigma}'(0) - i\int_0^t \hat{\hat{H}}_1'(t')\hat{\sigma}'(t')dt'. \tag{8.7}$$

Now substitute for $\hat{\sigma}'(t)$,

$$\hat{\sigma}'(t) = \hat{\sigma}'(0) - i\int_0^t \hat{\hat{H}}_1'(t')\hat{\sigma}'(0)dt' - \int_0^t\int_0^{t'} \hat{\hat{H}}_1'(t')\hat{\hat{H}}_1'(t'')\hat{\sigma}'(t'')dt''dt'. \tag{8.8}$$

We can continue this process,

$$\hat{\sigma}'(t) = \hat{\sigma}'(0) - i\int_0^t \hat{\hat{H}}_1'(t')\hat{\sigma}'(0)dt' - \int_0^t\int_0^{t'} \hat{\hat{H}}_1'(t')\hat{\hat{H}}_1'(t'')\hat{\sigma}'(0)dt''dt' + i\int_0^t\int_0^{t'}\int_0^{t''} \cdots, \tag{8.9}$$

but this is starting to get ugly, so we will just keep the first three terms (to be justified later).

Now take the ensemble average, representing the average effect of the perturbing Hamiltonian for the collection of spins. Let us assume that $\hat{\hat{H}}_1'(t)$ and $\hat{\sigma}'(0)$ are uncorrelated and $\overline{\hat{H}_1'(t)} = 0$, then

$$\int_0^t \overline{\hat{\hat{H}}_1'(t')\hat{\sigma}'(0)}dt' = 0, \tag{8.10}$$

Note, if $\overline{\hat{H}_1'(t')} \neq 0$, we can always incorporate any nonzero component into $\hat{H}_0$. This leads to

$$\hat{\sigma}'(t) - \hat{\sigma}'(0) = -\int_0^t\int_0^{t'} \overline{\hat{\hat{H}}_1'(t')\hat{\hat{H}}_1'(t'')\hat{\sigma}'(0)}dt''dt'. \tag{8.11}$$

Now select a time, $t = \Delta t$, very small, such that $\hat{\sigma}'(t) \cong \hat{\sigma}'(0)$. This assumption is the one that allows us to drop those higher-order terms. In essence, $\hat{\sigma}'(t)$ is assumed to vary slowly in time as compared to $\hat{H}_1'(t)$. In effect, we are conducting a 2nd order perturbation analysis. Now,

$$\Delta\hat{\sigma}' = \hat{\sigma}'(\Delta t) - \hat{\sigma}'(0) = -\int_0^{\Delta t}\int_0^{t'} \overline{\hat{\hat{H}}_1'(t')\hat{\hat{H}}_1'(t'')\hat{\sigma}'(0)}dt''dt'. \tag{8.12}$$

Define a new variable, $\tau = t' - t''$ and substitute,

$$\Delta\hat{\sigma}' = \hat{\sigma}'(\Delta t) - \hat{\sigma}'(0) = -\int_0^{\Delta t}\int_0^{t'} \overline{\hat{\hat{H}}_1'(t')\hat{\hat{H}}_1'(t' - \tau)\hat{\sigma}'(0)}d\tau dt'. \tag{8.13}$$

Equation (8.13) can be written more compactly by introducing a correlation superoperator

$$\hat{\hat{G}}(\tau) = \overline{\hat{\hat{H}}_1'(t')\hat{\hat{H}}_1'(t' - \tau)}. \tag{8.14}$$

Note that $\hat{\hat{G}}(\tau)$ is independent of t' and therefore the first integral equals Δt. Hence,

$$\Delta\hat{\sigma}' = -\Delta t \int_0^{\Delta t} \hat{\hat{G}}(\tau)\hat{\sigma}'(0)d\tau \cong -\Delta t \int_0^{\Delta t} \hat{\hat{G}}(\tau)d\tau\hat{\sigma}'(t). \tag{8.15}$$

We now need to make some additional important assumptions. First assume Δt sufficiently **small** that $\Delta\hat{\sigma}'/\Delta t \approx d\hat{\sigma}'/dt$, but, at the same time, we want Δt sufficiently **large** so that we can extend the integration to ∞,

$$\frac{d\hat{\sigma}'}{dt} = -\int_0^{\infty} \hat{\hat{G}}(\tau)d\tau\hat{\sigma}'(t). \tag{8.16}$$

The key question is, can we actually find such a Δt? Let us look at some numbers.

Consider $\Delta t = 10^{-6}$ s. Typically, nuclear magnetic resonance (NMR) relaxation times are on the order of milliseconds to seconds, hence $\Delta\hat{\sigma}'/\Delta t \approx d\hat{\sigma}'/dt$ holds. Now what about correlation time? Tissue water rotational correlation times due to molecular tumbling are on the order of 10^{-9} s, hence $\hat{\hat{G}}(\Delta t) \cong 0$ for $\Delta t \gg 10^{-9}$ s. Therefore, extending the upper limit in the integral of Eq. (8.16) from Δt to ∞ is also valid. In general, Redfield theory is valid for relaxation times much longer than the correlations times driving the relaxation processes. This holds true for most *in vivo* tissues.

We next need to find an explicit expression for $\hat{H}_1'(t) = e^{i\hat{\hat{H}}_0 t}\hat{H}_1(t)$, and the best choice is to express $\hat{H}_1(t)$ as a linear combination of eigenoperators of $\hat{\hat{H}}_0$. Let $\hat{A}_q$ be an eigenoperator of $\hat{\hat{H}}_0$, then

$$\hat{\hat{H}}_0\hat{A}_q = e_q\hat{A}_q, \tag{8.17}$$

where e_q is the scalar eigenvalue. For example, let $\hat{H}_0 = -\omega_I\hat{I}_z - \omega_S\hat{S}_z$, and consider the operators $\hat{I}_+ = \hat{I}_x + i\hat{I}_y$ and $\hat{I}_+\hat{S}_+$.

$$\hat{\hat{H}}_0\hat{I}_+ = -\omega_I i\hat{I}_y - \omega_I\hat{I}_x = -\omega_I\hat{I}_+, \tag{8.18}$$

and

$$\hat{\hat{H}}_0\hat{I}_+\hat{S}_+ = -(\omega_I + \omega_S)\hat{I}_+\hat{S}_+. \tag{8.19}$$

Hence, $\hat{I}_+$ and $\hat{I}_+\hat{S}_+$ are eigenoperators of $\hat{H}_0$ with eigenvalues $-\omega_I$ and $-(\omega_I + \omega_S)$, respectively. (See Exercise 4.8 for more examples of eigenoperators and their corresponding eigenvalues). Now let

$$\hat{H}_1(t) = \sum_q F_q(t)\hat{A}_q, \tag{8.20}$$

where $\hat{A}_q$ is an eigenoperator of $\hat{\hat{H}}_0$, and the functions $F_q(t)$ are random functions of time (typically dependent on molecular orientation).

We first note that for $\hat{H}_1(t)$ to represent a physical process, $\hat{H}_1(t)$ must be Hermitian; however, the $F_q(t)\hat{A}_q$s are, in general, complex. Hence, for every $F_q(t)\hat{A}_q$, the sum in Eq. (8.20) must also contain a term $F_q^*(t)\hat{A}_q^\dagger$, which we will denote as

$$F_{-q}(t)\hat{A}_{-q}. \tag{8.21}$$

Remembering that the $\widehat{A}_q$s are eigenoperators of $\widehat{\widehat{H}}_0$, namely, $\widehat{\widehat{H}}_0\widehat{A}_q = e_q\widehat{A}_q$, we note that

$$e^{i\widehat{\widehat{H}}_0 t}\widehat{A}_q = e^{ie_q t}\widehat{A}_q. \tag{8.22}$$

Combining these results yields

$$\widehat{H}'_1(t) = e^{i\widehat{\widehat{H}}_0 t}\widehat{H}_1(t) = \sum_q F_q(t)\widehat{A}_q e^{ie_q t}. \tag{8.23}$$

We are almost at the last step. Write the full expression for the correlation superoperator as

$$\widehat{\widehat{G}}(\tau) = \overline{\widehat{\widehat{H}}'_1(t')\widehat{\widehat{H}}'_1(t'-\tau)} = \sum_p \sum_q \overline{F_p(t')F_q(t'-\tau)}\widehat{\widehat{A}}_p\widehat{\widehat{A}}_q e^{ie_p t'} e^{ie_q(t'-\tau)}. \tag{8.24}$$

Using a secular approximation, one can show that only the terms for which $e_p = -e_q$ need to be kept. The other terms average out as they oscillate fast as compared to the relaxation rate of $\widehat{\sigma}'$. Thus,

$$\widehat{\widehat{G}}(\tau) = \sum_q \overline{F_{-q}(t')F_q(t'-\tau)}\widehat{\widehat{A}}_{-q}\widehat{\widehat{A}}_q e^{-ie_q\tau}. \tag{8.25}$$

Defining a set of correlation functions, $G_q(\tau) = \overline{F_{-q}(t')F_q(t'-\tau)}$, yields

$$\begin{aligned}
\frac{d\widehat{\sigma}'}{dt} &= -\int_0^\infty \widehat{\widehat{G}}(\tau)d\tau\widehat{\sigma}'(t) \\
&= -\int_0^\infty \sum_q G_q(\tau)e^{-ie_q\tau}\widehat{\widehat{A}}_{-q}\widehat{\widehat{A}}_q d\tau\widehat{\sigma}'(t) \\
&= -\int_0^\infty \sum_q G_q(\tau)e^{-ie_q\tau}d\tau\widehat{\widehat{A}}_{-q}\widehat{\widehat{A}}_q\widehat{\sigma}'(t).
\end{aligned} \tag{8.26}$$

Finally, define a set of spectral density functions,

$$J_q(e_q) = \int_0^\infty G_q(\tau)e^{-ie_q\tau}d\tau, \tag{8.27}$$

with $G_q(\tau) = G_q(0)e^{-\tau/\tau_c}$, where τ_c is correlation time characteristic of the perturbation.

As defined, each J_q is complex, however; in practice, the real part is much larger than the imaginary component. A more formal treatment shows the imaginary components basically cancel due to the combination of the terms: $\widehat{\widehat{A}}_{-q}\widehat{\widehat{A}}_q$ and $\widehat{\widehat{A}}_q\widehat{\widehat{A}}_{-q}$. Thus, we can write

$$J_q(\omega) = \int_0^\infty G_q(\tau)e^{-ie_q\tau}d\tau \approx \int_0^\infty G_q(\tau)\cos\omega\tau\, d\tau = G_q(0)\frac{\tau_c}{1 + \omega^2\tau_c^2} \tag{8.28}$$

To put it all together, define the relaxation superoperator as

$$\widehat{\widehat{\Gamma}} = \sum_q J_q(e_q)\widehat{\widehat{A}}_{-q}\widehat{\widehat{A}}_q \tag{8.29}$$

with

$$J_q(e_q) = \int_0^\infty \overline{F_{-q}(t')F_q(t'-\tau)}e^{-ie_q\tau}d\tau. \tag{8.30}$$

Substituting Eq. (8.29) back into Eq. (8.26) yields

$$\frac{d\widehat{\sigma}'}{dt} = -\widehat{\widehat{\Gamma}}\widehat{\sigma}'. \tag{8.31}$$

We can now add the Boltzmann correction for thermal equilibrium and finally switch back to the laboratory frame. Note that the relaxation superoperator is the same in both frames of reference, as befits relaxation times (see Exercise E.8.1). This results in what is known as the ***Master Equation of NMR***:

$$\frac{d\hat{\sigma}}{dt} = -i\hat{\hat{H}}_0\hat{\sigma} - \hat{\hat{\Gamma}}(\hat{\sigma} - \hat{\sigma}_B) \tag{8.32}$$

8.3 Calculating Relaxation Times

Rather than directly solving the master equation, we often just want to calculate the time dependences of particular coherences, e.g., $\overline{\langle \hat{I}_x \rangle}$, $\overline{\langle \hat{I}_y \rangle}$, or $\overline{\langle \hat{I}_z \rangle}$. To do so, first express the density operator in the product operator basis,

$$\hat{\sigma} = \sum_j \overline{\langle \hat{C}_j \rangle} \hat{C}_j \xrightarrow{\text{equivalently}} \vec{\sigma} = \begin{bmatrix} \overline{\langle \hat{E} \rangle} \\ \overline{\langle \hat{I}_x \rangle} \\ \overline{\langle \hat{S}_x \rangle} \\ \vdots \\ \overline{\langle 2\hat{I}_z\hat{S}_z \rangle} \end{bmatrix}, \tag{8.33}$$

where $\overline{\langle \hat{C}_j \rangle} = \text{Tr}(\hat{\sigma}\hat{C}_j)$, and $\vec{\sigma}$ is a "vector" in a 16-D coherence (Liouville) space for the two-spin case of $\hat{C}_j \in \left\{ \frac{1}{2}\hat{E}, \hat{I}_x, \hat{S}_x, \hat{I}_y, \hat{S}_y, \dots 2\hat{I}_z\hat{S}_z \right\}$. We can now rewrite the Master equation as a vector/matrix equation. In the matrix equation, $\hat{\hat{\Gamma}}$ translates to $\underline{R}$, where $\underline{R}$ is the relaxation "supermatrix" with elements

$$R_{jk} = \text{Tr}(\hat{C}_j\hat{\hat{\Gamma}}\hat{C}_k) = \langle \hat{C}_j|\hat{\hat{\Gamma}}|\hat{C}_k \rangle. \tag{8.34}$$

Here we have used a more compact notation, but hopefully not too confusing. $\langle \hat{C}_j|\hat{\hat{\Gamma}}|\hat{C}_k \rangle$ denotes calculating the Liouville space metric (see discussion of vector space metrics in Chapter 3) rather than denoting the expected value of an operator.

If we reorder $\vec{\sigma}$ to first list populations, then single-quantum terms, then double-quantum terms, we generate the picture for $\underline{R}$ shown in Figure 8.1, often referred to as the ***Redfield kite***. Using the secular approximation, this relaxation supermatrix is block diagonal. Cross-relaxation only occurs for coherences with degenerate or close to degenerate eigenvalues of $\hat{\hat{H}}_0$. The eigenvalues are the transition frequencies, *i.e.*, sums and differences of the system energy levels (see Exercise E.4.8).

To calculate relaxation times, rewrite the master equation in terms of the operator coefficients,

$$\frac{d}{dt}\overline{\langle \hat{C}_j \rangle} = \sum_k (-i\langle \hat{C}_j|\hat{\hat{H}}_0|\hat{C}_k \rangle\overline{\langle \hat{C}_k \rangle} - (\overline{\langle \hat{C}_k \rangle} - \overline{\langle \hat{C}_k \rangle}_B)\langle \hat{C}_j|\hat{\hat{\Gamma}}|\hat{C}_k \rangle), \tag{8.35}$$

where $\overline{\langle \hat{C}_k \rangle}_B$ correspond to the Boltzmann thermal equilibrium values for the various coherences. For example, longitudinal relaxation times are calculated by

$$\frac{1}{T_{1,I}} = \langle \hat{I}_z|\hat{\hat{\Gamma}}|\hat{I}_z \rangle, \frac{1}{T_{1,S}} = \langle \hat{S}_z|\hat{\hat{\Gamma}}|\hat{S}_z \rangle, \text{ and } \frac{1}{T_{1,cross}} = \langle \hat{I}_z|\hat{\hat{\Gamma}}|\hat{S}_z \rangle. \tag{8.36}$$

The transverse relaxation rate, T_2, (shown just for the I spin) is simply

$$\frac{1}{T_{2,I}} = \langle \hat{I}_x|\hat{\hat{\Gamma}}|\hat{I}_x \rangle = \langle \hat{I}_y|\hat{\hat{\Gamma}}|\hat{I}_y \rangle. \tag{8.37}$$

All that is left is to compute some, or more commonly, a bunch of commutators!

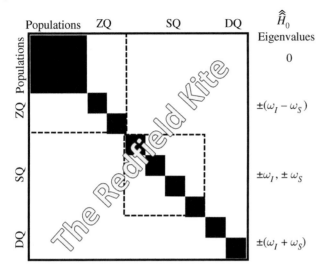

Figure 8.1 Structure of the Redfield kite relaxation supermatrix $\underline{R}$ where the terms have been grouped by populations, zero-quantum terms (ZQ), single-quantum terms (SQ), and double-quantum terms (DQ). The corresponding eigenvalues are listed on the right.

Let us investigate a simple example, relaxation due to a one-dimensional random perturbing field. Consider a Hamiltonian of the form: $\hat{H} = \hat{H}_0 + \hat{H}_1(t) = -\gamma B_0 \hat{I}_z - \gamma \Delta B(t)\hat{I}_z$ with $\langle \Delta B(t) \rangle = 0$ and $\langle \Delta B(t)\Delta B(t - \tau) \rangle = \langle \Delta B^2 \rangle e^{-|\tau|/\tau_c}$. Note, for this simplified case, the perturbation is limited to just the z-direction. Noting that $\hat{\hat{H}}_0 \hat{I}_z = 0 = 0 \cdot \hat{I}_z$, $\hat{I}_z$ is an eigenoperator of $\hat{\hat{H}}_0$ with eigenvalue $= 0$. Thus, $\hat{A}_0 = \hat{I}_z$, $F_0(t) = -\gamma \Delta B(t)$, and

$$J_0(\omega) = \gamma^2 \langle \Delta B^2 \rangle \frac{\tau_c}{1 + \omega^2 \tau_c^2} \equiv J(\omega). \tag{8.38}$$

The relaxation superoperator is simply

$$\hat{\hat{\Gamma}} = \sum_q J_q(e_q)\hat{\hat{A}}_{-q}\hat{\hat{A}}_q = J_0(0)\hat{\hat{I}}_z\hat{\hat{I}}_z, \tag{8.39}$$

and

$$\frac{1}{T_2} = \langle \hat{I}_x | \hat{\hat{\Gamma}} | \hat{I}_x \rangle = \gamma^2 \langle \Delta B^2 \rangle \tau_c \mathrm{Tr}(\hat{I}_x \hat{\hat{I}}_z \hat{\hat{I}}_z \hat{I}_x)$$

$$= \gamma^2 \langle \Delta B^2 \rangle \tau_c \mathrm{Tr}(i\hat{I}_x \hat{\hat{I}}_z \hat{I}_y) = \gamma^2 \langle \Delta B^2 \rangle \tau_c \mathrm{Tr}(\hat{I}_x \hat{I}_x). \tag{8.40}$$

Assuming we are working with normalized operators,

$$\frac{1}{T_2} = \gamma^2 \langle \Delta B^2 \rangle \tau_c. \tag{8.41}$$

An analogous calculation for T_1 yields, $1/T_1 = 0$, which is reasonable as there are no transverse field components in the perturbation. For the more general case (see derivation in Exercise E.8.2), the equations for T_1 and T_2 when the perturbation is isotropic in space, *i.e.*,

$$\Delta \vec{B}(t) = B_x(t)\vec{x} + B_y(t)\vec{y} + B_z(t)\vec{z} \ll B_0, \tag{8.42}$$

are

$$\frac{1}{T_1} = \langle \hat{I}_z | \hat{\hat{\Gamma}} | \hat{I}_z \rangle = 2\gamma^2 \langle \Delta B^2 \rangle J(\omega_0) = 2\gamma^2 \langle \Delta B^2 \rangle \frac{\tau_c}{1 + \omega_0^2 \tau_c^2}, \tag{8.43}$$

and

$$\frac{1}{T_2} = \langle \hat{I}_x | \widehat{\hat{\Gamma}} | \hat{I}_x \rangle = \langle \hat{I}_y | \widehat{\hat{\Gamma}} | \hat{I}_y \rangle = \gamma^2 \langle \Delta B^2 \rangle (\mathrm{J}(0) + \mathrm{J}(\omega_0)) = \gamma^2 \langle \Delta B^2 \rangle \left(\tau_c + \frac{\tau_c}{1 + \omega_0^2 \tau_c^2} \right). \tag{8.44}$$

8.4 Relaxation Mechanisms

8.4.1 Dipolar Coupling Revisited

The complete dipolar coupling Hamiltonian, given in Eq. (4.16), can be rewritten as

$$\hat{H}_D = -\frac{\mu_0 \gamma_I \gamma_S \hbar}{4\pi r^3} \sum_q F_q(t) \hat{A}_q, \tag{8.45}$$

where

$$\begin{aligned}
\hat{A}_0 &= \sqrt{\frac{1}{6}} \left(2\hat{I}_z \hat{S}_z - \frac{1}{2}\hat{I}_+ \hat{S}_- - \frac{1}{2}\hat{I}_- \hat{S}_+ \right) \\
\hat{A}_{\pm 1} &= \pm \frac{1}{2}(\hat{I}_\pm \hat{S}_z + \hat{I}_z \hat{S}_\pm) \\
\hat{A}_{\pm 2} &= \frac{1}{2}\hat{I}_\pm \hat{S}_\pm,
\end{aligned} \tag{8.46}$$

and the temporal functions are rank-2 spherical harmonics,

$$\begin{aligned}
F_0(t) &= \sqrt{\frac{3}{2}}(3\cos^2\theta - 1) \\
F_{\pm 1}(t) &= \pm 3 \sin\theta \cos\theta e^{\mp i\phi} \\
F_{\pm 2}(t) &= \frac{3}{2}\sin^2\theta e^{\mp 2i\phi}.
\end{aligned} \tag{8.47}$$

Table 8.1 lists the eigenoperators and associated eigenvalues for this perturbing Hamiltonian (see Exercise E.4.8). Together with $\overline{F_{-q}F_q} = \frac{6}{5}$, we can now compute $\widehat{\hat{\Gamma}}$.

Table 8.1 Eigenoperators and corresponding eigenvalues for the operators involved in dipolar coupling.

Eigenoperator	Eigenvalue
$\hat{I}_z \hat{S}_z$	0
$\hat{I}_+ \hat{S}_+$	$-(\omega_I + \omega_S)$
$\hat{I}_- \hat{S}_-$	$\omega_I + \omega_S$
$\hat{I}_+ \hat{S}_-$	$-(\omega_I - \omega_S)$
$\hat{I}_- \hat{S}_+$	$\omega_I - \omega_S$
$\hat{I}_+ \hat{S}_z$	$-\omega_I$
$\hat{I}_- \hat{S}_z$	ω_I
$\hat{I}_z \hat{S}_+$	$-\omega_S$
$\hat{I}_z \hat{S}_-$	ω_S

First consider the case of unlike spins. After much algebra, we arrive at the somewhat horrendous-looking equation for the relaxation superoperator of

$$
\widehat{\widehat{\Gamma}} = \frac{\mu_0^2}{16\pi^2} \frac{\gamma_I^2 \gamma_S^2 \hbar^2}{10r^6} \left\{ 2J(0)(\widehat{\widehat{I_z S_z}} \, \widehat{\widehat{I_z S_z}}) \right.
$$
$$
+ \left(\frac{1}{4}J(\omega_I - \omega_S) + \frac{3}{2}J(\omega_I + \omega_S) \right) (\widehat{\widehat{I_x S_x}} \, \widehat{\widehat{I_x S_x}} + \widehat{\widehat{I_y S_y}} \, \widehat{\widehat{I_y S_y}} + \widehat{\widehat{I_x S_y}} \, \widehat{\widehat{I_x S_y}} + \widehat{\widehat{I_y S_x}} \, \widehat{\widehat{I_y S_x}})
$$
$$
+ \frac{3}{2}J(\omega_I)(\widehat{\widehat{I_x S_z}} \, \widehat{\widehat{I_x S_z}} + \widehat{\widehat{I_y S_z}} \, \widehat{\widehat{I_y S_z}}) + \frac{3}{2}J(\omega_S)(\widehat{\widehat{I_z S_x}} \, \widehat{\widehat{I_z S_x}} + \widehat{\widehat{I_z S_y}} \, \widehat{\widehat{I_z S_y}})
$$
$$
\left. - \left(\frac{1}{4}J(\omega_I - \omega_S) - \frac{3}{2}J(\omega_I + \omega_S) \right) (\widehat{\widehat{I_x S_x}} \, \widehat{\widehat{I_y S_y}} + \widehat{\widehat{I_y S_y}} \, \widehat{\widehat{I_x S_x}} - \widehat{\widehat{I_x S_y}} \, \widehat{\widehat{I_y S_x}} - \widehat{\widehat{I_y S_x}} \, \widehat{\widehat{I_x S_y}}) \right\}, \quad (8.48)
$$

where we have used the notation that $\widehat{\widehat{I_z S_z}}$ is the superoperator of $\widehat{I_z}\widehat{S_z}$, in contrast to $\widehat{\widehat{I_z}}\widehat{\widehat{S_z}}$, the product of superoperators $\widehat{\widehat{I_z}}$ and $\widehat{\widehat{S_z}}$. In general, $\widehat{\widehat{I_z S_z}} \neq \widehat{\widehat{I_z}}\widehat{\widehat{S_z}}$.

Before calculating a bunch of commutators, we should note that there are multiple terms of the form $\widehat{\widehat{C_q}}\widehat{\widehat{C_q}}$, and this can make things easier. Namely, noting that all product operators cyclically commute,

$$
\widehat{\widehat{C_q}}\widehat{\widehat{C_q}}\widehat{C_p} = \begin{cases} 0 \text{ if } \widehat{\widehat{C_q}}\widehat{C_p} = 0 \\ \widehat{C_p} \text{ if } \widehat{\widehat{C_q}}\widehat{C_p} \neq 0 \end{cases}, \quad (8.49)
$$

and the terms of the form $\widehat{\widehat{C_q}}\widehat{\widehat{C_r}}$ give rise to cross-relaxation. For example

$$
\widehat{\widehat{I_x S_y}} \, \widehat{\widehat{I_y S_x}}\widehat{I_z} = \frac{1}{4}\widehat{S_z}. \quad (8.50)
$$

To calculate T_2, start with the dipolar coupling constant $q = \frac{\mu_0^2}{16\pi^2} \frac{\gamma_I^2 \gamma_S^2 \hbar^2}{10r^6}$. Then,

$$
\widehat{\widehat{\Gamma}}\widehat{I_x} = q \left(2J(0) + \frac{1}{2}J(\omega_I - \omega_S) + 3J(\omega_I + \omega_S) + \frac{3}{2}J(\omega_I) + 3J(\omega_S) \right) \widehat{I_x}. \quad (8.51)
$$

and

$$
\frac{1}{T_{2,I}} = \langle \widehat{I_x} | \widehat{\widehat{\Gamma}} | \widehat{I_x} \rangle = \frac{q}{2}(4J(0) + J(\omega_I - \omega_S) + 6J(\omega_I + \omega_S) + 3J(\omega_I) + 6J(\omega_S)). \quad (8.52)
$$

T_1 is found via

$$
\widehat{\widehat{\Gamma}}\widehat{I_z} = q[(J(\omega_I - \omega_S) + 6J(\omega_I + \omega_S) + 3J(\omega_I))\widehat{I_z} + (J(\omega_I - \omega_S) - 6J(\omega_I + \omega_S))\widehat{S_z}], \quad (8.53)
$$

and

$$
\frac{1}{T_{1,I}} = \langle \widehat{I_z} | \widehat{\widehat{\Gamma}} | \widehat{I_z} \rangle = q(J(\omega_I - \omega_S) + 6J(\omega_I + \omega_S) + 3J(\omega_I)). \quad (8.54)
$$

For completeness, the cross-relaxation term is given by

$$
\frac{1}{T_{1,IS}} = \langle \widehat{S_z} | \widehat{\widehat{\Gamma}} | \widehat{I_z} \rangle = q(J(\omega_I - \omega_S) - 6J(\omega_I + \omega_S)). \quad (8.55)
$$

The relaxation superoperator for the case where the two spins have the same (or nearly the same) chemical shift is even longer due to cross terms between $\widehat{I_z S_z}$ and $\widehat{I_\pm S_\mp}$, but ultimately yields

$$
\frac{1}{T_1} = \frac{3q}{2}(J(\omega_0) + 4J(2\omega_0)), \quad (8.56)
$$

and

$$\frac{1}{T_{2,I}} = \langle \hat{I}_x | \hat{\hat{\Gamma}} | \hat{I}_x \rangle = \frac{q}{4}(3J(0) + 5J(\omega_0) + 2J(2\omega_0)). \tag{8.57}$$

Note, now there is also transverse cross-relaxation between spins I and S,

$$\frac{1}{T_{2,IS}} = \langle \hat{S}_x | \hat{\hat{\Gamma}} | \hat{I}_x \rangle = \frac{q}{2}(2J(0) + 3J(\omega_0)). \tag{8.58}$$

This transverse cross-relaxation effect is exploited in some spin lock experiments (van de Ven 1996).

From this dipole coupling example, we can see that, starting with the Master Eq. (8.32), relaxation arises from perturbations having energy at the transition frequencies. If the eigenvalues of $\hat{H}_0$, (the energy levels of the system divided by $\hbar$) are $e_1, e_2, e_3...$, then the spectral density function is probed at the difference frequencies $e_i - e_j$. Cross-relaxation only occurs between coherences with the same transition frequencies. Although Redfield theory may seem much more complicated than the Solomon equations for dipolar relaxation, it is actually very useful. For example, T_1 and T_2 due to CSA or scalar relaxation of the 1st and 2nd kind are readily calculated.

8.4.2 Scalar Relaxation of the 1st Kind and 2nd Kind

Consider a J-coupled spin pair with the following Hamiltonian:

$$\hat{H} = \hat{H}_0 + \hat{H}_1 = -\omega_I \hat{I}_z - \omega_S \hat{S}_z + 2\pi J(\hat{I}_z \hat{S}_z + \hat{I}_x \hat{S}_x + \hat{I}_y \hat{S}_y). \tag{8.59}$$

We would normally expect a doublet from the I spin; however, chemical exchange by the S spin (we will refer to these exchanging spins as S_i) can become a relaxation mechanism (e.g., S is a group of unpaired electrons on a contrast agent). Under exchange, with an exchange time of τ_{ex}, the coupling constant between the I spin and a spin S_i becomes a random function of time. Rewriting the perturbing Hamiltonian:

$$\hat{H}_1(t) = A_i(t)\vec{\hat{I}} \cdot \vec{\hat{S}}, \tag{8.60}$$

where

$$\langle A_i^2 \rangle = \begin{cases} A^2 = 4\pi^2 J^2 \text{ if } I \text{ and } S_i \text{ are on the same molecule} \\ 0 \text{ otherwise} \end{cases}, \tag{8.61}$$

Then the probability of the I and S_i spins being on the same molecule at time $t + \tau$, given that they were together at time t, is given by

$$\langle A_i(t)A_i(t + \tau) \rangle = A^2 e^{-|\tau|/\tau_{ex}}. \tag{8.62}$$

Thus, we have $\hat{H}_0(t) = -\omega_I \hat{I}_z - \omega_S \hat{S}_z$ and $\hat{H}_1(t) = A_i(t)(\hat{I}_z \hat{S}_z + \hat{I}_x \hat{S}_x + \hat{I}_y \hat{S}_y)$. Written as a sum of eigenoperators of $\hat{\hat{H}}_0$, namely $\hat{I}_z \hat{S}_z$, $\hat{I}_+ \hat{S}_-$, and $\hat{I}_- \hat{S}_+$ with their corresponding eigenvalues of 0, $-(\omega_I - \omega_S)$, and $\omega_I - \omega_S$, the perturbing Hamiltonian becomes

$$\hat{H}_1(t) = A_i(t)\hat{I}_z \hat{S}_z + \frac{1}{2}A_i(t)\hat{I}_+ \hat{S}_- + \frac{1}{2}A_i(t)\hat{I}_- \hat{S}_+. \tag{8.63}$$

All we need now is the spectral density function, which we will denote $J_{ex}(\omega)$. Let P_i be the probability that spins I and S_i are on the same molecule, then

$$J_{ex}(\omega) = \sum_i P_i \int_0^\infty \langle A_i(t)A_i(t + \tau) \rangle e^{-i\omega\tau} d\tau. \tag{8.64}$$

Assume the I spin is always coupled to some S_i spin, *i.e.*, $\sum_i P_i = 1$, then

$$J_{ex}(\omega) = A^2 \frac{\tau_{ex}}{1 + \omega^2 \tau_{ex}^2}. \tag{8.65}$$

Hence,

$$\widehat{\widehat{\Gamma}} = A^2 J_{ex}(0)\widehat{\hat{I}_z \hat{S}_z}\, \widehat{\hat{I}_z \hat{S}_z} + \frac{1}{4} A^2 J_{ex}(\omega_I - \omega_S)\widehat{\hat{I}_+ \hat{S}_-}\, \widehat{\hat{I}_+ \hat{S}_-} + \frac{1}{4} A^2 J_{ex}(\omega_I - \omega_S)\widehat{\hat{I}_- \hat{S}_+}\, \widehat{\hat{I}_- \hat{S}_+}. \tag{8.66}$$

From which it follows that

$$\begin{aligned}
\frac{1}{T_{1,I}} &= \langle \hat{I}_z | \widehat{\widehat{\Gamma}} | \hat{I}_z \rangle \\
&= 2A^2 \frac{S(S+1)}{3} \frac{\tau_{ex}}{1 + (\omega_I - \omega_S)^2 \tau_{ex}^2} \\
&= \frac{8\pi^2 J^2 S(S+1)}{3} \frac{\tau_{ex}}{1 + (\omega_I - \omega_S)^2 \tau_{ex}^2}
\end{aligned} \tag{8.67}$$

Note, the $S(S+1)/3$ factor comes from $\mathrm{Tr}\left(\hat{S}_p^2\right) = S(S+1)/3$. Also,

$$\begin{aligned}
\frac{1}{T_{2,I}} &= \langle \hat{I}_x | \widehat{\widehat{\Gamma}} | \hat{I}_x \rangle = \langle \hat{I}_y | \widehat{\widehat{\Gamma}} | \hat{I}_y \rangle = \langle \hat{I}_+ | \widehat{\widehat{\Gamma}} | \hat{I}_+ \rangle \\
&= \frac{4\pi^2 J^2 S(S+1)}{3} \left(\tau_{ex} + \frac{\tau_{ex}}{1 + (\omega_I - \omega_S)^2 \tau_{ex}^2} \right).
\end{aligned} \tag{8.68}$$

These expressions for T_1 and T_2 have the same form for scalar relaxation of the 2nd kind with the correlation times replaced by the dipolar S-spin relaxation times $T_{1,S}$ and $T_{2,S}$ instead of τ_{ex}. Namely, consider the case where the T_1 relaxation time of the S spin is very short ($T_{1,S} \ll 1/J$). Typically, this is when the spin of the S nucleus is >1/2 and $T_{1,S}$ is dominated by quadrupolar coupling (to be discussed in more detail in Chapter 9). One way of analyzing this system is to assume the S spin is in continuous equilibrium with the lattice because of its short relaxation time. The perturbing Hamiltonian can then be rewritten as

$$\hat{H}_1(t) = S_z(t)\hat{I}_z + S_x(t)\hat{I}_x + S_y(t)\hat{I}_y \tag{8.69}$$

where $S_z(t)$, $S_x(t)$, and $S_y(t)$ are well modeled as stochastic functions with the following correlation functions

$$\langle S_z(t)S_z(t+\tau)\rangle = \frac{(2\pi J)^2 S(S+1))}{3} e^{-\tau/T_{1,S}} \tag{8.70}$$

and

$$\langle S_+(t)S_-(t+\tau)\rangle = \frac{(2\pi J)^2 S(S+1))}{6} e^{i\omega_S \tau} e^{-\tau/T_{2,S}} \tag{8.71}$$

One can then show (see Exercise E.8.4) that

$$\frac{1}{T_1} = \frac{8\pi^2 J^2 S(S+1)}{3} \left(\frac{T_{2,S}}{1 + (\omega_I - \omega_S)^2 T_{2,S}^2} \right), \tag{8.72}$$

and

$$\frac{1}{T_2} = \frac{4\pi^2 J^2 S(S+1)}{3} \left(T_{1,S} + \frac{T_{2,S}}{1 + (\omega_I - \omega_S)^2 T_{2,S}^2} \right). \tag{8.73}$$

8.4.3 Chemical Shift Anisotropy (CSA)

The Hamiltonian for a single-spin system in a magnetic field subject to both the isotropic part of the chemical shift shielding tensor, σ, and an anisotropic component, $\Delta\sigma$, is given by $\hat{H} = \hat{H}_0 + \hat{H}_1(t)$ where $\hat{H}_0 = -\gamma B_0(1 - \sigma)\hat{I}_z$, and

$$\hat{H}_1(t) = \gamma B_0 \Delta\sigma \left(\frac{1}{3}\sqrt{\frac{2}{3}} F_0(t)\hat{I}_z - \frac{1}{6}F_1(t)\hat{I}_+ - \frac{1}{6}F_{-1}(t)\hat{I}_- \right), \tag{8.74}$$

where, as before, $F_0(t) = \sqrt{\frac{3}{2}}(3\cos^2\theta - 1)$, and $F_{\pm 1}(t) = \pm 3\sin\theta\cos\theta\, e^{\mp i\phi}$ (Abragam 1983). For this case, one finds (see Exercise E.8.3)

$$\frac{1}{T_1} = \langle \hat{I}_z | \hat{\hat{\Gamma}} | \hat{I}_z \rangle = \frac{2}{15}(\Delta\sigma)^2\gamma^2 B_0^2 J(\omega_0), \tag{8.75}$$

and

$$\frac{1}{T_2} = \langle \hat{I}_x | \hat{\hat{\Gamma}} | \hat{I}_x \rangle = \frac{2}{15}(\Delta\sigma)^2\gamma^2 B_0^2 \left(\frac{2}{3} J(0) + \frac{1}{2} J(\omega_0) \right). \tag{8.76}$$

For hyperpolarized ^{13}C MRS experiments, ^{13}C-labeled substrates with the longest T_1s are most desirable (in order to extend the lifetime of the hyperpolarized signal). Dipolar coupling is minimized by labeling ^{13}C nuclei that are not directly bound to ^1H nuclei. A good example is [1-^{13}C]Pyr, where $T_1 \cong 60$ s at 3 T. Moving the label to the C_3-position (*i.e.*, the methyl group, [3-^{13}C]Pyr) results in a T_1 of less than 1 s. The difference is due to significant dipolar ^1H-^{13}C coupling for [3-^{13}C]Pyr versus minimal dipolar coupling in the case of [1-^{13}C]Pyr. In the absence of dipolar coupling, CSA becomes the dominant T_1 relaxation mechanism. An important note is that CSA relaxation scales as B_0^2. The optimum field strength for *in vivo* hyperpolarized ^{13}C experiments seems to be about 3 T, which represents a tradeoff between T_1 relaxation and chemical shift separation between metabolite resonances.

8.5 Relaxation in the Rotating Frame

T_1 probes the spectral density function at $J(\omega_0)$ while T_2 probes spectral density function at $J(0)$ and $J(\omega_0)$. In contrast, $T_{1\rho}$, "spin–lattice relaxation in the rotating frame," is often cited as an important tool for investigating slow fluctuations, typically 100 Hz to a few kHz, characteristic of multiple *in vivo* processes such as chemical exchange. But have not we been doing almost all our NMR calculations in the rotating frame? We explicitly showed that $\hat{\hat{\Gamma}}' = \hat{\hat{\Gamma}}$, where $\hat{\hat{\Gamma}}'$ and $\hat{\hat{\Gamma}}$ are the relaxation superoperators in the rotating and laboratory frames of references respectively (see Exercise E.8.1). The question thus arises: How does T_1 in the rotating frame differ from T_1? The answer is: it does not. Rather, "spin–lattice relaxation in the rotating frame" may not be the best name for $T_{1\rho}$. Perhaps $T_{1\rho}$ may be more accurately described as relaxation in the presence of a radiofrequency field.

 QUESTION: Is there also a $T_{2\rho}$?

8.5.1 Physics of $T_{1\rho}$

Let us derive the equations for $T_{1\rho}$ in the presence of a small isotropic perturbation. Consider a system of spin-$\frac{1}{2}$ particles subject to a large main magnetic field $B_0\vec{z}$ plus small independent randomly fluctuating magnetic fields, $\Delta\vec{B}(t) \ll B_0$, given by $\Delta\vec{B}(t) = B_x(t)\vec{x} + B_y(t)\vec{y} + B_z(t)\vec{z}$, where $\langle B_x^2 \rangle = \langle B_y^2 \rangle = \langle B_z^2 \rangle = \langle \Delta B^2 \rangle$. We wish to perform what is known as a spin-lock experiment and measure how much relaxation occurs during the application of a constant spin-lock Rf pulse (see Figure 8.2).

The general Hamiltonian in the laboratory frame is $\hat{H}_{lab} = \hat{H}_0 + \hat{H}_{Rf}(t) + \hat{H}_{\Delta B}(t)$, where $\hat{H}_{Rf}(t)$ is the applied Rf pulse and $\hat{H}_{\Delta B}(t)$ is the perturbation. In the presence of an on-resonance spin-lock Rf field, the Hamiltonian becomes

$$\hat{H}_{lab}(t) = -\gamma B_0\hat{I}_z - \gamma B_1(\hat{I}_x \cos\omega_0 t - \hat{I}_y \sin\omega_0 t) - \gamma B_z(t)\hat{I}_z - \gamma B_x(t)\hat{I}_x - \gamma B_y(t)\hat{I}_y$$
$$= -\omega_0\hat{I}_z - \omega_1(\hat{I}_x \cos\omega_0 t - \hat{I}_y \sin\omega_0 t) - \gamma B_z(t)\hat{I}_z - \gamma B_x(t)\hat{I}_x - \gamma B_y(t)\hat{I}_y. \tag{8.77}$$

Let us now switch to the interaction frame of reference, *i.e.*, $\hat{H}' = e^{i\hat{H}_0 t}\hat{H}_{lab}$. After a little algebra and fun with commutators, we get

$$\hat{H}'(t) = -\gamma B_1\hat{I}_x - \gamma B_z(t)\hat{I}_z - \gamma e^{i\hat{H}_0 t}(B_x(t)\hat{I}_x + B_y(t)\hat{I}_y). \tag{8.78}$$

NOTA BENE: Deriving Eq. (8.78) is great practice for working with product operators.

It is easier at this point if we switch from the product operators to the eigenoperators of $\hat{H}_0 = -\omega_0\hat{I}_z$, which are $\hat{I}_z$, $\hat{I}_+ = \hat{I}_x + i\hat{I}_y$, and $\hat{I}_- = \hat{I}_x - i\hat{I}_y$, with eigenvalues 0, $-\omega_0$, and ω_0, respectively. Substituting yields

$$\hat{H}'(t) = -\omega_1\hat{I}_x - \gamma B_z(t)\hat{I}_z - \frac{1}{2}\gamma(B_x(t) - iB_y(t))e^{-i\omega_0 t}\hat{I}_+ - \frac{1}{2}\gamma(B_x(t) + iB_y(t))e^{i\omega_0 t}\hat{I}_-. \tag{8.79}$$

First note that this Hamiltonian looks very similar to what we used to compute $1/T_1$. Namely, compare $\hat{H}(t) = -\omega_0\hat{I}_z + \hat{H}_{pert}(t)$ versus $\hat{H}'(t) = -\omega_1\hat{I}_x + \hat{H}'_{pert}(t)$. Proceed by switching to a doubly rotating frame, *i.e.*, rotating about the $\hat{I}_x$ axis at frequency $\gamma B_1 = \omega_1$ (the spin-lock amplitude).

$$\sigma''(t) = e^{-i\omega_1\hat{I}_x}\hat{\sigma}'(t) = e^{-i\omega_1\hat{I}_x}e^{-i\omega_0\hat{I}_z}\hat{\sigma}_{lab}(t) \tag{8.80}$$

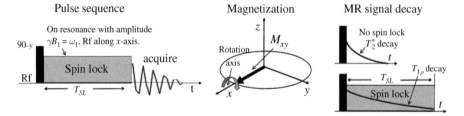

Figure 8.2 A simple spin-lock experiment for measuring relaxation during the application of a constant Rf pulse. Note, the spin lock Rf pulse is along the *x*-axis, on-resonance, and has amplitude $\gamma B_1 = \omega_1$. T_{SL}, spin lock time.

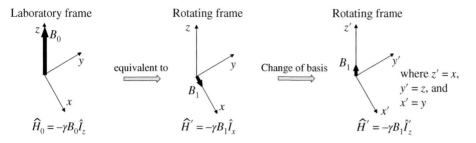

Figure 8.3 Mathematically, relaxation in the rotating frame looks analogous to T_1 relaxation with an appropriate change of variables.

which yields

$$\hat{H}''(t) = -\gamma B_z(t)e^{-i\omega_1 t\hat{\hat{I}}_x}\hat{I}_z - \frac{1}{2}\gamma(B_x(t) - iB_y(t))e^{-i\omega_0 t}e^{-i\omega_1 t\hat{\hat{I}}_x}\hat{I}_+ - \frac{1}{2}\gamma(B_x(t) + iB_y(t))e^{i\omega_0 t}e^{-i\omega_1 t\hat{\hat{I}}_x}\hat{I}_-.$$

(8.81)

However, computing the corresponding commutators is not so straightforward as $\hat{I}_z, \hat{I}_+$, and $\hat{I}_-$ are not eigenoperators of $\hat{\hat{I}}_x$. The solution is to make a simple change of coordinates as shown in Figure 8.3.

In this rotating frame of reference, the new "longitudinal" axis is $z' = x$, and the new "transverse" axes are $x' = y$ and $y' = z$. Now, calling $T_{1\rho}$ the "spin–lattice relaxation time in the rotating frame" makes a little more sense, as we are looking for

$$\frac{1}{T_{1\rho}} = \frac{d}{dt}\overline{\left\langle \hat{I}_{z'} \right\rangle}.$$

(8.82)

Let us proceed by computing the corresponding relaxation superoperator. Like before, we want to express the Hamiltonian in terms of the eigenoperators and eigenvalues of $\hat{\hat{I}}_{z'}$.

In this case, these are given by $\hat{\hat{I}}_{z'}\hat{I}_{z'} = 0 \cdot \hat{I}_{z'}$, $\hat{\hat{I}}_{z'}\hat{I}_{+'} = \hat{I}_{+'}$, and $\hat{\hat{I}}_{z'}\hat{I}_{-'} = -\hat{I}_{-'}$, where $\hat{I}_{+'} = \hat{I}_{x'} + i\hat{I}_{y'}$ and $\hat{I}_{-'} = \hat{I}_{x'} - i\hat{I}_{y'}$. Substituting for the prior operators using $\hat{I}_z = \frac{i}{2}\left(\hat{I}_{+'} - \hat{I}_{-'}\right)$, $\hat{I}_+ = \hat{I}_{z'} + \frac{i}{2}\left(\hat{I}_{+'} + \hat{I}_{-'}\right)$, and $\hat{I}_- = \hat{I}_{z'} - \frac{i}{2}\left(\hat{I}_{+'} + \hat{I}_{-'}\right)$, we have

$$\hat{H}''(t) = \gamma B_z(t)e^{-i\omega_1 t\hat{\hat{I}}_{z'}}\frac{i}{2}\left(\hat{I}_{+'} - \hat{I}_{-'}\right) - \frac{\gamma}{2}(B_x(t) - iB_y(t))e^{-i\omega_0 t}e^{-i\omega_1 t\hat{\hat{I}}_{z'}}\left(\hat{I}_{z'} + \frac{i}{2}\left(\hat{I}_{+'} + \hat{I}_{-'}\right)\right)$$
$$- \frac{\gamma}{2}(B_x(t) + iB_y(t))e^{i\omega_0 t}e^{-i\omega_1 t\hat{\hat{I}}_{z'}}\left(\hat{I}_{z'} - \frac{i}{2}\left(\hat{I}_{+'} + \hat{I}_{-'}\right)\right).$$

(8.83)

Taking the commutators and grouping terms yields

$$\hat{H}''(t) = -\frac{\gamma}{2}\left((B_x(t) - iB_y(t))e^{-i\omega_0 t} + (B_x(t) + iB_y(t))e^{i\omega_0 t}\right)\hat{I}_{z'}$$
$$+ \frac{i\gamma}{2}\left(\left(B_z(t)e^{-i\omega_1 t} - \frac{1}{2}(B_x(t) - iB_y(t))\right)e^{-i(\omega_0+\omega_1)t} + \frac{1}{2}(B_x(t) + iB_y(t))e^{-i(-\omega_0+\omega_1)t}\right)\hat{I}_{+'}$$
$$- \frac{i\gamma}{2}\left(\left(B_z(t)e^{i\omega_1 t} + \frac{1}{2}(B_x(t) - iB_y(t))\right)e^{-i(\omega_0-\omega_1)t} - \frac{1}{2}(B_x(t) + iB_y(t))e^{-i(-\omega_0-\omega_1)t}\right)\hat{I}_{-'}.$$

(8.84)

Although Eq. (8.84) looks horrendous, it is written in precisely the form we are looking for, namely

$$\hat{H}(t) = F_0(t)A_0 + F_1(t)A_1 + F_{-1}(t)A_{-1},$$

(8.85)

with $A_0 = \hat{I}_{z'}$ and $A_{\pm 1} = \hat{I}_{\pm'}$. The relaxation superoperator in this doubly rotating frame is

$$\hat{\hat{\Gamma}}'' = \sum_{q=-1}^{1} \int_{t=0}^{\infty} \int_{t'=0}^{\infty} \overline{\langle F_q(t) F_{-q}(t') \rangle} \hat{\hat{A}}_q \hat{\hat{A}}_{-q} \, dt' \, dt''. \tag{8.86}$$

To proceed, we now need to find the expressions for the correlation functions and corresponding spectral densities. Consider the simple model of an isotropic perturbation $\Delta \vec{B}(t) = B_x(t)\vec{x} + B_y(t)\vec{y} + B_z(t)\vec{z}$ where $\Delta \vec{B}(t) \ll B_0$, and having the correlation function

$$\langle B_p(t) B_q(t - \tau) \rangle = \begin{cases} 0 \text{ if } p \neq q \\ \langle \Delta B^2 \rangle e^{-|\tau|/\tau_{ex}} \text{ if } p = q \end{cases} \tag{8.87}$$

for $p, q \in \{x, y, z\}$ and τ_{ex} is the exchange rate. The corresponding correlation function is $G(\tau) = \langle \Delta B^2 \rangle e^{-|\tau|/\tau_{ex}}$. This results in a spectral density function of the form

$$\frac{J(\omega)}{\langle \Delta B^2 \rangle} = \frac{1}{\langle \Delta B^2 \rangle} \int_0^{\infty} G(\tau) e^{-i\omega\tau} \, d\tau = \frac{\tau_{ex}}{1 + \omega^2 \tau_{ex}^2}. \tag{8.88}$$

Next, make the usual secular approximations and, after much algebra, we arrive at:

$$\hat{\hat{\Gamma}}'' = \frac{\gamma^2}{4} \left(\langle B_x^2 \rangle + \langle B_y^2 \rangle \right) J(\omega_0) \hat{\hat{I}}_{z'} \hat{\hat{I}}_{z'} + \frac{\gamma^2}{4} \left(\langle B_z^2 \rangle J(\omega_1) + \frac{1}{4} \left(\langle B_x^2 \rangle + \langle B_y^2 \rangle \right) J(\omega_0 + \omega_1) \right.$$
$$\left. + \frac{1}{4} \left(\langle B_x^2 \rangle + \langle B_y^2 \rangle \right) J(\omega_0 - \omega_1) \right) \left(\hat{\hat{I}}_{+'} \hat{\hat{I}}_{-'} + \hat{\hat{I}}_{-'} \hat{\hat{I}}_{+'} \right). \tag{8.89}$$

Assuming $\omega_0 \gg \omega_1$, this simplifies to

$$\hat{\hat{\Gamma}}'' \approx \frac{\gamma^2}{2} \langle \Delta B^2 \rangle J(\omega_0) \hat{\hat{I}}_{z'} \hat{\hat{I}}_{z'} + \frac{\gamma^2}{2} \langle \Delta B^2 \rangle (J(\omega_1) + J(\omega_0)) \left(\hat{\hat{I}}_{x'} \hat{\hat{I}}_{x'} + \hat{\hat{I}}_{y'} \hat{\hat{I}}_{y'} \right). \tag{8.90}$$

In the end, $1/T_{1\rho}$ is simply the "longitudinal" relaxation rate in this doubly rotating frame and is given by the deceptively simple equation

$$\frac{1}{T_{1\rho}} = \left\langle \hat{I}_{z'} | \hat{\hat{\Gamma}}'' | \hat{I}_{z'} \right\rangle = \gamma^2 \langle \Delta B^2 \rangle (J(\omega_1) + J(\omega_0)). \tag{8.91}$$

Here we note, instead of probing the spectral density function at $\omega = 0$ as is seen with T_2, the spectral density function is probed at $\omega = \omega_1$, and the correlation time is given by the exchange time. These are key features we can now use to investigate some interesting *in vivo* processes.

In summary, $T_{1\rho}$ differs from T_2 in that for a random isotropic perturbation it probes the density function at $J(\omega_1) + J(\omega_0)$ rather than $J(0) + J(\omega_0)$, and the user selects the spin-lock amplitude γB_1 of interest, typically 0.1–3 kHz. In the limit of $\omega_1 = 0$, $1/T_{1\rho}$ just equals $1/T_2$ (which is expected since this case simply become relaxation along the $\hat{I}_x$ axis),

$$\lim_{\omega_1 \to 0} \frac{1}{T_{1\rho}} = \gamma^2 \langle \Delta B^2 \rangle (J(0) + J(\omega_0)) = \frac{1}{T_2}. \tag{8.92}$$

In general, the study of the change in relaxation rates versus magnetic field (called **dispersion**) yields further insights into different relaxation processes.

8.5.2 The Spin-Lock Experiment

There are a variety of magnetic resonance imaging (MRI) approaches for generating $T_{1\rho}$ contrast (see Figure 8.4). These generally fall into two broad measurement strategies. The first is to fix ω_1

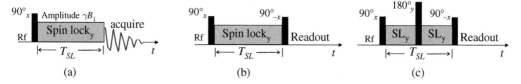

Figure 8.4 Three different spin-lock strategies. (a) Simplest implementation, (b) implementation often favored in fast imaging applications, and (c) a spin-lock method for improved robustness to B_0 inhomogeneities.

and collect a series of data sets in which the length of the spin lock pulse (T_{SL}) is incremented. In combination with imaging gradients, this method is used to generate spatial maps of $T_{1\rho}$. The second is to fix T_{SL} and collect data with increasing values of ω_1. This is a way of generating a $T_{1\rho}$-dispersion curve.

The power of $T_{1\rho}$ is as follows. Rapid processes (*e.g.*, with correlation times near $\tau_c = 1/\omega_0$) can be studied using magnets at different fields strengths. However, investigating slower processes by lowering B_0 results in an unacceptably low SNR. $T_{1\rho}$ imaging allows the study of these slower processes without this dramatic SNR loss.

8.5.3 Choosing the Optimum Spin-Lock Frequency

The following logic can often be found in the MRI literature: "Because $T_{1\rho}$ probes the density function at $J(\gamma B_1)$, $T_{1\rho}$ is maximally sensitive to fluctuations at the spin lock frequency of $\omega_1 = \gamma B_1$. Hence, to best assess a chemical exchange process with exchange rate k_{ex}, one should use $T_{1\rho}$ with a spin-lock frequency of $\omega_1 = k_{ex}$." Is this true? Is $\omega_1 = k_{ex}$ the optimum and, if so, in what sense? Let us find out.

Multiple physical processes can drive $T_{1\rho}$ (as well as T_2 and T_1) relaxation. These include molecular tumbling in combination with dipolar coupling, chemical exchange processes, dynamic scalar coupling interactions, and diffusion effects. Moreover, equations describing $T_{1\rho}$ relaxation differ among these sources, and the relative values of $1/\omega_1$ versus the characteristic time constants associated with these different relaxation processes are critically important.

Spectral density functions quantify the amount of energy available from a stochastic process at a given frequency and are the primary mathematical tools used for analyzing MRI relaxation processes. Let us analyze the spectral density functions and associated T_2 and $T_{1\rho}$ relaxation times for a simplified two-pool model of tissue water. Namely, consider a system of water protons where all spins are subject to dipolar coupling and molecular tumbling and a fraction of the spins, f_{ex}, are also subject to relaxation via a chemical exchange process characterized by an exchange rate k_{ex} and frequency shift between sites of δ.

For this two-pool model, the total magnetization decay rates are given by the combination of dipolar coupling and chemical exchange effects. Namely,

$$\frac{1}{T_{1\rho}} = \frac{1}{T_{1\rho,D}} + f_{ex}\frac{1}{T_{1\rho,ex}} \tag{8.93}$$

and

$$\frac{1}{T_2} = \frac{1}{T_{2,D}} + f_{ex}\frac{1}{T_{2,ex}}. \tag{8.94}$$

Using spectral density function for dipolar coupling (Solomon 1955; Gilani and Sepponen 2016), the equations for T_2 and $T_{1\rho}$ are given by

$$\frac{1}{T_{1\rho,D}} = K(3J_D(\omega_1) + 5J_D(\omega_0) + 2J_D(2\omega_0)) = K\left(\frac{3\tau_c}{1 + \omega_1^2\tau_c^2} + \frac{5\tau_c}{1 + \omega_0^2\tau_c^2} + \frac{2\tau_c}{1 + 4\omega_0^2\tau_c^2}\right)$$

(8.95)

and

$$\frac{1}{T_{2,D}} = K(3J_D(0) + 5J_D(\omega_0) + 2J_D(2\omega_0)) = K\left(3\tau_c + \frac{5\tau_c}{1 + \omega_0^2\tau_c^2} + \frac{2\tau_c}{1 + 4\omega_0^2\tau_c^2}\right),$$

(8.96)

where

$$K = \frac{3}{40}\left(\frac{\mu_0}{4\pi}\right)^2\frac{\gamma^4\hbar^2}{r^6}.$$

(8.97)

However, Eq. (8.93) is typically not particularly useful. For most *in vivo* experiments $\omega_1\tau_c \ll 1$ (a typical value for *in vivo* water is $\tau_c = 10^{-9}$ s) and hence, for dipolar coupling, $T_{1\rho,D} \approx T_2$.

The case for chemical exchange is more interesting. In this case the spins are subject to a random perturbation in ΔB_z (secular approximation), *i.e.*, just a modulation of chemical shift. Namely, $\Delta\vec{B}(t) = \Delta B_z(t)\vec{z}$ where $\Delta\vec{B}(t) \ll B_0$, and having the correlation function

$$\langle\Delta B_z(t)\Delta B_z(t-\tau)\rangle = \begin{cases} 0 & \text{if spin at site A} \\ \delta^2 e^{-|\tau|/\tau_{ex}} & \text{if spin at site B} \end{cases}.$$

(8.98)

Before deriving the final expression for $T_{1\rho}$ under chemical exchange, there is one additional adjustment to be made. Namely, if the spin-lock pulse is on resonance for large water pool, then it is off-resonance for the exchanging spins. The net result is we need to replace ω_1 with the effective field $\omega_{eff} = \sqrt{\delta^2 + \omega_1^2}$, *i.e.*,

$$|B_{eff}| = \frac{1}{\gamma}\omega_{eff} = \frac{1}{\gamma}\sqrt{\delta^2 + \omega_1^2},$$

(8.99)

In this case, $T_{1\rho}$ probes the spectral density at ω_{eff} and is given by the expression (Cobb et al. 2011),

$$\frac{1}{T_{1\rho,ex}} = J_{ex}(\omega_{eff}) = \frac{\delta^2 k_{ex}}{k_{ex}^2 + \omega_{eff}^2} = \frac{\delta^2 k_{ex}}{k_{ex}^2 + \delta^2 + \omega_1^2}.$$

(8.100)

To get a sense of scale, let $\omega_1 = 500$ Hz, and $\delta = 127$ Hz (1 ppm at 3 T), then $\omega_{eff} = \sqrt{\delta^2 + \omega_1^2} = 516$ Hz (as a side note, compare the above expression with Eq. (7.95)). Representative spectral density curves are plotted in Figure 8.5.

Under the assumption that $1/T_2$ and $1/T_{1\rho}$ are dominated by spectral density terms evaluated at $J(0)$ and $J(\omega_{eff})$ respectively, then the corresponding transverse magnetization rates, including both molecular tumbling-driven dipolar-coupling modulations and chemical exchange effects, are given by

$$\frac{1}{T_2} \approx 3K\tau_c + \alpha\frac{\delta^2}{k_{ex}}$$

(8.101)

and

$$\frac{1}{T_{1\rho}} \approx 3K\tau_c + \alpha\frac{\delta^2 k_{ex}}{k_{ex}^2 + \omega_{eff}^2}.$$

(8.102)

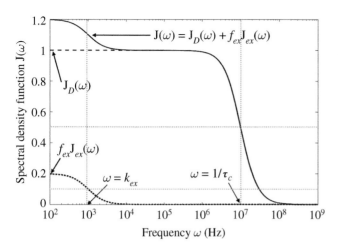

Figure 8.5 Spectral density functions for a representative tissue containing a mixture of water molecules for which all of the molecules are subject to dipolar coupling with a rotational correlation time τ_c and f_{ex} fraction of the molecules are subject to both dipolar coupling and chemical exchange with a partner having chemical shift difference δ and exchange rate $= k_{ex}$. T_2, $T_{1\rho}$, and T_1 are largely determined by the values of the spectral density function $J(\omega)$ evaluated at $\omega = 0$, $\omega = \omega_{eff} = \sqrt{\delta^2 + \omega_1^2}$, and $\omega = \omega_0$ respectively. Note, $J_{ex}(\omega)$ and $J_D(\omega)$ are decreased by 50% at $\omega = k_{ex}$ and $\omega = 1/\tau_c$, respectively.

T_2-weighted acquisitions aways contain maximal information regarding chemical exchange-driven relaxation. But this is not necessarily the case for $T_{1\rho}$ where the dependence of the chemical exchange fraction parameter f_{ex}, is a function of the spin-lock frequency. If $\omega_{eff} \ll k_{ex}$, both T_2 and $T_{1\rho}$ contain chemical exchange information. However, if $\omega_{eff} \gg k_{ex}$, then $T_{1\rho}$ is independent of α and, hence, contains NO chemical exchange information. In other words, for high spin-lock frequencies, the dephasing effects of chemical exchange are fully refocused. Using the tissue model described above, data acquired with $\omega_{eff} \gg k_{ex}$ would contain no information regarding chemical exchange.

A chemical exchange process with a characteristic exchange rate of k_{ex} does not have excess energy at $\omega = k_{ex}$. There is no positive or negative peak in the corresponding spectral density function at $J(k_{ex})$. As compared to T_2, $T_{1\rho}$ data acquired at a spin-lock frequency of ω_1 **suppresses** rather than enhances contribution from processes with rotational or chemical exchange rates near or below ω_1. Setting the spin-lock frequency to $\omega_1 = k_{ex}$ (or more accurately setting $\omega_1 = \omega_{eff}$ for a chemical exchange process) simply results in $T_{1\rho}$ information for which the contributions from the exchange process have been **reduced** by 50%, i.e., $J(\omega_{eff}) = J(0)/2$.

A spin-lock pulse is fundamentally a refocusing mechanism. Locking the spins along an applied Rf field results in the elimination of dephasing arising from dynamic processes that are slow compared to the ω_1. This has the effect of **removing** such relatively slow processes from acting as relaxation mechanisms.

The "refocusing" nature of $T_{1\rho}$ experiments can be further appreciated via comparisons to other well-known MRS and MRI acquisition methods, with a representative set of such acquisitions shown in Figure 8.6. First, compare a simple spin-lock acquisition (Figure 8.6a) with proton-decoupled ^{13}C MRS (Figure 8.6b). In the case of decoupling, the non-random nature of J coupling results in discrete peak splitting rather than a general spectral broadening (i.e., T_2 decay). Nonetheless, a series of 180° Rf pulses removes this splitting (i.e., refocuses J coupling) if the spacing between the 180° is short compared to $1/J$, analogous to a spin-lock acquisition with $\omega_1 = 1/\tau_{180}$.

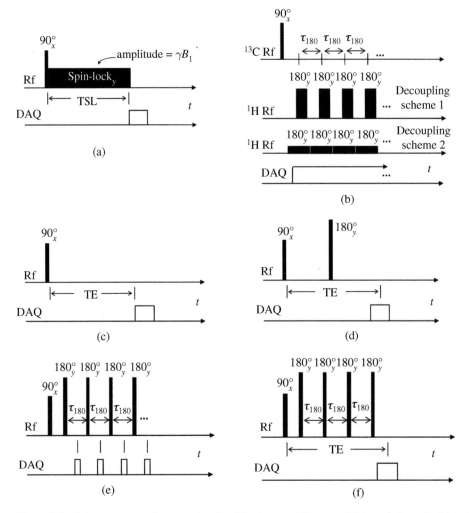

Figure 8.6 Pulse sequence diagrams showing Rf pulses and data acquisition windows for (a) a simple spin-lock implementation, (b) ^{1}H decoupling of ^{13}C spectra, and (c) gradient-echo, (d) spin-echo, (e) fast spin-echo, and (f) CPMG acquisitions.

Figure 8.6b further depicts three different heteronuclear decoupling schemes with increasing broad but lower amplitude 180 s. The third scheme, with a continuous decoupling pulse minimizes the Rf power requirements. In practice, decoupling schemes usually employ phase modulation (*e.g.*, alternating the sign of the 180 s in a 180–180–180–180…scheme) to increase the spectral bandwidth performance. This raises an important issue for comparing Figure 8.6a spin-lock acquisition where transverse magnetization is decaying with time constant $T_{1\rho}$ to the acquisitions shown in Figure 8.6b–f. A long continuous spin-lock pulse of amplitude ω_1 has a much narrower spectral bandwidth as compared, for example, to a series of 180° pulses. This represents a spin-lock frequency tradeoff between spectral bandwidth and Rf power.

The simplistic spin-lock implementation shown in Figure 8.6a is quite sensitive to off-resonance effects and, hence, will likely perform poorly in regions with significant B_0 inhomogeneities. There have been some proposals for more robust spin-lock implementations (Pala et al. 2023); however, imagers might follow the lead of spectroscopists whereby decoupling spectral bandwidths can be increased with minimum increases in Rf power using phase-modulated decoupling, for example via composite pulses or other broadband decoupling schemes (Freeman and Kupce 1997; Freeman and Hurd 1997). The same techniques are also used in homonuclear spin-lock MRS experiments such as TOCSY (Kirschstein et al. 2008).

One effect we have yet to mention is that due to residual dipolar coupling (RDI). RDI results in a nuclear spin Hamiltonian term analogous to J coupling. Given all spins typically do not experience precisely the same value of RDI, the result is a broadening of the peaks rather than the discrete splitting seen with J coupling. A spin-lock pulse will refocus RDI if $\omega_1 \gg \gamma \text{RDI}$. This effect has been observed in cartilage where RDI is reported to be the order of 1 kHz) (Borthakur et al. 2006; Akella et al. 2004).

Next compare a gradient-echo (Figure 8.6c) to a spin-echo (Figure 8.6d) acquisition. The analogy here is T_2^* is to T_2 as T_2 is to $T_{1\rho}$. For on-resonance spins, the spin-echo acquisition is comparable to a spin-lock acquisition with $\omega_1 = 1/TE$ (modulo the noted differences being the spectral bandwidth differences between a long continuous Rf verses a short 180° pulse). All signal loss occurring on a time-scale short compared to $1/TE$ is refocused. Choosing to study static susceptibility effects using only spin echoes, which refocuses these effects, would likely be a poor choice. Both the fast-spin echo (FSE) and Carr–Purcell–Meiboom–Gill (CPMG) pulse sequences shown in Figure 8.6e and f are analogous to $T_{1\rho}$ studies with $\omega_1 = 1/\tau_{180}$.

The important issue is finding the best acquisition or set of acquisitions to assess *in vivo* chemical exchange processes, and the answer varies depending on the goals of the experiment. If the goal is to eliminate a source of transverse magnetization decay associated with a dynamic process with a characteristic exchange rate k_{ex} or, for example, to suppress RDI effects, choosing $\omega_1 \gg k_{ex}$ would be ideal.

In contrast, to isolate contributions from chemical exchange one needs to compare $T_{1\rho}$ at two or more different spin-lock frequencies. This is consistent with studies suggesting using $T_{1\rho}$-dispersion data rather than $T_{1\rho}$ data acquired at a single value of ω_1 (Duvvuri et al. 2001; Elsayed et al. 2023; Keenan et al. 2015; Regatte et al. 2003; Wang et al. 2015), while others have suggested analyzing maps of $T_{1\rho}/T_2$ ratios (Keenan et al. 2015). However, the full dispersion curve is unnecessary unless the goal is to calculate k_{ex} (Cobb et al. 2011; Wang et al. 2015). Similarly, $T_{1\rho}$ and T_2 maps (which each require at least two acquisitions with different values of the spin-lock duration TSL) are also overkill. Perhaps the simplest metric, first suggested by Regatte et al. (2003), is to compute the ratio of $T_{1\rho}$-weighted images acquired at two different values of ω_1. One promising approach would be to acquire a $T_{1\rho}$-weighted image, I_1, with $\omega_1 = 0$ (equivalent to a T_2-weighted image) and a $T_{1\rho}$-weighted image, I_2, acquired at the largest ω_1 allowable within Rf power deposition constraints. Let

$$I = M_0 \left(1 - e^{-TR/T_1}\right) e^{-TSL/T_{1\rho}}, \tag{8.103}$$

then, Eqs. (8.101) and (8.102) yield

$$\ln\left(\frac{I_1}{I_2}\right) = f_{ex}\frac{\delta^2}{k_{ex}} TSL \left(1 - \frac{1}{1 + \left(\frac{\omega_{eff}}{k_{ex}}\right)^2}\right) = f_{ex}\frac{\delta^2}{k_{ex}} TSL \left(1 - \frac{k_{ex}^2}{k_{ex}^2 + \omega_{eff}^2}\right). \tag{8.104}$$

From Eq. (8.104), the sensitivity of this metric to changes in the chemical exchange fraction α is maximized for $\omega_{eff} \gg k_{ex}$. Of course, for lower spin-lock frequencies the sensitivity decreases, with $\ln(I_1/I_2)$ being completely independent of α for $\omega_{eff} \ll k_{ex}$. Figure 8.7 is a plot of the relative sensitivity of $\ln(I_1/I_2)$, as given by the derivative with respect to α, as a function of ω_{eff}. Going a step further, we can also estimate the optimum spin-lock duration TSL. This proposed metric

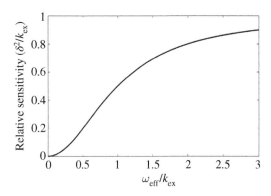

Figure 8.7 Relative sensitivity of $\ln(I_1/I_2)$ to the fraction of spins α engaged in chemical exchange versus $\omega_{eff} = \sqrt{\delta^2 + \omega_1^2}$. The tissue is modeled as having all water spins subject to relaxation via dipolar coupling and a fraction α of the spins also subject to chemical exchange with an exchange rate of k_{ex} and a frequency shift difference of δ. I_1 is a T_2-weighted image, I_2 is a $T_{1\rho}$-weighted image with spin-lock frequency ω_1.

in Eq. (8.104) is essentially a two-point estimate of the decay rate for an exponential decaying function with time constant $\tau = T_{1\rho} - T_2$. The optimum sampling times for two-point measurements of an exponentially decaying function were derived by Jones et al. (1996) and correspond to $TSL_{opt} \approx 1.11 \times (T_{1\rho} - T_2)$.

When multiple processes involved, each with its own characteristic time constants, e.g., RDI and proteoglycan hydroxyl chemical exchange in cartilage (Borthakur et al. 1999), the best choices for ω_1 might be to bracket the characteristic exchange rate of the targeted process of interest (Regatte et al. 2003).

To study changes in k_{ex}, e.g., exchange rate changes as a function of pH (Spear and Gore 2016) as compared to changes in the exchanging fraction f_{ex}, the optimum sensitivity is achieved by setting ω_1 equal to the inflection point in the $T_{1\rho}$ dispersion curve given by $\omega_{ip} = \frac{1}{3}\sqrt{k_{ex}^2 + \delta^2}$ (Wang et al. 2015). This is slightly different from, perhaps, the natural choice of $\omega_{eff} = k_{ex}$ (i.e., $\omega_1 = \sqrt{k_{ex}^2 - \delta^2}$).

We finally note that there is a B_0 field dependence for $T_{1\rho}$ studies of chemical exchange. The chemical shift parameter, δ^2, scales as B_0^2, suggesting higher fields might be better for assessing chemical exchange effects. Unfortunately, the Rf power deposition of the spin-lock pulse also scales as B_0^2, limiting maximum *in vivo* spin-lock frequencies. Similarly, conducting studies at lower magnetic fields would ease SAR limitations, but at the cost of reduced sensitivity to chemical exchange.

In summary, while $T_{1\rho}$ acquisitions do indeed probe the spectral density function at $J(\omega_1)$, the effects of chemical exchange on transverse magnetization relaxation are partially or fully refocused using spin-lock pulse sequences, depending on the choice of the spin-lock frequency. At a given B_0 field strength, unlike T_1 or T_2, tissues do not have a single $T_{1\rho}$ value. $T_{1\rho}$ is a function of the spin-lock frequency, thus the chosen ω_1 for a given experiment must be specified and ideally analyzed in relation to the characteristic exchange rate (or peak broadening) of the processes of interest.

8.5.4 Rf Power Considerations

Whether the optimum ω_1, as chosen based on the characteristic correlation rates of the associated relaxation process of interest, is achievable in *vivo* is a separate question. Typically, the amount of Rf energy dissipated in the subject is a critical limitation, both in the choice of the maximum B_1 and T_{SL}. More precisely, the specific absorption rate (SAR) is the rate of energy absorption by tissues during MRI as measured in watts per kilogram (W/kg). In the United States, the Food and Drug Administration (FDA) limits the amount of energy absorbed during the body over the course of a single MRI examination to 1 °C/kg, which in practice typically implies the body cannot be exposed to greater than 4 W/kg (ICNIRP 2004). SAR increases as both B_0^2 (*i.e.*, worse on high-field scanners) and B_1^2 (limiting the temporal frequencies that can be investigated). SAR is further influenced by the size and shape of the patient (increased SAR with obesity), and the number of Rf pulses per unit time (*i.e.*, worse for sequences with lots of 180° pulses such as fast spin echo).

Consequently, several approaches for measuring $T_{1\rho}$ at reduced SAR are under active investigation. One approach is to use a lower-field scanner. However, while SAR scales with B_0^2, studies of chemical exchange are limited by the size of the perturbation similarly scaling as the square of the chemical shift difference δ (see Eq. (8.96)). A second approach is to just apply the spin-lock pulse during acquisition of central k-space lines (*i.e.*, with a fast spin echo acquisition).

Spin-lock acquisitions can also reduce the magic-angle effects that arise from residual dipolar interactions (RDI). However, SAR restrictions typically limit maximum spin-lock frequencies for *in vivo* studies. Thus, there is a reduction in magic-angle effects, but they rarely disappear entirely.

8.5.5 Adiabatic Spin-Lock

The $T_{1\rho}$ described in the previous sections corresponds to the special case whereby the spin-lock pulse has constant amplitude, is on-resonance, and points along the x or y axis in the rotating frame of reference. To emphasize this condition, the term "on-resonance $T_{1\rho}$" is sometimes used.

More generally, the spin-lock pulse can point along any spatial axis, and both its direction and amplitude can be time-varying and need not be restricted to the on-resonance case. The most common choice is to implement a spin-lock using an adiabatic Rf pulse, such as a frequency-modulated hyperbolic secant (Michaeli et al. 2006). In this case, there is no initial 90° Rf pulse to tip the spins into the transverse plane. Rather, the magnetization typically starts along the z-axis and, if the rate of change of the spin-lock B_1 pulse axis is sufficiently slow, remains "locked" along the direction of this applied field. Such adiabatic pulses have the advantage that the magnetization vector remains highly insensitive to changes in the B_1 amplitude; a property highly useful for minimizing the effects of spatially varying Rf inhomogeneities or exchange rates (Gilani and Sepponen 2016).

In this frequency-swept case, the direction of spin-lock field, B_{eff}, is the vector sum of $(\omega_0 - \omega_{Rf})/\gamma$ and, the spin-lock pulse frequency is swept across the resonance frequency. Relaxation of the magnetization along the axis defined by the now time-varying spin-lock pulse is known as "$T_{1\rho}$-adiabatic," while relaxation transverse to the spin-lock pulse is called "$T_{2\rho}$-adiabatic." In the case of chemical exchange, generalizing Eq. (8.100) yields

$$\frac{1}{T_{1\rho,ex}}(t) = f_{ex}\delta^2 \sin^2\alpha(t)\frac{\delta^2 k_{ex}}{k_{ex}^2 + \delta^2 + \omega_1^2} \tag{8.105}$$

and

$$\frac{1}{T_{2\rho,ex}}(t) = f_{ex}\delta^2 \cos^2\alpha(t)\frac{1}{k_{ex}}, \tag{8.106}$$

where $\alpha(t)$ is the time-varying angle between B_0 and the applied spin-lock pulse. Note these relaxation rates are now also time-varying.

Further modification of adiabatic $T_{1\rho}$ methods include adding both amplitude and frequency modulation to the spin-lock pulse (Liimatainen et al. 2010). In general, T_1, T_2, $T_{1\rho}$, $T_{1\rho}$-adiabatic, and $T_{2\rho}$-adiabatic all differ and can provide unique insights into in vivo tissue properties. An excellent review of $T_{1\rho}$ methods can be found in Gilani and Sepponen (2016).

8.5.6 Applications

Given that $T_{1\rho}$ is a relaxation mechanism suitable for studying chemical exchange processes, it may serve as a biomarker for alterations of these processes in tissues. In clinical MRI research, the most ubiquitous use of $T_{1\rho}$ contrast has been for the study of cartilage and osteoarthritis (OA; Choi and Gold 2011) (to be discussed in more detail in Chapter 10), although there have been numerous other proposed applications. Here we provide just a partial list. Liver fibrosis involves the accumulation of collagen, proteoglycans, and other macromolecules in the extracellular matrix, and thus $T_{1\rho}$ relaxation could be a candidate biomarker for liver fibrosis (Takayama et al. 2015). Degeneration of lumbar discs has been studied with $T_{1\rho}$ (Brayda-Bruno et al. 2014). $T_{1\rho}$ is potentially sensitive to high-intensity focused ultrasound (HIFU)-induced protein denaturation with significant changes in tumor $T_{1\rho}$ observed after HIFU treatment, with the changes reported to be more pronounced than changes in T_2 (Hectors et al. 2015). $T_{1\rho}$-mapping has also been used for the study of patients with chronic myocardial infarction (van Oorschot et al. 2014), proposed as a biomarker in the diagnosis of Alzheimer's disease (Haris et al. 2015), and quantitative $T_{1\rho}$ mapping may provide a useful marker for assessing disease progression in Huntington disease (Wassef et al. 2015).

8.6 Illustrative Redfield Theory Examples

This section discusses two additional applications arising from hyperpolarized ^{13}C (HP-^{13}C) experiments. Although HP-^{13}C is not widely used, it provides an excellent opportunity to highlight the utility of Redfield theory for studying multiple relaxation processes given that the ^{13}C-labeling pattern is typically chosen to minimize dipolar coupling effects.

8.6.1 Hyperpolarized ^{13}C-urea

The first example comes from a paper by Shang et al., entitled "Handheld Electromagnet Carrier for Transfer of Hyperpolarized Carbon-13 Samples" (Shang et al. 2016). The authors state that "Some HP ^{13}C substrates can lose polarization extremely quickly in low magnetic fields when they are transferred between the polarizer and the MR scanner, reducing the SNR." The questions are: Which substrates? Why is this a low-field effect? And how low is "low"?

The targeted ^{13}C-labled substrate for this study was urea, which is of considerable interest for measuring *in vivo* perfusion. This is an interesting molecule from an MR perspective in that scalar coupling between fast-relaxing spin-1 quadrupolar-coupled ^{14}N and spin-½ ^{13}C nuclei results in rapid loss of polarization at low field.

Starting from Eq. (8.72) for scalar relaxation of the 2nd kind, in the case of $[^{13}$C-^{14}N$_2$]urea, T_1 of the ^{13}C nucleus is given by

$$\frac{1}{T_1} = \frac{1}{T_{1,0}} + \frac{32\pi^2 J^2}{3}\left(\frac{T_{2,N}}{1 + (\gamma_C - \gamma_N)^2 B_0^2 T_{2,N}^2}\right), \tag{8.107}$$

where the $1/T_{1,0}$ is due to relaxation from other mechanisms (such as dipolar coupling) and the extra factor of 2 due to their being two ^{14}N nuclei. Let us start with some numbers. $J = 14.5$ Hz, $T_{1,0} = 78$ s (at 3 T), $T_{2,N} = 2 \times 10^{-4}$ s, $\gamma_N = 19.331 \times 10^6$ rad/s/T, and $\gamma_C = 67.762 \times 10^6$ rad/s/T. To help visualize this equation, T_1 is plotted as a function of B_0 in Figure 8.8.

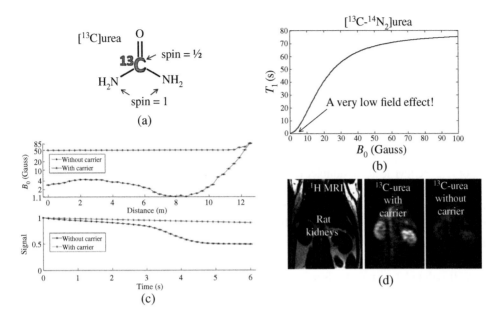

Figure 8.8 Scalar relaxation of the 2nd kind in an HP-^{13}C experiment. (a) [^{13}C-^{14}N$_2$]urea (b) T_1 of ^{13}C as a function of field strength for [^{13}C-^{14}N$_2$]urea. (c) *In vitro* and (d) *in vivo* data with and without the use of an electromagnetic carrier to keep the magnetic field of the sample above 50 G. Source: (c) and (d) Adapted with permission from Shang et al. (2016) Figures 1 and 5, respectively.

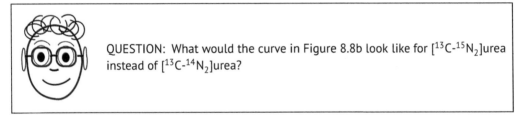

QUESTION: What would the curve in Figure 8.8b look like for [^{13}C-^{15}N$_2$]urea instead of [^{13}C-^{14}N$_2$]urea?

For conventional MR experiments, such a low-field effect would be of little consequence given that B_0 fields are typically ≥ 0.5 T, resulting in resonant frequencies in the MHz range. However, in the special case of HP-^{13}C imaging, the polarized sample must be transported from the polarizer to the patient lying in the scanner. During this transportation process, the sample typically passes through low-field regions, even as low as the earth's magnetic field ($\sim$0.5 G, where 1 T = 10,000 G). Although the transport is performed as quickly as possible, passing though such a low magnetic field causes significant T_1-decay signal loss due to scalar relaxation of the 2nd kind. The use of a magnetic carrier, such as proposed by Shang et al., even one having a field of only 50 G, prevents this signal loss. Note, safety concerns just dictate switching off the device upon entering the fringe field of the scanner!

8.6.2 Hyperpolarized ^{13}C-Pyr

The second example comes from papers by Lau et al. (2013) and Datta and Spielman (2017) (Figure 8.9).

Figure 8.9 Pyruvate labeled with ^{13}C in the C$_1$ and C$_2$ positions.

QUESTION: Which carbon, if labeled, would have the shortest T_1?

The key observation is that any sample of singly labeled pyruvate at the C_1 position, [1-^{13}C]Pyr, has 1% doubly labeled [1,2-^{13}C]Pyr due to the natural abundance of ^{13}C, and the carbon–carbon J coupling of [1,2-^{13}C]Pyr leads to doublet resonances. Under conventional thermal equilibrium conditions, these doublets are symmetric. However, there is a noted asymmetry in the hyperpolarized state arising from a non-negligible contribution of a $2\hat{I}_z\hat{S}_z$ coherence since the high temperature approximation is no longer valid at the close to 1.0 K temperature of the sample in the polarizer. Namely, the Hamiltonian is $\hat{H}_0 = -\omega_I\hat{I}_z - \omega_S\hat{S}_z + 2\pi J(\hat{\vec{I}} \cdot \hat{\vec{S}})$, and the spin density operator is $\hat{\sigma}_0 \approx \frac{1}{4}\hat{E} + \frac{1}{2}P_C\hat{I}_z + \frac{1}{2}P_C\hat{S}_z + \frac{1}{2}P_C^2 2\hat{I}_z\hat{S}_z$, where P_C is the carbon polarization. As shown in Figure 8.10a, this doublet asymmetry is a function of the polarization and decays over time after the sample leaves the polarizer.

Consider an experiment using hyperpolarized [1,2-^{13}C]Pyr, where the initial carbon polarization is P_C and the flip angle used for repeated measurements in the scanner of the C_1- and C_2-carbons is ϕ. Experimental results are shown in Figure 8.10b. Both the C_1 and the C_2 doublets

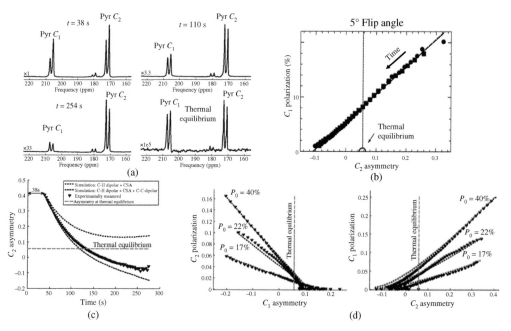

Figure 8.10 ^{13}C spectra for hyperpolarized [1,2-^{13}C]Pyr. (a) Doublet asymmetry occurs in the hyperpolarized case, and (b) the experimental data plotting the relationship between doublet asymmetry and polarization (18 ml of 37 °C blood injected with 2 ml of 80 mM pre-polarized Pyr solution) changes over time. (c) Good model fits to the decay of the C_2 asymmetry parameter require considering CSA and dipolar coupling effects simultaneously (C_1 asymmetry fits are similar). (d) Experimental *in vitro* data for hyperpolarized [1,2-^{13}C]Pyr and corresponding fits for C_1 and C_2 asymmetry parameters versus time for different levels of initial polarization. Source: (b) Adapted with permission from Lau et al. (2013) Figure 4.

are asymmetric primarily due to the $2\hat{I}_z\hat{S}_z$ term generating an antiphase signal (with a minor contribution from residual strong coupling effects). Lau et al., defined an asymmetry parameter a_C = (inner peak – outer peak)/(inner peak + outer peak), for which one can show that at $t = 0$, a_{C_2} is given by $a_{C_2}(0) \approx P_C \cos\phi + \sin\theta$, where $\sin\theta$ is the residual strong coupling parameter for [1,2-^{13}C]Pyr at 3 T and equals 0.056.

Given that polarization of the C_1 and C_2 Pyr peaks decay with T_1s ~50 s *in vitro* at 3 T, one might expect a_{C_2} to exponentially decay with time constant T_1 towards a thermal equilibrium value of 0.056. Namely, $a_{C_2}(t) = P_C e^{-t/T_1} \cos\phi + \sin\theta$. However, as shown in Figure 8.10b, the observed data does not fit this proposed equation with $a_{C_2}(t)$ appearing to reach a very different thermal equilibrium value! Also, $a_{C_2}(t)$ does not asymptotically approach the equilibrium value but rather exhibits a decay curve seemingly independent of the ultimate thermal equilibrium value (Figure 8.10c). Writing $a_{C_2}(t)$ in terms of the coherences yields

$$a_{C_2}(t) = \frac{\overline{\langle 2\hat{I}_z\hat{S}_z\rangle}(t)}{\overline{\langle \hat{S}_z\rangle}(t)} \cos\phi + \frac{1}{2}\frac{\overline{(\langle \hat{I}_z\rangle(t) + \langle \hat{S}_z\rangle(t))}}{\overline{\langle \hat{S}_z\rangle}(t)} \sin\theta. \tag{8.108}$$

We now need to find the relaxation rates of the I_z, S_z and $2I_zS_z$ coherences, and the primary mechanisms to consider are: (i) C_1–C_2 dipolar coupling, (ii) C_1–H, C_2–H dipolar coupling, (iii) chemical shift anisotropy (CSA) for both C_1 and C_2, and (iv) the combination of all of these.

In general, the relaxation matrix for the coherences of interest can be written as

$$\frac{d}{dt}\begin{bmatrix} \overline{\langle \hat{I}_z\rangle}(t) \\ \overline{\langle \hat{S}_z\rangle}(t) \\ \overline{\langle 2\hat{I}_z\hat{S}_z\rangle}(t) \end{bmatrix} = -\begin{bmatrix} R_{11} & R_{12} & R_{13} \\ R_{21} & R_{22} & R_{23} \\ R_{31} & R_{32} & R_{33} \end{bmatrix}\begin{bmatrix} \overline{\langle \hat{I}_z\rangle}(t) - I_z^{eq} \\ \overline{\langle \hat{S}_z\rangle}(t) - S_z^{eq} \\ \overline{\langle 2\hat{I}_z\hat{S}_z\rangle}(t) - 2I_zS_z^{eq} \end{bmatrix}. \tag{8.109}$$

Let's use Redfield relaxation theory to find analytic expressions for the R_{ij}s. We will use literature values for the needed parameters. Namely, $\tau_c = 3 \times 10^{-12}$ s, $r = 1.1$ Å, $T_{1,CH} = 70$ s, $T_{1,CSA} = 200$ s, where the notations "$T_{1,IS}$" and "$T_{1,CSA}$" denote T_1 relaxation due to dipolar coupling between I and S nuclei and chemical shift anisotropy respectively.

From these numbers, one can readily show that C_1–C_2 dipolar coupling is not a dominant source of relaxation, as confirmed by experimental T_1 values for C_1 in [1,2-^{13}C]Pyr being similar to [1-^{13}C]Pyr. Add C-H dipolar coupling, noting that this is a major source of relaxation as ^{1}H is a stronger magnet than ^{13}C, and the $2\hat{I}_z\hat{S}_z$ coherence decays at twice the rate of $\hat{I}_z$ or $\hat{S}_z$. There is also no field dependence for small molecules. Next include CSA for which there is a moderate effect at 3 T, and the effect is field dependent as T_1 values for both the C_1 and C_2 carbons decrease with increasing B_0. However, as shown in Figure 8.8c, we still fail to track the observed relaxation behavior.

This mystery can be solved by including what are known as ***interference effects***. When we include the various relaxation mechanisms occurring simultaneously, some new terms can emerge that capture the effects from the interaction of the different relaxations. First consider the case of X protons dipolar-coupled to C_1 and Y protons dipolar-coupled to C_2. The relevant Hamiltonians are

$$\hat{H}_{D-IX} = -\frac{\gamma_C\gamma_H\hbar}{r_{IX}^3}\sum_{q=-2}^{2}F_q\hat{A}_q \text{ and } \hat{H}_{D-SY} = -\frac{\gamma_C\gamma_H\hbar}{r_{SY}^3}\sum_{q=-2}^{2}F_q\hat{B}_q, \tag{8.110}$$

where

$$
\hat{A}_0 = \sqrt{\frac{1}{6}} \left(2\hat{I}_z\hat{X}_z - \frac{1}{2}\hat{I}_+\hat{X}_- - \frac{1}{2}\hat{I}_-\hat{X}_+ \right)
$$
$$
\hat{A}_{\pm 1} = \pm\frac{1}{2} \left(\hat{I}_{\pm}\hat{X}_z + \hat{I}_z\hat{X}_{\pm} \right)
$$
$$
\hat{A}_{\pm 2} = \frac{1}{2}\hat{I}_{\pm}\hat{X}_{\pm}, \tag{8.111}
$$

$$
\hat{B}_0 = \sqrt{\frac{1}{6}} \left(2\hat{S}_z\hat{Y}_z - \frac{1}{2}\hat{S}_+\hat{Y}_- - \frac{1}{2}\hat{S}_-\hat{Y}_+ \right)
$$
$$
\hat{B}_{\pm 1} = \pm\frac{1}{2} \left(\hat{S}_{\pm}\hat{Y}_z + \hat{S}_z\hat{Y}_{\pm} \right)
$$
$$
\hat{B}_{\pm 2} = \frac{1}{2}\hat{S}_{\pm}\hat{Y}_{\pm}, \tag{8.112}
$$

and

$$
F_0 = \sqrt{\frac{3}{2}}(3\cos^2\theta - 1)
$$
$$
F_{\pm 1} = \pm 3\sin\theta\cos\theta\, e^{\mp i\phi}
$$
$$
F_{\pm 2} = \frac{3}{2}\sin^2\theta\, e^{\mp 2i\phi}. \tag{8.113}
$$

The relaxation superoperator is $\hat{\hat{\Gamma}} = \sum_q J_q(\omega_q)(\hat{\hat{A}}_{-q} + \hat{\hat{B}}_{-q})(\hat{\hat{A}}_q + \hat{\hat{B}}_q)$ but $[\hat{A}_q, \hat{B}_q] = 0$ for all q. Hence, $\hat{\hat{\Gamma}} = \hat{\hat{\Gamma}}_A + \hat{\hat{\Gamma}}_B = \sum_q J_q(\omega_q)\hat{\hat{A}}_{-q}\hat{\hat{A}}_q + \sum_q J_q(\omega_q)\hat{\hat{B}}_{-q}\hat{\hat{B}}_q$, and we can just independently sum the resulting relaxation rates.

But what about CSA and C_1–C_2 dipolar coupling? The relevant Hamiltonians for these relaxation mechanisms are

$$
\hat{H}_{D-IS} = -\frac{\gamma_C^2\hbar}{r_{IS}^3} \sum_{q=-2}^{2} F_q\hat{A}_q, \tag{8.114}
$$

and

$$
\hat{H}_{CSA1} = \gamma_C B_0\Delta\sigma \sum_{q=-1}^{1} F_q\hat{B}_q \text{ and } \hat{H}_{CSA2} = \gamma_C B_0\Delta\sigma \sum_{q=-1}^{1} F_q\hat{C}_q, \tag{8.115}
$$

where $\Delta\sigma$ is the anisotropic component of the chemical shift tensor,

$$
\hat{A}_0 = \sqrt{\frac{1}{6}} \left(2\hat{I}_z\hat{S}_z - \frac{1}{2}\hat{I}_+\hat{S}_- - \frac{1}{2}\hat{I}_-\hat{S}_+ \right)
$$
$$
\hat{A}_{\pm 1} = \pm\frac{1}{2} \left(\hat{I}_{\pm}\hat{S}_z + \hat{I}_z\hat{S}_{\pm} \right)
$$
$$
\hat{A}_{\pm 2} = \frac{1}{2}\hat{I}_{\pm}\hat{S}_{\pm}, \tag{8.116}
$$

$$
\hat{B}_0 = \frac{1}{3}\sqrt{\frac{3}{2}}\hat{I}_z \text{ and } \hat{B}_{\pm 1} = -\frac{1}{6}\hat{I}_{\pm}, \tag{8.117}
$$

and

$$
\hat{C}_0 = \frac{1}{3}\sqrt{\frac{3}{2}}\hat{S}_z \text{ and } \hat{C}_{\pm 1} = -\frac{1}{6}\hat{S}_{\pm}. \tag{8.118}
$$

Note, the functions F_0, $F_{\pm 1}$, and $F_{\pm 2}$ are the same as given in Eq. (8.113). In this case $\hat{A}_q$, $\hat{B}_q$, and $\hat{C}_q$ do not all commute, hence, when calculating $\hat{\bar{\Gamma}} = \sum_q J_q(\omega_q)(\hat{A}_{-q} + \hat{B}_{-q} + \hat{C}_{-q})(\hat{A}_q + \hat{B}_q + \hat{C}_q)$, we get new cross relaxation terms such as

$$R_{13} = R_{31} = \langle \hat{I}_z | \hat{\bar{\Gamma}} | 2\hat{I}_z\hat{S}_z \rangle = \frac{2}{5} q_{CC,CSA_1} J(\omega_C) \equiv \frac{1}{T_{I_z,I_zS_z-cross}}$$

$$R_{23} = R_{32} = \langle \hat{S}_z | \hat{\bar{\Gamma}} | 2\hat{I}_z\hat{S}_z \rangle = \frac{2}{5} q_{CC,CSA_2} J(\omega_C) \equiv \frac{1}{T_{S_z,I_zS_z-cross}}, \quad (8.119)$$

where

$$q_{CC,CSA_1} = \left(\frac{\mu_0}{4\pi}\right) \frac{\gamma_C^3 \hbar}{r_{CC}^3} B_0 \Delta\sigma_1 \text{ and } q_{CC,CSA_2} = \frac{\mu_0}{4\pi} \frac{\gamma_C^3 \hbar}{r_{CC}^3} B_0 \Delta\sigma_2. \quad (8.120)$$

Assuming extreme narrowing, i.e. $J(0) \approx J(\omega_C) \approx J(2\omega_C)$ the direct relaxation terms can be written as

$$R_{11} \approx \frac{1}{T_{1,CSA_1}} + \frac{1}{T_{1,HC_1}} + \frac{1}{T_{1,CC}},$$

$$R_{22} \approx \frac{1}{T_{1,CSA_2}} + \frac{1}{T_{1,HC_2}} + \frac{1}{T_{1,CC}},$$

$$R_{33} \approx \frac{1}{T_{1,CSA_1}} + \frac{1}{T_{1,CSA_2}} + \frac{1}{T_{1,HC_1}} + \frac{1}{T_{1,HC_2}} + \frac{3}{5T_{1,CC}}, \quad (8.121)$$

and the cross-relaxation terms are

$$R_{12} = R_{21} \approx \frac{1}{2T_{1,CC}},$$

$$R_{13} = R_{31} \approx \sqrt{\frac{6}{5(T_{1,CC}T_{1,CSA_1})}}$$

$$R_{23} = R_{32} \approx \sqrt{\frac{6}{5(T_{1,CC}T_{1,CSA_2})}}. \quad (8.122)$$

Adding these dipolar/CSA interference terms yields much improved fits to the observed data (see Figure 8.10c and d). Hence, despite C–C coupling having a negligible direct effect, cross-relaxation, via a CSA interference effect, is indeed important and needed to explain the observed data. Note, in principle, thermal equilibrium for this asymmetry parameter is not reached until ~8 minutes, by which time, the SNR of the decay curve is too low to accurately measure.

The conclusions from this example are that sometimes simple experiments can yield quite confusing and unexpected results. In this case, an adequate explanation of the observed data required the use of multiple MR relaxation mechanisms. The observed peak asymmetry was primarily driven by $2\hat{I}_z\hat{S}_z$ direct relaxation, the C–C dipolar coupling/CSA interference effect, and residual strong coupling effects. In this experiment, the use of an asymmetry metric to estimate instantaneous polarization requires both knowledge of the initial polarization and the time history of the sample.

8.7 Summary

This chapter provided a more mathematical approach for analyzing MR relaxation processes based on Redfield theory. The theory is valid for most liquid-state systems, and, hence, directly applicable to many *in vivo* tissues. Formulas for T_1, T_2, and $T_{1\rho}$ relaxation times were derived in

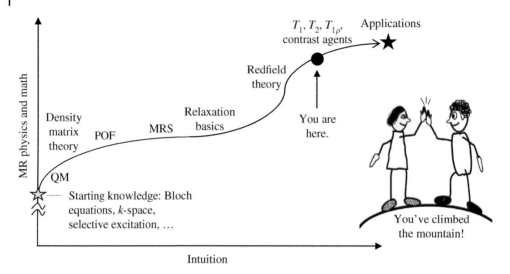

Figure 8.11 The roadmap at the end of Chapter 8.

terms of the spectral density function, the key mathematical quantity for including the influences of local time-vary magnetic fields as presented by processes such as molecular tumbling and chemical exchange. We have now completed the most challenging mathematic parts of this text (Figure 8.11).

Exercises

E.8.1 **The Interaction Frame**

Starting with the Liouville-Von Neumann equation $d\hat{\sigma}/dt = -i\hat{\hat{H}}\hat{\sigma}$, where $\hat{\hat{H}}$ is the superoperator associated with the Hamiltonian $\hat{H}(t) = \hat{H}_0 + \hat{H}_1(t)$, we wish to switch to a frame of reference rotating about $\hat{H}_0$, namely,

$$\hat{\sigma}'(t) = e^{i\hat{\hat{H}}_0 t}\hat{\sigma}(t) \text{ and } \hat{H}'(t) = e^{i\hat{\hat{H}}_0 t}\hat{H}(t).$$

a) Show $d\hat{\sigma}'/dt = -i\hat{\hat{H}}_1'\hat{\sigma}'$.

b) Given, $d\hat{\sigma}'/dt = -i\hat{\hat{\Gamma}}(\hat{\sigma}' - \hat{\sigma}_B)$, where $\hat{\hat{\Gamma}}$ is the relaxation superoperator and $\hat{\sigma}_B$ is the Boltzmann thermal equilibrium spin density operator, show $d\hat{\sigma}/dt = -i\hat{\hat{H}}_0\hat{\sigma} - \hat{\hat{\Gamma}}(\hat{\sigma} - \hat{\sigma}_B)$.

E.8.2 **Relaxation by Random Fields**

Consider a system of spin-½ particles subject to a large main magnetic field $B_0\vec{z}$ plus a small isotropic randomly fluctuating magnetic field, $\Delta\vec{B}(t) \ll B_0$, given by $\Delta\vec{B}(t) = B_x(t)\vec{x} + B_y(t)\vec{y} + B_z(t)\vec{z}$, where $\langle B_x^2 \rangle = \langle B_y^2 \rangle = \langle B_z^2 \rangle = \langle \Delta B^2 \rangle$. Assuming an exponential correlation time for the perturbation fields of τ_c, use Redfield theory to show...

$$\frac{1}{T_1} = 2\gamma^2\langle\Delta B^2\rangle\frac{\tau_c}{1+\omega_0^2\tau_c^2} \text{ and } \frac{1}{T_2} = \gamma^2\langle\Delta B^2\rangle\left(\tau_c + \frac{\tau_c}{1+\omega_0^2\tau_c^2}\right).$$

E.8.3 Chemical Shift Anisotropy

The Hamiltonian for a single-spin system in a magnetic field subject to both the isotropic part of the chemical shift shielding tensor, σ, and an anisotropic component, $\Delta\sigma$, is given by $\hat{H} = \hat{H}_0 + \hat{H}_1(t)$, where

$$\hat{H}_0 = -\gamma B_0(1 - \sigma)\hat{I}_z,$$

$$\hat{H}_1(t) = \gamma B_0 \Delta\sigma \left(\frac{1}{3}\sqrt{\frac{2}{3}}F_0(t)\hat{I}_z - \frac{1}{6}F_1(t)\hat{I}_+ - \frac{1}{6}F_{-1}(t)\hat{I}_- \right),$$

$$F_0(t) = \sqrt{\frac{3}{2}}(3\cos^2\theta - 1),$$

$$F_{\pm 1}(t) = \pm 3\sin\theta\cos\theta e^{\mp i\phi}.$$

a) Find $1/T_1$ and $1/T_2$.
b) For hyperpolarized ^{13}C MRS studies, CSA is the dominant relaxation mechanism. Why?

E.8.4 Scalar Relaxation of the 2nd Kind

Consider a system of J-coupled spins with $\hat{H} = -\omega_I\hat{I}_z - \omega_S\hat{S}_z + 2\pi J(\hat{I}_z\hat{S}_z + \hat{I}_x\hat{S}_x + \hat{I}_y\hat{S}_y)$. In this case, the T_1 relaxation time of the S spin is very short ($T_{1,S} \ll 1/J$). One way of analyzing this system is to assume the S spin is in continuous equilibrium with the lattice because of its short relaxation time. By assuming the S spin is part of the lattice, the perturbing Hamiltonian can be rewritten as $\hat{H}_1 = S_z(t)\hat{I}_z + S_x(t)\hat{I}_x + S_y(t)\hat{I}_y$, where $S_z(t)$, $S_x(t)$, and $S_y(t)$ are well modeled as stochastic functions with the following correlation functions. Namely,

$$\langle S_z(t)S_z(t+\tau) \rangle = \frac{(2\pi J)^2 S(S+1)}{3}e^{-\tau/T_{1,S}},$$

and

$$\langle (S_x(t) + iS_y(t))(S_x(t+\tau) - iS_y(t+\tau)) \rangle = \langle S_+(t)S_-(t+\tau) \rangle = \frac{(2\pi J)^2 S(S+1)}{6}e^{i\omega_S\tau}e^{-\tau/T_{2,S}}.$$

Show

$$\frac{1}{T_1} = \frac{2(2\pi J)^2 S(S+1)}{3}\frac{T_{2,S}}{1 + (\omega_I - \omega_S)^2 T_{2,S}^2}$$

and

$$\frac{1}{T_2} = \frac{(2\pi J)^2 S(S+1)}{3}\left(T_{1,S} + \frac{T_{2,S}}{1 + (\omega_I - \omega_S)^2 T_{2,S}^2} \right),$$

where S = spin of the unpaired electron system or nucleus. Note, the $S(S+1)/3$ factor comes from $\text{Tr}\left(\hat{S}_p^2\right) = S(S+1)/3$, where p is a product operator.

E.8.5 The ABCs of $T_{1\rho}$

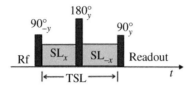

A research group at a major university is interested in studying early changes in cartilage due to OA. From published literature, they conclude that one of the earliest signs of OA is loss of glycosaminoglycans (GAG), and $T_{1\rho}$ might be a good MRI contrast mechanism to look for this effect. The assumption is that, in the presence of GAG, there is a chemical exchange between the GAG hydroxyl groups (–OH) and the bulk water with exchange rate k_{ex}. As an additional constraint, the maximum spin-lock amplitude that they can safely achieve *in vivo* on their 3 T scanner is $\gamma B_1 = 500$ Hz. Fortunately, a graduate student has coded up the spin-lock pulse sequence module shown above combined with a fast-imaging readout. Three students, A, B, and C, came up with the following proposals for the best approach to examine this hypothesized loss of GAG.

Student A's proposal: "$T_{1\rho}$ is sensitive to chemical exchange. Therefore, I propose that, for each subject, we acquire images with $\gamma B_1 = 500$ Hz and have the length of the spin lock (T_{SL}) incrementally extended to generate a series of images from which we can compute a spatial map of $T_{1\rho}$. Changes in GAG content should be apparent in the $T_{1\rho}$ maps."

Student B's proposal: "A's proposal is totally worthless. Based on our assumptions, A's $T_{1\rho}$ maps will be completely independent of GAG content! The better approach is to acquire just two images/subject in which T_{SL} is fixed and amplitude of the spin-lock pulse is switched between 0 and 500 Hz. This will result in a T_2-weighted image and a $T_{1\rho}$-weighted image. A map of the ratio of these two images is the best way to look for any changes in GAG content. I even have a method for figuring out the optimal T_{SL}."

Student C's proposal: "$T_{1\rho}$ is too hard to study; let's move on to something else."
Which proposal do you support and why?

Historical Notes

Alfred G. Redfield, a pioneering physicist and biochemist, made exceptional contributions to the field of NMR. He obtained his Ph.D. at the University of Illinois in 1953. He then entered Harvard for a fellowship studying NMR of solids.

In 1955, Redfield joined the IBM Watson Laboratory at Columbia University, continuing his research on the properties of normal and superconducting metals at very low temperatures. He also held a faculty appointment in the physics department at Columbia.

A significant achievement during Redfield's time at IBM was the development of the High-Resolution Pulsed Nuclear Magnetic Resonance (NMR) Spectrometer, earning him the prestigious IBM Outstanding Contribution Award in 1970. This cutting-edge device significantly enhanced the capabilities of NMR spectroscopy and opened new possibilities for applications.

In 1972, he joined the faculty at Brandeis University as a tenured professor in the physics department. Redfield's contributions to NMR extended beyond technical innovations. His ground-breaking theoretical work in Redfield relaxation theory, applied to statistical systems, had profound implications across the physical sciences. The work initially appeared in a research publication of Bell Labs but was later formally published in the monograph series "Advances in Magnetic Resonance."

Despite his brilliance, Redfield remained humble about his achievements, modestly responding to questions about his theory with, "Well, it was just a better way of writing down what everybody already knew (Lynch 2019)." This humility and dedication to his work endeared him to his peers and students alike.

(Source: photo from https://commons.wikimedia.org/wiki/File:Alfred_G._Redfield_2000s.jpg, licensed under the Creative Commons Attribution-Share Alike 4.0 International.)

9

MRI Contrast Agents

By exploiting the high tissue-water content, and consequently large number of ^{1}H nuclei, MRI is able to overcome its fundamentally low sensitivity as compared to other medical imaging modalities such as positron emission tomography (PET). We have also discussed in the previous chapters the multiple sources of MRI contrast that provide additional information content. However, the development of MRI contrast agents, exogenous substances used to modify the detected MR signal, is largely driven by a need to better differentiate pathological processes such as blood--brain barrier breakdown or inflammation.

For a typical spin-echo acquisition, the signal intensity is proportional to $M_0 \left(1 - e^{-TR/T_1} \right) e^{-TE/T_2}$, and the idea of using paramagnetic salts to shorten water relaxation times goes all the way back to Bloch et al. (1946). For *in vivo* use, MRI contrast agents must be both biocompatible pharmaceuticals and function as NMR relaxation probes, typically shortening T_1, T_2, or T_2^*. As a rule of thumb, a 10--20% change in the targeted relaxation time is required for robust *in vivo* detection. Further considerations include *in vivo* stability, excretability, and lack of toxicity (both acute and chronic).

Within these constraints, the contrast agents being currently used clinically, or under active research development, can be divided into roughly three classifications: (i) T_1 shortening agents, primarily utilizing gadolinium (Gd), (ii) T_2 and/or T_2^* shortening agents, largely superparamagnetic iron oxide (SPIO) nanoparticles, and (iii) PARACEST agents. In this chapter, we will first review the physics underlying these major classes of contrast agents, followed by an overview of their current clinical use or potential.

9.1 Paramagnetic Relaxation Enhancement

As listed in Table 9.1, different materials have varying magnetic properties.

Paramagnetic materials affect both T_1 and T_2, and the addition of these to a solution increases both $1/T_1$ and $1/T_2$ relaxation rates, with both the diamagnetic and paramagnetic contributions being additive,

$$\frac{1}{T_i} = \frac{1}{T_{i,d}} + \frac{1}{T_{i,p}} \text{ for } i = 1, 2. \tag{9.1}$$

Solvent relaxation rates are generally linearly proportional to the concentration of the paramagnetic species, $[M]$,

$$\frac{1}{T_i} = \frac{1}{T_{i,d}} + [M]R_i, \tag{9.2}$$

where R_i is called the ***relaxivity***.

Fundamentals of In Vivo Magnetic Resonance: Spin Physics, Relaxation Theory, and Contrast Mechanisms, First Edition. Daniel M. Spielman and Keshav Datta.
© 2024 John Wiley & Sons, Inc. Published 2024 by John Wiley & Sons, Inc.
Companion website: www.wiley.com/go/Spielman

Table 9.1 Material magnetic properties. Materials become magnetized to widely varying degrees in the presence of an applied magnetic field. In contrast to diamagnetic, paramagnetic, and superparamagnetic materials, ferromagnetic materials maintain their magnetization even after the initial applied magnetic field is removed.

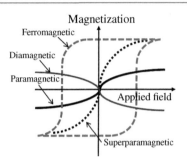

Magnetic property	Polarization vs applied field	Magnetic susceptibility	Typical materials
Diamagnetism	Opposite	−10ppm	Water, fat, most biological tissues
Paramagnetism	Same	+1	Simple salts and chelates of metals (Gd, Fe, Mn, Cu), molecular O_2
Superparamagnetism	Same	+5000	Ferritin, hemosiderin, SPIO contrast agents
Ferromagnetism	Same	>10,000	Iron, steel

In the presence of paramagnetic ions, nuclear spins see the lattice as the combined electron spin system along with the other molecular degrees of freedom. To reduce *in vivo* toxicity (to be discussed in more detail shortly), paramagnetic ions are typically used in a chelated, or chemically protected, form as shown in Figure 9.1. The water interactions, which can be divided into both ***inner-*** and ***2nd-sphere*** effects, include chemical exchange as well as both scalar and dipolar coupling, with the resulting relaxation rates given by

$$\frac{1}{T_{i,p}} = \left(\frac{1}{T_i}\right)_{inner} + \left(\frac{1}{T_i}\right)_{2nd} + \left(\frac{1}{T_i}\right)_{outer}. \qquad (9.3)$$

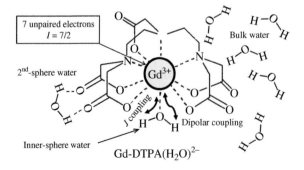

Figure 9.1 Inner-sphere, 2nd-field, and bulk water interactions with a chelated paramagnetic ion.

Of particular importance for this discussion is a parameter known as the inner-sphere ***paramagnetic relaxation enhancement*** (PRE). Driven by the chemical exchange contributions, the excess spin–lattice relaxation rate for the ligand (in this case water) due to the paramagnetic material is given by

$$\left(\frac{1}{T_i}\right)_{inner} = \frac{f_M}{\tau_M + T_{1,M}}, \tag{9.4}$$

where M is the ligand in the paramagnetic complex, f_M is the molar fraction of the bound ligand nuclei, and τ_M is the lifetime in the ligand complex. For paramagnetic complexes, the relaxivity is typically defined as the PRE normalized to 1 mM.

9.1.1 Solomon–Bloembergen–Morgan Theory

A more formal analysis of these chelated paramagnetic systems is known as Solomon–Bloembergen–Morgan (SBM) theory (Kowalewski and Mäler 2006), in which the spin–lattice relaxation rate for bound nuclei is a combination of both the scalar and dipole–dipole couplings between the nuclear and electron spins. Hence,

$$\frac{1}{T_{1,M}} = \left(\frac{1}{T_{1,M}}\right)_{sc} + \left(\frac{1}{T_{1,M}}\right)_{dd}. \tag{9.5}$$

Following the derivations of T_1 relaxation times given in Eqs. (8.54) and (8.67),

$$\frac{1}{T_{1M}} = \frac{2}{3}A_{sc}^2 S(S+1)\frac{\tau_{e2}}{1+(\omega_S-\omega_I)^2\tau_{e2}^2} + \frac{2}{15}\left(\frac{\mu_0}{4\pi}\right)^2\frac{\gamma_I^2\gamma_S^2\hbar^2}{r_{IS}^6}S(S+1)$$

$$\times \left[\frac{\tau_{c2}}{1+(\omega_S-\omega_I)^2\tau_{c2}^2} + \frac{3\tau_{c1}}{1+\omega_I^2\tau_{c1}^2} + \frac{6\tau_{c2}}{1+(\omega_S+\omega_I)^2\tau_{c2}^2}\right]. \tag{9.6}$$

where the first term in this sum is from scalar coupling and the second the dipolar coupling contribution. Noting that $\omega_S = \gamma_e B_0$ is much larger than $\omega_I = \gamma_H B_0$, this equation reduces to

$$\frac{1}{T_{1M}} = \frac{2}{3}A_{sc}^2 S(S+1)\frac{\tau_{e2}}{1+\omega_s^2\tau_{e2}^2} + \frac{2}{15}b_{IS}^2 S(S+1)\left[\frac{7\tau_{c2}}{1+\omega_S^2\tau_{c2}^2} + \frac{3\tau_{c1}}{1+\omega_I^2\tau_{c1}^2}\right]. \tag{9.7}$$

The two contributions to the dipolar coupling component are typically referred to as the "7-term" and the "3-term." The time constants are, however, quite interesting. Namely,

$$\frac{1}{\tau_{e2}} = \frac{1}{\tau_M} + \frac{1}{T_{2,e}}, \quad \frac{1}{\tau_{e1}} = \frac{1}{\tau_M} + \frac{1}{T_{1,e}}, \quad \text{and} \quad \frac{1}{\tau_{cj}} = \frac{1}{\tau_R} + \frac{1}{\tau_M} + \frac{1}{T_{je}}; \quad j = 1, 2, \tag{9.8}$$

where τ_M is the water exchange time, $T_{1,e}$ and $T_{2,e}$ are the electron T_1 and T_2 relaxation times, and τ_R is the rotational correlation time of the complex.

Typically, the scalar term is small compared to the dipolar coupling term (valid for Gd^{3+} but not necessarily true for Mn^{2+}). If the rotational correlation time dominates the dipolar coupling term, then the field dependence of the PRE is shown in Figure 9.2. Note this type of plot known as an ***NMR dispersion*** (NMRD) curve.

The plot shown in Figure 9.2 is the behavior typically observed when constructing MRI phantoms using e.g., $MnCl_2$ or $CuSO_4$. But what about $T_{1,e}$ and $T_{2,e}$? Are not these relaxation times themselves field dependent? The answer is yes, and the complete SBM theory includes the field dependence of the electron relaxation times. Calculating ESR relaxation rates is typically quite complicated. In the case of the paramagnetic complexes used as MR contrast agents, electron relaxation rates

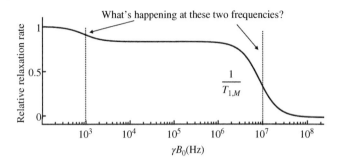

Figure 9.2 Representative NMR dispersion (NMRD) curves.

are dominated by Zero-field Splitting (ZFS), the electron spin equivalent of nuclear quadrupolar coupling (Kowalewski and Mäler 2006). Namely,

$$\frac{1}{T_{1,e}} = \frac{\Delta_t^2}{5}\left(\frac{\tau_v}{1+\omega_S^2\tau_v^2} + \frac{4\tau_v}{1+4\omega_S^2\tau_v^2}\right) \tag{9.9}$$

and

$$\frac{1}{T_{2,e}} = \frac{\Delta_t^2}{10}\left(3\tau_v + \frac{5\tau_v}{1+\omega_S^2\tau_v^2} + \frac{2\tau_v}{1+4\omega_S^2\tau_v^2}\right), \tag{9.10}$$

where Δ_t is the trace of transient ZFS tensor and the correlation time for ZFS modulations, τ_v, is typically $\sim 10^{-9}$ s.

It is worth comparing these equations to the relaxation times arising from quadrupolar coupling. Namely, any nucleus with spin $S > \frac{1}{2}$ has an electrical quadrupolar moment due to its nonuniform charge distribution. This electrical quadrupole moment interacts with local electric field (E-field) gradients. Static E-field gradients result in shifts of the resonance frequencies of the observed peaks, while dynamic (time-varying) E-field gradients are a very effective relaxation mechanism, typically significantly stronger than dipolar coupling or the other relaxation mechanisms we have discussed. The quadrupolar coupling contribution to spin–lattice relaxation is

$$\frac{1}{T_{1,Q}} \cong \frac{3\pi}{100}\frac{2S+3}{S^2(2S-1)}\left(\frac{e^2qQ}{\hbar}\right)^2(J(\omega_s)+4J(4\omega_s)), \tag{9.11}$$

where Q is the quadrupolar coupling constant and the parameter q is the electric field gradient contribution.

Hence, the full SBM equations for the T_1 relaxation times of water in the presence of a chelated paramagnetic ion are

$$\frac{1}{T_{1M}} = \frac{2}{3}A_{sc}^2 S(S+1)\frac{\tau_{e2}}{1+\omega_S^2\tau_{e2}^2} + \frac{2}{15}b_{IS}^2 S(S+1)\left[\frac{7\tau_{c2}}{1+\omega_S^2\tau_{c2}^2} + \frac{3\tau_{c1}}{1+\omega_I^2\tau_{c1}^2}\right], \tag{9.12}$$

$$\frac{1}{\tau_{e2}} = \frac{1}{\tau_M} + \frac{1}{T_{2,e}}, \quad \frac{1}{\tau_{e1}} = \frac{1}{\tau_M} + \frac{1}{T_{1,e}}, \quad \text{and} \quad \frac{1}{\tau_{cj}} = \frac{1}{\tau_R} + \frac{1}{\tau_M} + \frac{1}{T_{je}}; \quad j=1,2, \tag{9.13}$$

where

$$\frac{1}{T_{1,e}} = \frac{\Delta_t^2}{5}\left(\frac{\tau_v}{1+\omega_S^2\tau_v^2} + \frac{4\tau_v}{1+4\omega_S^2\tau_v^2}\right), \tag{9.14}$$

and

$$\frac{1}{T_{2,e}} = \frac{\Delta_t^2}{10}\left(3\tau_v + \frac{5\tau_v}{1+\omega_S^2\tau_v^2} + \frac{2\tau_v}{1+4\omega_S^2\tau_v^2}\right). \tag{9.15}$$

The SBM theory works reasonably well, but there are multiple extensions and modifications (Kowalewski and Mäler 2006). For completeness, the equations for the $T_{2,e}$ relaxation time are given by

$$\frac{1}{T_{2M}} = \frac{2}{3}A_{sc}^2 S(S+1)\left(\tau_{e1} + \frac{\tau_{e2}}{1+\omega_S^2\tau_{e2}^2}\right) + \frac{2}{15}b_{IS}^2 S(S+1)\left[4\tau_{c1} + \frac{3\tau_{c1}}{1+\omega_I^2\tau_{c1}^2} + \frac{13\tau_{c2}}{1+\omega_S^2\tau_{c2}^2}\right]. \tag{9.16}$$

If the field-dependence of the electron relaxation rates is included, the NMRD curves can yield much more interesting relaxivity behavior. An excellent example is the NMRD curve for Gd-DTPA bound to albumin as shown in Figure 9.3b. For this contrast agent, choice of the scanner field strength can be very important.

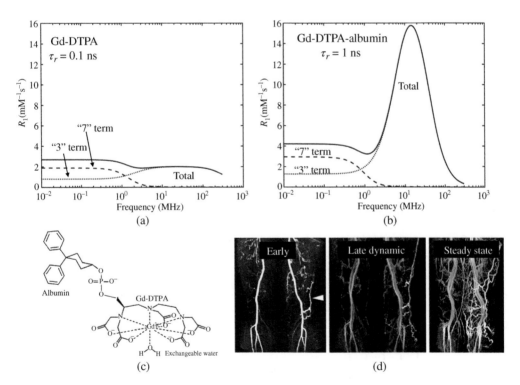

Figure 9.3 NMRD curves for (a) Gd-DTPA and (b) Gd-DTPA bound to albumin, which increases the rotational correlation time. (c) Chemical structure of Gd-DTPA-albumin with a representative example as an intravascular MRI contrast agent. (d) Example lower limb 3D contrast-enhanced magnetic resonance angiography (CE-MRA) using Gd-DTPA-albumin. Source: (d) Adapted with permission from Lewis et al. (2012) Figure 3.

9.1.2 Gd³⁺-Based T_1 Contrast Agents

The reason Gd, and to a lesser degree manganese (Mn), are the most common ions used for T_1 MRI contrast agents can be found in Table 9.2. The magnetic moment of these ions arises from unpaired electrons and contains contributions from both electron spin and orbital angular momentum components. A key parameter is the electron T_1, which is strongly influenced by the electron configuration. For highly symmetric electron configurations, the electric fields and gradients largely cancel, lengthening the T_1. In contrast, with low symmetry, electric field gradients enhance quadrupolar relaxation. In addition to the magnetic moment, T_1 relaxation is driven by field fluctuations at the Larmor frequency. Hence, at the most common field strengths used for clinical scanners (0.5–7 T), Gd^{+3} and Mn^{+2} are the most favorable ions for driving T_1 relaxation.

As an example, compare a gadolinium ion, Gd^{3+}, with dysprosium, Dy^{3+}. Gd^{3+} has 7 unpaired e⁻s, a magnetic moment $\mu = 7.9$, and $T_{1e} = $ ~1 ns, ideal parameters for use in a T_1 contrast agent. In contrast, Dy^{3+} has 5 unpaired e⁻s and a larger net magnetic moment compared to Gd^{3+} of $\mu = 10.6$. However, with $T_{1e} = $ ~10^{-4} ns, this ion provides little to no T_1 relaxivity enhancement.

QUESTION: Does Dy^{3+} have any use as a contrast agent?

An important question that we have not addressed is whether a compound such as Gd-DTPA also shortens T_2? The answer is yes, but the relaxation rates are additive. *In vivo* tissue T_2s are typically considerably shorter than T_1s. On a percentage basis, T_1 agents such as Gd-DTPA increase $1/T_1$ much more than $1/T_2$. Hence, for a typical MRI sequence, the signal is usually enhanced, unless the Gd-DTPA concentration gets very high (for example, as seen in the renal pelvis as the kidney filters and concentrates the Gd-DTPA from the bloodstream). Beyond use as a direct imaging contrast agent, there are also multiple innovative *in vivo* Gd-DTPA targeting strategies including labeling of nanoparticles or even agents where the Gd ion is only exposed to water after some enzymatic reaction (Wahsner et al. 2019).

Table 9.2 Common ions used in MRI contrast agents along with their MR properties.

Ion	Spin	Electron configuration	Magnetic moment (μB)	Electron T_1 (ns)	ΔR_1 @ 0.5 T (L/(mmol⁻ˢ))	ΔR_2 @ 0.5 T (L/(mmol⁻ˢ))
²⁴Cr³⁺	3/2	↑ ↑ ↑ — —	3.9	$10^{-1}-1$	4.36	10.10
²⁵Mn²⁺	5/2	↑ ↑ ↑ ↑ ↑	5.9	$1-10$	7.52	41.60
²⁶Fe³⁺	5/2	↑ ↑ ↑ ↑ ↑	5.9	$10^{-1}-1$	8.37	12.80
²⁹Cu²⁺	1/2	↑↓ ↑↓ ↑↓ ↑↓ ↑	1.7	10^{-1}	0.83	0.98
⁶³Eu³⁺	3	↑↓ ↑ ↑ ↑ ↑ ↑ ↑ ↑	3.4	$10^{-4}-10^{-3}$	0.38	0.41
⁶⁴Gd³⁺	7/2	↑ ↑ ↑ ↑ ↑ ↑ ↑	7.9	$1-10$	12.10	15.00
⁶⁶Dy³⁺	5/2	↑↓ ↑↓ ↑ ↑ ↑ ↑ ↑	10.6	$10^{-4}-10^{-3}$	0.56	0.56

9.2 T_2 and T_2^* Contrast Agents

The experimentally observed loss of transverse magnetization is typically greater than $1/T_2$, and denoted by the relaxation time T_2^*. T_2^* relaxation includes loss of transverse signals due to both spin–spin interactions and macroscopic magnetic field inhomogeneities,

$$\frac{1}{T_2^*} = \frac{1}{T_2} + \gamma \Delta B/2 \tag{9.17}$$

Spin–spin interactions, driven for example by molecular tumbling, are not reversible on the time scale of the Rf pulses used in typical NMR pulse sequences and hence contribute to irreversible T_2 relaxation. However static macroscopic field inhomogeneities, as well as some slower time-varying contributions, are indeed reversible, for example, using a spin echo sequence. For the rest of the discussion in this section, we are going to ignore this difference between T_2 and T_2^*, except to acknowledge the observed effects will differ on spin-echo versus gradient-echo acquisitions.

9.2.1 T_2, Diffusion, and Outer-Sphere Relaxation

Among the lanthanides, Dy^{3+} has the largest magnetic moment, and when chelated in the appropriate complexes, such Dy^{3+} agents can be very effective T_2 shortening agents. However, we are now going to instead focus on contrast agents where the Dy^{3+} or more commonly iron Fe^{+3} are confined into nanoparticles. Consider water molecules diffusing past a superparamagnetic center. Starting with Fick's law of diffusion, one can derive an exponentially decaying temporal correlation function with correlation time

$$\tau_D = d^2/D$$

where d is the closest approach of the water molecules and the nanoparticle and D is the water diffusion coefficient (Kowalewski and Mäler 2006). As a result, we obtain a familiar result that $1/T_2 \propto [M_s]$, the molar concentration of the contrast agent.

9.2.2 SPIOs and USPIOs

Ferrimagnetic (Fe^{3+}) or ferromagnetic (Fe^{2+}) materials are magnetized in an external magnetic field. When clustered together, for example in very small crystallites (1–20 nm), these materials can form a single magnetic domain. This is like paramagnetism, but instead of individual atoms influenced by the magnetic field, magnetic moment of the entire crystallite aligns with the applied field. Agglomerated iron oxide can be formed into superparamagnetic iron oxide nanoparticles (SPIOs), which typically consist of an iron oxide core of (3–20 nm) stabilized with a monomer or polymer coating. Having hydrodynamic diameters >50 nm, these materials are filtered by the liver, but also taken up by macrophages; and hence are attractive MRI contrast agents due to potential high sensitivity and low toxicity.

Going smaller, ultra-small superparamagnetic iron oxides (USPIOs) have iron oxide cores of 3–20 nm, but are coated to preserve their individuality and have hydrodynamic diameters <50 nm. These agents remain in the blood pool longer than SPIOs (good for angiography), tend to be cleared by the kidney, and are usable as both T_1 and T_2 contrast agents. USPIOs and SPIOs are also commonly chosen for molecular imaging applications (Bashir et al. 2015) (see Figure 9.4).

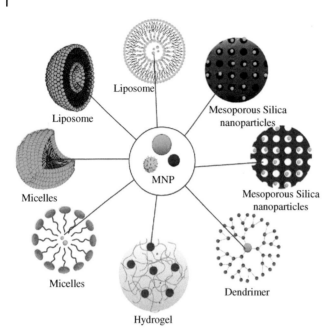

Figure 9.4 Example magnetic nanoparticle arrangements with polymers and other materials. Source: Reproduced with permission from Anik et al. (2021) Figure 5.

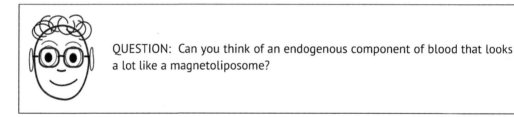

QUESTION: Can you think of an endogenous component of blood that looks a lot like a magnetoliposome?

9.3 PARACEST Contrast Agents

The third broad class of MRI contrast agents is based on the chemical exchange saturation transfer (CEST) mechanism (see discussion in Chapter 7). CEST contrast can arise from multiple types of proton exchange groups, so molecules endogenous to tissues as well as exogenous agents are potential imaging candidates. A large chemical shift difference between an exchanging proton site and the solvent water is highly advantageous for CEST contrast, in that this allows selective Rf excitation without directly saturating the water signal to be imaged.

Lanthanide-based agents of this type are referred to as PARACEST agents. A good CEST contrast agent must exhibit slow to intermediate exchange with a chemical shift, $\Delta\omega$, being significantly larger than the exchange rate, k_{ex}. A large $\Delta\omega$ permits shorter proton lifetimes and thus larger CEST effects. Among the lanthanides, $\Delta\omega$ is largest for Tm^{3+} or Dy^{3+}; however, Eu^{3+} is typically preferred for PARACEST agents as water exchange is found to be slowest in Eu^{3+} complexes and the long relaxation times of Eu^{3+} are also favorable (Sherry and Wu 2013).

The development of novel PARACEST agents is an active area of research as they provide molecular-level information having great potential to predict therapeutic effects or provide earlier assessments of therapeutic effects (Pagel 2011). Specifically, PARACEST agents can be designed with a variety of chemical groups (*e.g.*, amide, amine, hydroxyl, or phosphates) that directly

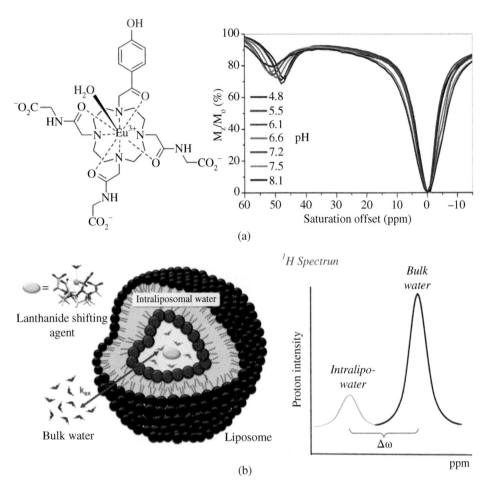

Figure 9.5 Representative PARACEST contrast agents. (a) PARACEST agent for reporting pH and associated CEST spectra recorded at 9.4 T and 303 K in plasma. (b) Simplified scheme for a LipoCEST contrast agent. Source: (a) Adapted with permission from Wu et al. (2016) Figures 1 and 3, and (b) adapted with permission from Wahsner et al. (2019) Figure 25, Copyright 2019 American Chemical Society.

interact with metabolites, enzymes, or biological ions, and hence potentially provide highly specific metabolic information. The major impediment to clinical translation is relatively poor sensitivity. Although water exchange improves detection sensitivity over conventional ^{1}H-MRS by ~100–1000-fold, most PARACEST agents remain much less sensitive than the T_1 and T_2 contrast agents previously discussed (Pagel 2011). Figure 9.5 shows two representative PARACEST agents.

9.4 Contrast Agents in the Clinic

There is a long history of MR contrast agent development, and the vast majority of agents have not entered routine clinical use. However, of those that have, these agents have had a profound clinical impact, particularly for MRI of oncology and inflammatory processes. Most clinical MR contrast agents work via enhancing T_1 recovery or reducing T_2^*, with the agents often classified by the tissues they target; examples being intravascular versus extravascular agents, blood pool agents, and hepatobiliary agents (see Figure 9.6).

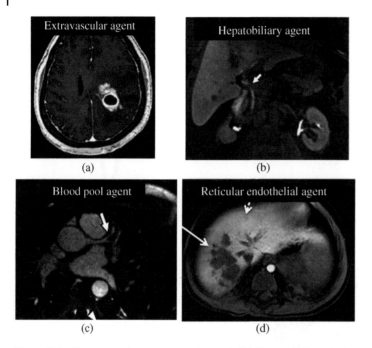

Figure 9.6 Representative contrast-enhanced MRI. Source: Adapted with permission from (a) Hirschler et al. (2023) Figure 3, (b) Poetter-Lang et al. (2023) Figure 3, (c) Wagner et al. (2011), and (d) Ayyala et al. (2017) Figure 3.

9.4.1 Gd-Based Agents

The most commonly used clinical MRI contrast agents reduce T_1 using the gadolinium ion, Gd^{+3}, which has 7 unpaired electrons in combination with a favorable electron T_1 (see Table 9.2). Alone, Gd^{+3} is highly toxic, hence contrast agent preparations use chelation to molecules such as diethylenetriamine pentaacetate (DTPA) to shield the body from gadolinium toxicity as well as to reduce retention times. T_1 relaxivity is provided by water transiently bonding to a coordination site. As shown in Figure 9.7a, gadolinium-based contrast agents (GBCAs) can be divided according to the structure of the chelating ligand (macrocyclic or linear) as well as the net charge on chelated complex (ionic or non-ionic). In general, macrocyclic agents are more stable than linear ones; the macrocyclic encirclement of the Gd^{3+} ion by the chelating ligand creates a stronger molecular cage, reducing the likelihood of dissociation. Some of the most commonly used clinical agents are shown in Figure 9.7b, and an excellent review of Gd-based MRI contrast agents can be found in Li and Meade (2019). In addition, there are blood pool imaging agents such as Ablavar (considered safe, but currently not commercially available), where binding Gd-DTPA to albumin not only increases the T_1 relaxivity but also extends the lifetime in the blood pool.

However, there are safety concerns with gadolinium-based contrast agents (GBCAs). These agents are engineered to be highly water-soluble and readily eliminated, intact, from the body after intravenous injection. Unfortunately, in practice, some small fraction of the injected dose is not eliminated, and the residual gadolinium can cause toxic effects. The first of these, described in 2000, is nephrogenic systemic fibrosis (NSF). Symptoms include increased collagen deposition resulting in thickening and hardening of skin of extremities, potentially resulting in immobility or deformity, with additional toxic effects also noted in the lungs, heart, diaphragm, esophagus, and skeletal muscle (Le Fur and Caravan 2019). Although the exact mechanism behind NSF is

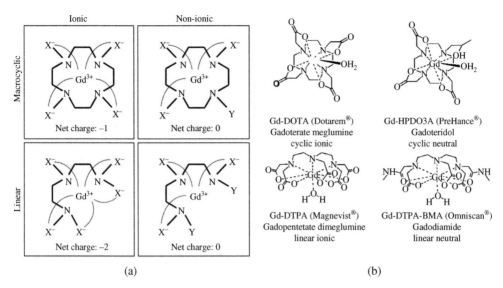

Figure 9.7 Structure and commercial names of common gadolinium-based contrast agents (GBCA). (a) GBCAs can be divided according to the structure of the chelating ligand (macrocyclic versus linear) and according to the net charge (ionic versus non-ionic). (b) Trade and generic names of four representative GBCAs. Source: (a) and (b) Adapted with permission from Fraum et al. (2017) Figure 1 and Aime and Caravan (2009) Figure 3, respectively.

not fully understood, the working hypothesis is that the delayed clearance of gadolinium contrast agents allows Gd^{3+} disassociation resulting in Gd^{3+} deposits in tissue. The increased stability of the macrocyclic over linear agents has led to the macrocyclic agents being the predominant choice for clinical use.

The primary risk of gadolinium exposure is in patients with acute or chronic kidney failure/injury. NSF is uncommon with a reported incidence in patients of 0.1–4% (Prince et al. 2008). Unfortunately, the clinical course is quite variable and there are no consistently successful treatments. A second finding of currently unknown clinical significance is that of Gd brain deposition following contrast-enhanced MRI (Le Fur and Caravan 2019).

Overall, these concerns have led to avoidance in dialysis patients or those with renal failure, reduced doses, and a search for alternative contrast agents. One alternative is to consider Mn^{+3}-based agents, given the favorable MR properties of this ion in combination with the theory that manganese, which is naturally present in trace quantities in the body, may prove less toxic than gadolinium. Currently, the potential utility of such agents and their safety profiles have yet to be fully assessed (Daksh et al. 2022). However, probably the most active search for improved clinical contrast agents involves the use of SPIOs.

9.4.2 Iron-Based Agents

The USPIO ferumoxytol (Feraheme) is currently available in the United States for off-label use (*i.e.*, the practice of prescribing a drug for a different purpose than originally approved by the FDA). Ferumoxytol is superparamagnetic iron oxide coated with polyglucose sorbitol carboxymethyl ether, having an overall colloidal particle size of 17–31 nm. In low dosages it is an effective T_1 agent and is also used as a T_2^* agent in higher concentrations. Figure 9.8 illustrates the use of this highly effective MRI contrast agent for both shortening the T_1 of the blood pool and enhancing susceptibility effects. Figure 9.9 illustrates its use for tumor imaging. An excellent review of the

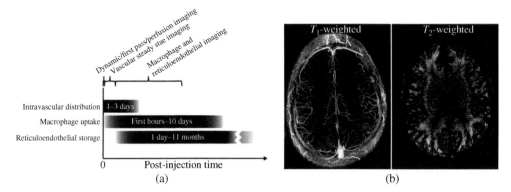

Figure 9.8 Ferumoxytol-enhanced imaging of the brain. (a) Schematic summarizing the distribution and key imaging timing for ferumoxytol. (b) T_1- and T_2-weighted gradient-recalled echo images of a normal volunteer 1–2 hours immediately following an i.v. administration of ferumoxytol showing T_1-enhancement of the blood pool and showing strong T_2^*-weighting, respectively. Source: (a) Adapted with permission from Bashir et al. (2015) Figure 1, and (b) images courtesy of Michael Moseley, PhD.

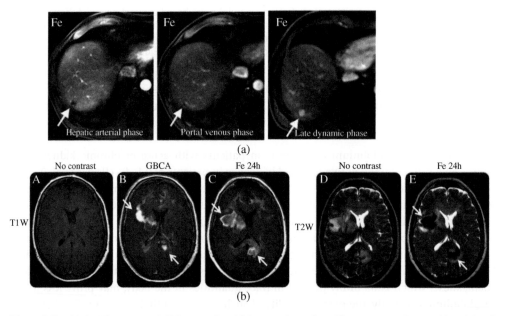

Figure 9.9 Clinical ferumoxytol (Fe) examples. (a) Images from dynamic contrast-enhance T1-weighted MRI examinations of the liver in a patient with a hepatic hemangioma. Peripheral nodular enhancement is apparent in both the hepatic arterial and portal venous phases, with complete fill-in of the lesion in the late dynamic phase, diagnostic of a hepatic hemangioma. (b) Images from a patient with primary central nervous system (CNS) lymphoma. (A, B) Precontrast and postcontrast gadolinium-enhanced T_1-weighted images demonstrate multifocal, enhancing, deep white matter lesions (white arrows). Enhancement is heterogeneous and poorly defined. (C) A T_1-weighted image obtained 24 hours after intravenous ferumoxytol administration demonstrates more confluent and extensive enhancement, likely related to phagocytic cell uptake, in a similar pattern to the gadolinium-enhanced images. (D, E) T_2-weighted images obtained before and 24 hours after intravenous ferumoxytol administration demonstrate several areas of confluent, focal, strong signal loss due to ferumoxytol uptake. Stereotactic needle biopsy of the right frontal T_2-hypointense lesion confirmed the suspected diagnosis of diffuse large B-cell lymphoma. Source: (a) Adapted with permission from Bashir et al. (2015) Figure 7, and (b) reproduced from Farrell et al. (2013).

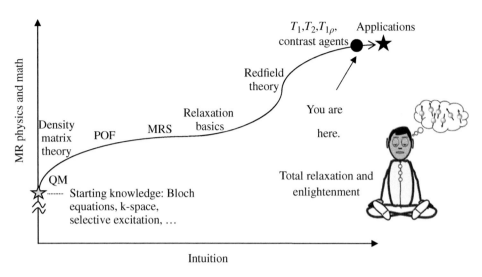

Figure 9.10 The roadmap at the end of Chapter 9.

clinical use of ferumoxytol can be found in an article by Daldrup-Link (2017). Safety issues with SPIOs include long retention times (*in vivo* retention half-life ~20 hours), but most importantly, a ~10-fold higher rate of adverse reactions compared to gadolinium agents, with anaphylaxis being the primary concern.

9.5 Summary

The development and use of MR contrast agents is a very active area of research, with significant efforts focused on agents for specific anatomical regions and/or pathologies. In addition to providing examples of current agents in clinical use, the mathematical foundation presented in this chapter also enables one to predict behavior of these agents at different field strengths. In addition to interesting physics, biochemistry, and physiology, there is a continued search for agents with improved safety profiles. However, even for new agents deemed both safe and effective, economic considerations such as market size and insurance reimbursements also need to be considered. The net result is that, although many new MR contrast agents are developed, very few reach widespread clinical use. Now that we have studied *in vivo* relaxation mechanisms, we will investigate some applications in the next chapter (Figure 9.10).

Exercises

E.9.1 NMRD Curves

Plot T_1 relaxivity NMRD curves for Gd-DTPA and Gd-DTPA bound to serum albumin.

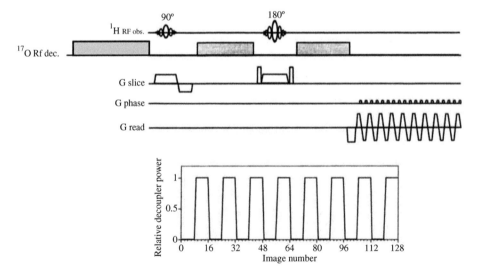

Albumin

Gd-DTPA

Gd-DTPA-albumin

H H Exchangeable water

H H Exchangeable water

E.9.2 ^{17}O Imaging

This problem investigates using ^{17}O with decoupling as an MRI contrast agent to study focal cerebral ishemia and is adapted with permission from de Crespigny et al. (2000). A research team decides to measure *in vivo* levels of $H_2{}^{17}O$ (^{17}O is a spin 5/2 nuclei with a natural abundance of 0.037%) using an indirect detection approach in which serial T_2-weighted echo planar images are acquired with the decoupler power alternately on and off every eighth image. The pulse sequence diagram and acquisition scheme are shown below.

The following *in vitro* and *in vivo* data were obtained. In particular, the data below show: (Left) signal time courses during serial decoupling experiments for four tubes of water and one tube of acetone. Baseline T_2-weighted echo-planar imaging (EPI) of the tubes is shown on the far left. Next to these are maps of the correlation coefficient between the signal time course for each pixel and the decoupler power waveform. The plots (on the right) are from rectangular ROIs indicated on the T_2-weighted images.

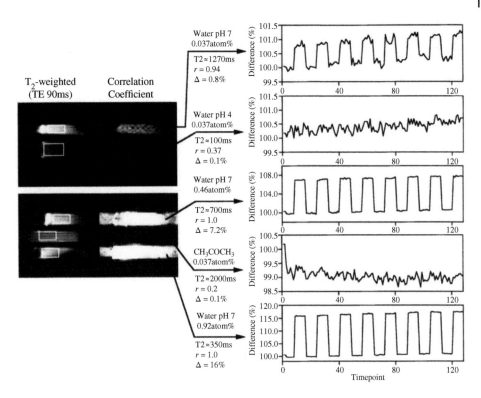

In vivo data from normal and ischemic rat brain are provided below.

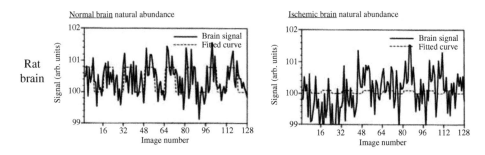

Give a theoretical explanation for the observed data. In particular,

a) Why does image signal intensity increase when the decoupler in on and why does the effect increase with increasing ^{17}O fractional enrichment?

b) Why does the effect disappear at low pH or with the use of acetone instead of water?

c) Give an explanation for the differential response between normal and ischemic brain.

d) Discuss potential applications for $H_2{}^{17}O$ imaging.

e) A graduate student proposes to improve the sensitivity of the method by turning on the decoupler prior to the 90° excitation in order to exploit the Nuclear Overhauser effect (NOE). Will this change the sensitivity and by what factor?

10

In vivo Examples

In this chapter, we have chosen three representative tissues, namely, brain, blood and cartilage, to highlight MR relaxation time differences throughout the human body and provide examples of how these variations can be used to enhance image contrast and extract clinically useful *in vivo* information.

10.1 Relaxation Properties of the Brain

In addition to its vasculature, the brain is primarily composed of gray matter, white matter, and cerebrospinal fluid (CSF). Each of these tissues has unique MR relaxation properties, enabling neuroscientists and physicians to investigate not only anatomy, but also features such as blood–brain barrier integrity, blood flow and perfusion, white matter tracts and other brain microstructure, iron deposition, and neuronal activation (see discussion in Section 10.2.3). In this section, we will discuss some of the primary MRI brain imaging tools currently available.

10.1.1 Morphological Imaging

Brain morphometry refers to the measurement of the size, volume, and shape characteristics of different brain structures and tissue types; proton density-, T_1-, and T_2-weighted imaging are standard acquisitions in most brain imaging protocols. As shown in Figure 10.1, of the three major brain tissues, CSF is the most liquid-like, white matter the most solid-like, and gray matter is in between.

Beyond the basic tissue types, the blood–brain barrier (BBB) relies on tight junctions between neighboring blood vessel endothelial cells to govern the passage of blood-borne components into and out of the neural microenvironment. As such, the BBB is essential for normal brain function and its disruption plays a key role in the onset and/or progression of multiple neurological disorders. This has driven considerable interest in the *in vivo* assessment of BBB integrity, and MRI using Gd-based contrast agents is the primary imaging method of choice (Veksler et al. 2014).

10.1.2 Perfusion Imaging

MRI of local brain perfusion, to image the blood flow through brain's capillary network, is a clinically invaluable tool for assessing multiple pathologies including stroke, brain tumors, and neurodegenerative diseases. One approach, known as arterial spin labeling (ASL), uses magnetically labeled water in arterial blood (*e.g.*, labeled via an Rf inversion pulse) as a freely diffusible flow

Fundamentals of In Vivo Magnetic Resonance: Spin Physics, Relaxation Theory, and Contrast Mechanisms, First Edition. Daniel M. Spielman and Keshav Datta.
© 2024 John Wiley & Sons, Inc. Published 2024 by John Wiley & Sons, Inc.
Companion website: www.wiley.com/go/Spielman

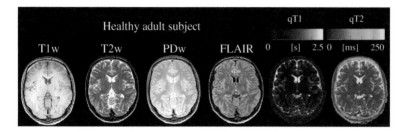

Figure 10.1 MRI of the human brain comparing T_1-weighted (T1w), T_2-weighted (T2w), proton density-weighted (PDw), and fluid-attenuated inversion-recovery (FLAIR) images, and quantitative T_1 (qT1) and T_2 (qT2) maps from a healthy adult. Source: Adapted with permission from Kleinloog et al. (2023) Figure 1.

tracer (Haller et al. 2016). The second method is to dynamically acquire images following the bolus injection of an exogenous, intravascular, nondiffusible contrast agent such as Gd-DTPA. This approach is generally referred to as dynamic contrast-enhanced (DCE) MRI (Essig et al. 2013).

10.1.3 Diffusion-weighted Imaging (DWI)

As we have discussed, time varying magnetic fields experienced by the spins drive MR relaxation. However, static field inhomogeneities can also cause local dephasing if the spins are moving, particularly via diffusion. The thermally driven random motion of molecules is known as Brownian motion, and was famously analyzed by Einstein in 1905 (Einstein 1905b). Due to cellular membranes and other *in vivo* structures, water molecules typically undergo hindered diffusion, with the degree of hindrance providing important biological and microstructural information. For example, extracellular water is generally freer to diffuse than intracellular, and diffusion along white matter tracts is faster than that across these tracts. Mathematically, *in vivo* water diffusion is characterized by a local diffusion tensor, typically assumed to be spherical for isotropic diffusion processes and ellipsoidal in the anisotropic case (see Figure 10.2a).

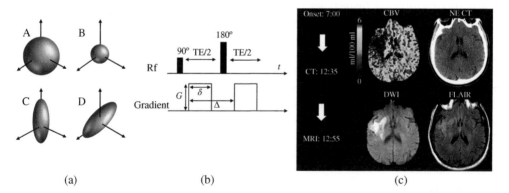

Figure 10.2 DWI. (a) Visualization of spin displacement front due to diffusion. (A) Isotropic diffusion. (B) Restricted diffusion. (C, D) Anisotropic diffusion. (b) A simple pulsed-gradient spin-echo pulse sequence to impart sensitivity to diffusion, typically quantified by the b-value, *i.e.,* $b = (\gamma G\delta)^2(\Delta - \delta/3)$. (c) CT versus DWI comparison of ischemic stroke imaged $\sim$5.5 hours after symptom onset. Source: (c) Adapted with permission from Broocks et al. (2020) Figure 1.

MRI can measure water diffusion on a time scale of a few tens of milliseconds. However, this diffusion process only results in signal dephasing in the presence of spatial magnetic field inhomogeneities. To enhance this signal loss, large "diffusion" gradients are used. Most commonly, DWI is performed using a spin–echo acquisition where the echo is surrounded by large symmetric gradient pulses (see Figure 10.2b). MR signals from moving spins are dephased, while signals from stationary spins are refocused, and the observed signal reduction is related to the local water ***apparent diffusion coefficient*** (ADC) with the MR-sensitivity, a function of the gradient amplitudes and timings, typically characterized by a parameter known as the ***b-value.*** Other metrics used in DWI include the ***fractional anisotropy*** (FA) and ***mean diffusivity*** (MD). Typically, the key to successful DWI studies is the reduction of artifacts from bulk motion, and single-shot sequences, such as echo planar imaging (EPI), are often employed for fast imaging to reduce these artifacts.

The imaging of ischemic stroke is one of the most important applications of DWI. Lacking oxygen and the glucose-derived energy normally supplied by blood flow, homeostatic ion pumps break down, and the affected cells swell, resulting in a DWI-detectable shift of water from the extra to the more hindered intracellular space (see Figure 10.2c).

An important extension of DWI, known as ***diffusion tensor imaging*** (DTI), measures diffusion along multiple spatial axes (see Figure 10.3). Taking advantage of faster water diffusion parallel as compared to perpendicular to white matter tracts, an important DTI application is brain tractography, also known as fiber tracking.

In summary, quantitative diffusion imaging techniques enable the characterization of microstructural properties of the human brain *in vivo* and are widely used in both neuroscience and clinical applications. Together with methods such as fMRI (to be discussed in Section 10.2), DWI and DTI have become important tools for the understanding of the spatiotemporal interaction of normal brain function and adaptive processes such as brain plasticity (Martinez-Heras et al. 2021).

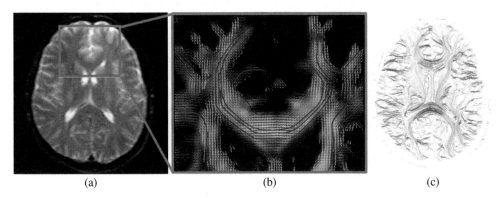

(a)   (b)   (c)

Figure 10.3 Diffusion tensor imaging (DTI) of the adult human brain. (a) T_2-weighted MRI, (b) Vector field derived from the spatial orientation of the major eigenvector of the diffusion tensor at each pixel is shown superimposed on a fractional anisotropy (FA) map to better reflect the major white matter tracts. All eigenvectors with underlying FA values below 0.2 were disregarded. The orientation of the major eigenvector was used to select the appropriate color from a RGB red, green, and blue color sphere (*i.e.,* red, left–right axis; green, anterior–posterior axis; blue, cephalo–caudal axis). These colors were assigned to facilitate 3D visualization of the vector field. By manually selecting seed points, fibers were selectively traced (bold red lines). (c) Using regularly placed seed points, fiber maps can be projected into the whole image plane. Source: Adapted with permission from Bammer et al. (2002) Figure 6.

10.1.4 Imaging Myelin

A layered phospholipid structure, myelin speeds neurotransmission while serving as a protective neuronal sheath. This myelin sheath consists of extracellular myelin water and lipid bilayers containing macromolecules. Furthermore, myelination plays a critical role in brain development and demyelinating processes are central to multiple brain pathologies, such as multiple sclerosis (Lee et al. 2021).

MRI is the primary method for *in vivo* myelin imaging, and most of the multiple MRI methods developed to interrogate this critical tissue and evaluate therapies are based on the unique anatomical compartmentalization shown in Figure 10.4. Myelin is typically modeled as containing two pools of spins, free water (which can be further divided into intracellular and extracellular water), and hydrogen bound to macromolecules and lipids (van der Weijden et al. 2022).

Magnetization transfer contrast (MTC), allowing the indirect detection of myelin-associated macromolecules, was the first myelin-targeted brain MRI technique and probably remains the most popular. However, other techniques, including targeting the myelin water located between lipid bilayers, measuring the nonaqueous protons of the phospholipid bilayer using ultrashort echo-time (UTE) techniques in combination with suppression of longer T_2 components, mapping the effects of the myelin sheaths on water diffusion, and imaging myelin-induced susceptibility effects (to be described in more detail in Section 10.1.5) have all shown promise.

10.1.5 Susceptibility-weighted Imaging (SWI)

Dephasing from local magnetic field inhomogeneities is the primary source of T_2^* contrast, and SWI pulse sequences utilize post-acquisition image processing via a combination of spatial filtering and image phase multiplication to enhance the contrast in T_2^*-weighted images (Haacke et al. 2009). SWI is used to target local field variations generated by tissues containing materials such as blood products, iron, calcium, and the deoxyhemoglobin in venous blood (Haller et al. 2021).

In general, all MR images have contributions from both the M_x and M_y components of transverse magnetization. Typically, magnitude images, $\sqrt{M_x^2 + M_y^2}$, are used for clinical interpretation, but, particularly for gradient-echo acquisitions, the image phase can also contain useful information. High-pass-filtering the segmented phase image can suppress artifacts arising from air/tissue interfaces and motion-induced phase errors to reveal more interesting structure. These filtered phase images can then be used to create spatial masks highlighting either positive phase shifts from paramagnetic (*e.g.*, deoxyhemoglobin) or negative shifts from diamagnetic materials (*e.g.*,

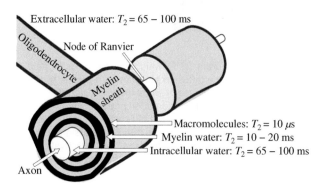

Figure 10.4 Central nervous system structures relevant for myelin MRI.

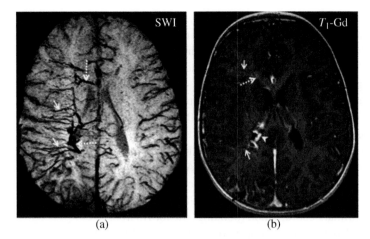

Figure 10.5 Susceptibility-weighted imaging (SWI). Representative images from a 2-year patient with Sturge–Weber Syndrome: (a) SWI (minimum intensity projection) and (b) T_1-weighted postgadolinium (T_1-Gd). The transmedullary veins (white solid arrow), and the connecting periventricular veins (white dashed arrow), can be seen only as mild enhancements in the right hemisphere on the T_1-Gd image. The T_1-Gd image also shows the enlarged choroid plexus (white arrowhead) and leptomeningeal abnormality (black solid arrow). In comparison, the SWI image shows a clear network of transmedullary veins and periventricular veins with much larger extent and higher contrast than that on T_1-Gd. However, the enlarged choroid plexus and leptomeningeal abnormality are not clearly visualized in SWI images. Source: Adapted with permission from Hu et al. (2008) Figure 3.

calcifications). Although the precise algorithms are proprietary to the individual scanner manufacturers, the basic idea is to multiply the acquired magnitude images with this processed phase mask. Susceptibility-shifted objects will then display local phase changes as a function of their relative shift to B_0. To further enhance the visibility of veins, SWI data are often displayed using a minimum-intensity projection over selected volumes of tissue. More recently, SWI has been extended via additional data processing steps to generate quantitative susceptibility maps (QSM), where image intensity is directly proportional to the local magnetic susceptibility (Haller et al. 2021). Given unique contrast, SWI and QSM are now used for multiple clinical applications, particularly for the detection of microbleeds or local iron deposition (see Figure 10.5).

10.2 Relaxation Properties of Blood

MRI of blood has an important role in both basic science investigations, *e.g.,* mapping neuronal activation in the brain using functional MRI (fMRI), and multiple clinical applications including studies of cardiac function, the vasculature, and hemorrhage. The ^{1}H MR signals from blood are dominated by signals from water protons, and these signals are strongly dependent on red blood cells and the fundamental role of hemoglobin (Hb) as an oxygen carrier.

10.2.1 Hemoglobin and Red Blood Cells

Let us start with a very basic question. What are the *in vivo* T_1 and T_2 relaxation times of water in human blood at a common clinical field strength such as 3 T? The answer is that for fully oxygenated blood, $T_1 \approx 1550$ ms and $T_2 \approx 165$ ms. But why do we need to specify that the blood is fully oxygenated?

In general, the blood relaxation time depends on multiple physical parameters, including field strength, temperature, the integrity of the erythrocytes (red blood cells), hematocrit, and the chemical state of the Hb in the red blood cells. Focusing on normal physiological parameters at typical imaging field strengths (*e.g.,* 37 °C, intact red blood cells, ~45% hematocrit, and 0.5–4 T), the most interesting parameter is the oxygen saturation level.

Hb, the oxygen-carrying molecule in human blood, contains four iron atoms that reversibly bind to oxygen (O_2), H_2O, and other small molecules. The primary Hb derivatives are oxyhemoglobin (oxy-Hb), deoxyhemoglobin (deoxy-Hb), methemoglobin (met-Hb), hemichromes, ferritin, and hemosiderin, and each of these derivatives has very different MR properties that determine how they affect the relaxation behavior of nearby water molecules. As shown in Figure 10.6, the Hb molecule consists of two alpha (α) and two beta (β) subunits, each containing an iron-containing heme group to which O_2 may bind. Each heme group consists of an iron (Fe) ion surrounded by a heterocyclic porphyrin ring, the primary coordination (bonding) site for Hb iron.

Oxy-Hb is diamagnetic, while deoxy-Hb is paramagnetic (four unpaired e⁻s). However, conformational changes in the absence of oxygen binding, blocks access to water in deoxy-Hb (see Figure 10.7). Without this direct binding, water-e⁻ dipole–dipole interactions are too weak to significantly contribute to T_1 relaxation. However, the unpaired electrons in deoxy-Hb do produce large magnetic susceptibility gradients (Figure 10.8a), and these local field distortions dephase nearby H_2O molecules, shortening T_2 and T_2^* relaxation times.

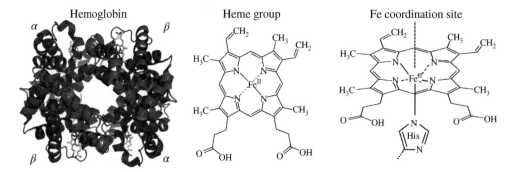

Figure 10.6 The chemical structure of hemoglobin.

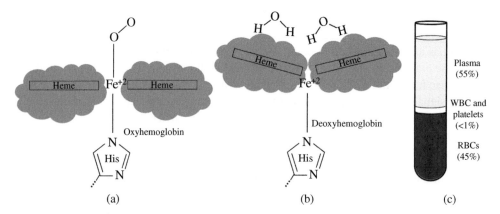

Figure 10.7 Blood constituents. (a) Oxyhemoglobin, (b) deoxyhemoglobin, and (c) relative contribution of primary blood components. WBC, white blood cells; RBC, red blood cells.

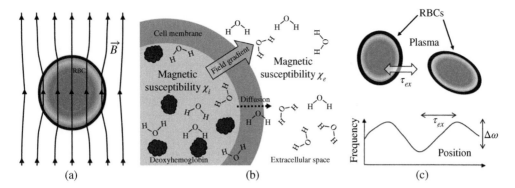

Figure 10.8 Red blood cells with deoxy-Hb exhibit (a) magnetic field gradients, (b) water diffusion across the cell membrane, and (c) time varying effective fields due to exchange of water into and out of the erythrocytes.

However, these susceptibility effects of deoxy-Hb depend on compartmentalization with both blood vessels and red blood cells (see Figure 10.8b). The largest constituent of blood is plasma, which is 95% water plus some dissolved proteins. The more solid elements are primarily red blood cells, which contain the Hb (see Figure 10.7c). The red blood cell blood fraction is known as the hematocrit and is typically 35–50% range in adults (somewhat higher in males versus females). Arterial blood contains a ~95/5 oxy/deoxy-Hb ratio, while the ratio in venous blood is closer to ~70/30 oxy/deoxy-Hb.

Let us start by modeling a red blood cell containing deoxy-Hb as having a bulk susceptibility shift relative to the surrounding plasma. In the deoxygenated state, the diffusion of water between plasma and erythrocytes results in dephasing and T_2 shortening. A more general model also includes diffusion through gradients. Hence,

$$\frac{1}{T_{2,blood}} = \frac{1}{T_{2,ox}} + \frac{1}{T_{2,deox}}, \tag{10.1}$$

where $T_{2,ox}$ and $T_{2,deox}$ are the T_2s of blood at 100% O_2 sat and fully deoxygenated, respectively. Including the deoxygenation effect, the exchange across the erythrocyte membrane, and diffusion through the surrounding magnetic field gradients, the net result is $\tau_{ex} \cong 8$–$10\,ms$ with $\Delta\omega = 1.0\,ppm$ as shown in Figure 10.8c.

10.2.2 MRI Blood Oximetry

A Carr–Purcell–Meiboom–Gill (CPMG) multi-spin-echo pulse sequence is commonly used for measuring blood T_2. Using the Luz–Meiboom model of relaxation, the T_2 of blood is given by:

$$\frac{1}{T_{2,blood}} = \frac{1}{T_{2,ox}} + f_a(1 - f_a)\left[\left(1 - \frac{\%HbO_2}{100}\right)\alpha\omega_{0/2\pi}\right]^2 \tau_{ex}\left(1 - \frac{2\tau_{ex}}{\tau_{180}}\tanh\frac{\tau_{180}}{2\tau_{ex}}\right), \tag{10.2}$$

where f_a is the fraction of protons at one of the sites under exchange, τ_{180} is the spacing between echoes, and α is a dimensionless constant involving both the susceptibility of deoxy-Hb and erythrocyte geometry (Wright et al. 1991a). This equation can be rewritten in the form:

$$\frac{1}{T_{2,blood}} = \frac{1}{T_{2,ox}} + K(\tau_{180}, \omega_0)\left(1 - \frac{\%HbO_2}{100}\right)^2. \tag{10.3}$$

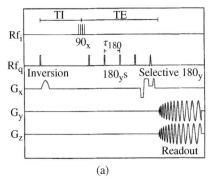

Subject	$T_{2.0}$ (ms)	τ_{180} (ms)	Aorta		Superior vena cava	
			T_{2b} (ms)	%HbO$_2$	T_{2b} (ms)	%HbO$_2$
1	224	6	223	97	185	74
		24	220	96	138	74
2	243	12	242	97	175	75
		24	230	93	155	76
3	214	12	213	97	154	73
		24	196	90	126	72
4	196	24	194	97	139	78
5	277	24	274	97	171	77

(a)   (b)

Figure 10.9 *In vivo* MR blood oximetry. (a) Short TI inversion recovery (STIR) pulse sequence for the *in vivo* estimation of T_{2b}, and corresponding blood oxygenation level. TI, inversion time; RF$_i$ and RF$_q$ are the in-phase and quadrature components of the Rf field, and τ_{180} is the separation time between echoes. (b) Representative %HbO$_2$ estimates from *in vivo* measurements of T_{2b} in healthy human adults. Source: Adapted with permission from Wright et al. (1991a) Figure 1 and Table 2.

Referring to Figure 10.9a, if time between 180 s, τ_{180}, is short compared to τ_{ex} the dephasing due to the presence of deoxy-Hb will be refocused. The parameter K is then a function of τ_{180}, τ_{ex}, and hematocrit. After obtaining a blood sample and measuring K for each subject, Wright et al. (1991a) used Eq. (10.2) to estimate *in vivo* oxygenation levels in the major blood vessels of normal control subjects using a 1.5 T MRI scanner. For maximum lipid suppression, their imaging sequence began with a short τ inversion recovery (STIR) sequence ($TI = 120$ ms), followed by a frequency-selective 90° excitation pulse exciting only the water protons. Data were then acquired using a fast single-shot spiral readout; representative data are shown in Figure 10.9b.

This *in vivo* MR blood oximetry approach can also be extended to other applications. For example, combining blood oxygenation from major blood vessels with measurements of blood flow, *e.g.*, in the major artery and vein feeding and draining a target organ, can be used to estimate organ-specific oxygen utilization (Li et al. 1998). A second application of the basic pulse sequence shown in Figure 10.9a is to adjust τ_{180} to enable flow-independent angiography, a contrast-free method for imaging deep vein thromboses (Wright et al. 1991b).

10.2.3 Functional Magnetic Resonance Imaging (fMRI)

Studies originally published by Ogawa et al. (1990) demonstrated large field inhomogeneities generated by deoxy-Hb within red blood cells. Magnetic field gradients within the diffusion distance of water shorten blood T_2, and the field variations around deoxygenated erythrocytes extend well beyond the boundary of the vessels, resulting in T_2^* shortening. This effect was coined blood oxygen level dependent (BOLD) contrast, the size of which depends on the relative amounts of oxy- versus deoxy-hemoglobin present (see Figure 10.10).

Imaging the BOLD effect has become a fundamental tool for noninvasive mapping of brain neuronal activation. Namely, within the brain, local blood flow and the associated oxygenation levels modulate with neuronal activation. As shown in Figure 10.11, active neurons require increased energy. This local metabolic demand increases O_2 consumption, increasing deoxy-Hb levels, resulting in a decrease in the BOLD signal (*i.e.*, T_2^* shortens). A few tens of ms later, local blood flow increases, overcompensating for the immediate metabolic needs and deoxy-Hb content decreases, increasing the BOLD signal. This locally increased BOLD signal is what is targeted in fMRI studies.

Anoxic mouse brain

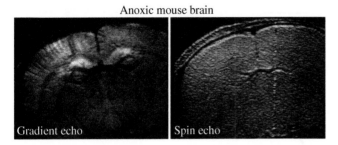

Figure 10.10   MRI signal loss in the mouse brain due to the accumulation of deoxy-Hb. Source: Adapted with permission from Ogawa et al. (1990) Figure 3.

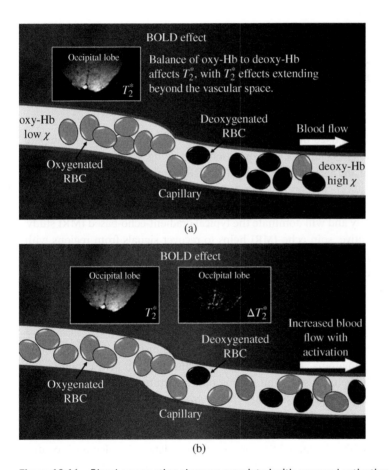

Figure 10.11   Blood oxygenation changes associated with neuronal activation.

One consideration is whether fMRI is better performed using gradient- versus spin-echo pulse sequences. Consider a single spherical magnetic inhomogeneity of radius R. The relative values of the diffusion coefficient D, the effective diffusion time τ_D, size of the inhomogeneous region, and local frequency shift parameter $\Delta\omega$ combine to determine the overall effect on T_2 relaxivity. Whereas previously we considered the effects of changing $\Delta\omega$ (e.g., SPIOs versus Dy-DTPA), now

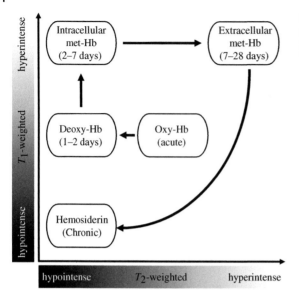

Figure 10.12 Schematic of the chemical changes seen in a brain hematoma over time and the associated appearance on T_1- and T_2-weighted MRI.

let us also look at size effects. For a fixed diffusion coefficient, D, τ_D scales as R^2. Now, compare a spin-echo (SE) versus a gradient-echo (GRE) acquisition. Assuming a Gaussian local field distribution, one can show that the larger effective R in a GRE acquisition leads to increased transverse magnetization relaxation. In contrast, for spin echo imaging, relaxivity occurring during a long τ_D (*i.e.*, large R) is refocused. Hence, for GRE images, large vessels are more efficient at increasing T_2 relaxivity and will dominate the typical gradient-echo-based fMRI study. Although significantly less sensitive, spin-echo-fMRI helps to recover signals from regions with poor homogeneity and better highlights small vessels, potentially better localizing the capillary bed associated with the activated neurons.

10.2.4 MRI of Hemorrhage

Hemorrhage is defined as the escape of blood from vessels into surrounding tissues, and a collection of clotted blood is known as a hematoma. The appearance of hemorrhage on MRI is multifactorial, depending on both the Hb chemical state and microscopic structure of the hematoma (see Figure 10.12). Furthermore, hemorrhage constituents change over time. Based on these effects, critical clinical information can be gleaned from MRI, with a more detailed overview and examples shown below.

Let us now look at the source of the changing MRI contrast in more detail. The electronic configuration of iron differs among the various Hb species. However, with respect to magnetic properties, the most important parameter is the number of unpaired electrons in the entire molecule, with the valence state of the ferrous (Fe^{+2}) or ferric (Fe^{+3}) iron less significant. Specifically, magnetic susceptibility, χ, is proportional to $N(N+2)$, where N is the number of unpaired electrons. The T_2 relaxation rate, in turn, is proportional to χ^2. Furthermore, the number of unpaired electrons per iron atom in oxy-Hb, deoxy-Hb, and met-Hb, and per molecule for ferritin and hemosiderin significantly differ, with the chemical structure and associated magnetic properties of these compounds given in Figure 10.13.

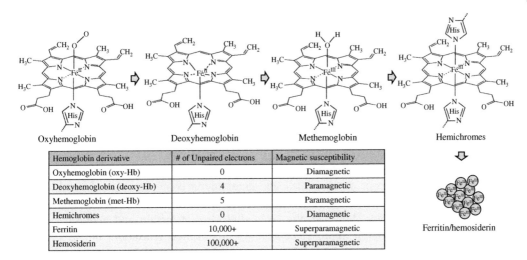

Hemoglobin derivative	# of Unpaired electrons	Magnetic susceptibility
Oxyhemoglobin (oxy-Hb)	0	Diamagnetic
Deoxyhemoglobin (deoxy-Hb)	4	Paramagnetic
Methemoglobin (met-Hb)	5	Paramagnetic
Hemichromes	0	Diamagnetic
Ferritin	10,000+	Superparamagnetic
Hemosiderin	100,000+	Superparamagnetic

Ferritin/hemosiderin

Figure 10.13 Magnetic properties of hemoglobin.

Although both Fe^{+2} and Fe^{+3} are paramagnetic, the impact of Hb as an MR contrast is also critically dependent on water accessibility. As previously mentioned, conformation changes going from oxy- to deoxyhemoglobin block access to water. However, the Fe^{+2} to Fe^{+3} transition, that occurs when deoxyhemoglobin is metabolized to methemoglobin (via enzymes such as methemoglobin reductase), causes additional conformational changes that unblock water access. Consequently, methemoglobin exhibits inner sphere relaxation, significantly shortening water T_1 values.

Let us now investigate the appearance of hemorrhage on various MRI pulse sequences. MRI of hyperacute hemorrhage (<12 hours) is characterized by oxy-Hb. The hematoma at this stage consists primarily of clotting erythrocytes with oxy-Hb being the predominating form of Hb . Edema may also form around the primary lesion. Deoxy-Hb only starts to form around the periphery. As a result, the core of the hematoma is isointense to white matter on T_1-weighted imaging, with mild hyperintensity on T_2 sequences. The initial increase of deoxy-Hb in the rim can be visualized as hyperintense on T_2^*-weighted images. Diffusion imaging will also show restricted water mobility due to a shift of water from the extracellular to intercellular space.

Over the next 12 hours–2 days, MRI of the hemorrhage is largely characterized by the formation of deoxy-Hb, as clotted erythrocytes accumulate in the periphery and spread inward. The increased susceptibility due to the paramagnetic deoxy-Hb shortens both T_2 and T_2^* relaxation times. In addition, the conformational changes of Hb associated with the loss of oxygen binding block water molecule access to the paramagnetic Fe^{2+}, resulting in significant T_2 changes with minimal alterations in T_1.

In the subacute phase (2 days–1 week), MRI contrast is dominated by the formation of met-Hb. The transition of the iron from Fe^{2+} to Fe^{3+} causes sufficient disruption of the Hb subunits to allow the binding of water, promoting inner sphere T_1 relaxation. Simultaneously, the paramagnetic properties of met-Hb shorten T_2 and T_2^* relaxation times. After several days, the periphery of the lesion typically starts to accumulate ferritin and hemosiderin.

Between one week and two months, the erythrocytes begin to break down, resulting in more homogeneous magnetic fields. While inner sphere relaxation of the intracellular met-Hb results in short T_1, T_2, and T_2^* values, the lysing of the red blood cells leads to the release of met-Hb into the extracellular space, resulting in a more uniform met-Hb distribution. This minimizes local susceptibility effects and increases the signal intensity on T_2-weighted imaging sequences. Due to the accumulation of extracellular met-Hb, the center of the hematoma becomes hyperintense on both T_1- and T_2-weighted imaging sequences. The continued accumulation of ferritin and hemosiderin in the lesion periphery decreases signal intensity on T_2^*-weighted images as well as DWI.

As the lesion transitions into the chronic stage, met-Hb is oxidatively denatured to form hemichromes, which are weakly diamagnetic and have minimal relaxation effects on the MR signal. These hemichromes subsequently break down, releasing free iron scavenged by macrophages and collected first as ferritin and then hemosiderin. Specifically, ferritin is a protein shell packed with hundreds to thousands of iron particles, and hemosiderin is an aggregate of hundreds to thousands of ferritin particles plus amorphous proteins and lipids. Both forms of iron cause marked T_2 and T_2^* shortening, however, the Fe binding sites of ferritin and hemosiderin are sequestered, prohibiting the close approach of water for inner sphere relaxation, and, as a result, there is only minimal shortening of T_1. After a few months, the original hematoma generally collapses, and the surrounding edema disappears, while the Hb molecules continue to degrade with the released iron typically deposited in the surrounding tissues. The result is that the central regions of old hematomas are water-like, with long T_1 and T_2 relaxation times and large ADC values. Figure 10.14 shows representative MR images of brain hemorrhage acquired at various time points post-injury.

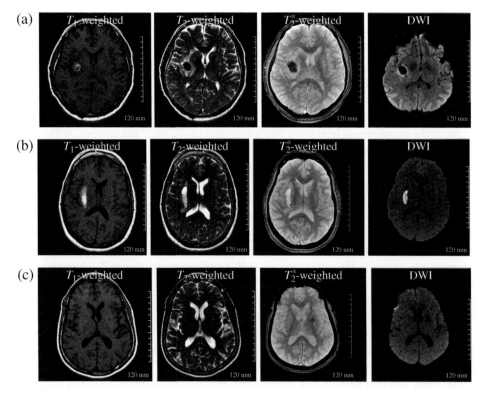

Figure 10.14 MRI appearance of brain hemorrhage in the (a) acute to early subacute, (b) late subacute, and (c) chronic stages. Source: Images courtesy of Dr. M. Ivy.

10.3   Relaxation Properties of Cartilage

Articular cartilage (hereafter, in this chapter, we will just use the term "cartilage") is another tissue with unique structure and function. The presence of multiple relaxation parameters and associated MRI contrast mechanisms make cartilage an excellent choice to illustrate how these mechanisms influence *in vivo* imaging results.

Cartilage provides a smooth lubricated surface for low-friction articulation of the joints and facilitates the transmission of physical loads. To perform these tasks, cartilage is structured into multiple zones with the cartilage extracellular matrix (ECM) composed primarily of type II collagen network and an interlocking mesh of fibrous proteins and proteoglycans (PGs). Chondrocytes are the only cells found in healthy cartilage. Only 2–4 mm thick, there are no blood vessels or nerves within the cartilage, and the overall composition is approximately 65–80% water, 10–29% collagen, and 10–15% proteoglycans (see Figure 10.15).

The collagen matrix gives cartilage its form and tensile strength, and the orientation and alignment of collagen fibers vary according to the depth from the articular surface as well as regionally within the joint. Exchangeable –OH and –NH protons on the GAG side chains of the proteoglycans provide a negative net charge that attracts positive counter-ions and water molecules. The strong electrostatic repulsive force between these proteoglycans results in swelling, which provides the needed interstitial fluid pressure for load bearing (see Figure 10.16).

From the medical perspective, MRI can detect multiple morphologic changes including cartilage loss, changes in fibrillation, GAG content, and osteophyte formation. At the cellular level, chondrocytes, the only cells found in healthy cartilage, secrete the materials needed to maintain and sustain the cartilage. They respond to outside stimuli and tissue damage, and are also responsible for degenerative conditions, such as osteoarthritis (OA), the most prevalent chronic disease of the elderly. OA is an important cause of disability in our society, with increasing incidence not only in the United States but in many other parts of the world (Choi and Gold 2011). Early signs of cartilage degeneration in OA are marked by changes in hydration driven by proteoglycan loss as well as some thinning and disruption of collagen although it has been suggested neither the content nor the type of collagen is altered at the earliest stages (Li and Majumdar 2013). Later changes

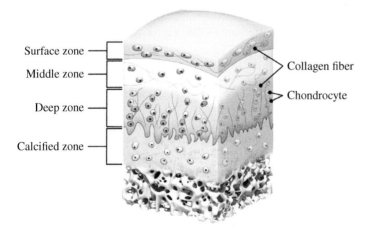

Figure 10.15   The zonal structure of articular cartilage. Cartilage is mostly acellular and avascular with the extracellular matrix consisting of water, collagen, and proteoglycans having negatively charged glycosaminoglycan (GAG) side chains. Source: Adapted with permission from Yu et al. (2023) Figure 1.

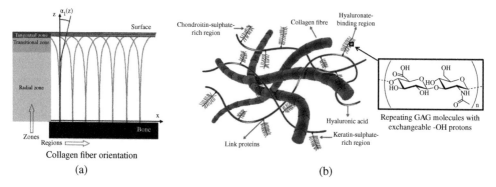

Figure 10.16 Cartilage anatomic details. (a) Predominant orientation of the cartilage collagen fibers by zones, and (b) structure relationships among collagen fibers, proteoglycans, and molecules. Source: (a) Adapted with permission from Gründer (2006) Figure 2 and (b) Adapted with permission from Ryan et al. (2015), copyright 2015 American Chemical Society.

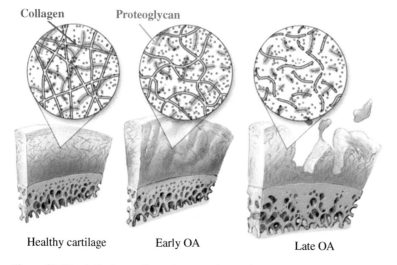

Figure 10.17 Articular cartilage degeneration during osteoarthritis (OA). Source: Adapted with permission from Li and Majumdar (2013) Figure 1.

include increased dehydration, further loss of proteoglycans, and GAG molecules, collagen disruption, extensive fibrillation, and thinning (see Figure 10.17).

The MR relaxation mechanisms in cartilage are thought to be primarily driven by three independent processes. The first is dipole–dipole coupling. Water molecules near macromolecules, such as collagen fibers, are more restricted in their tumbling motion (slower rotational correlation times) than water molecules further away. The second is chemical exchange. Hydrogen atoms are frequently involved in chemical exchange processes, wherein they are physically transferred from one molecule to another. In cartilage, there is natural chemical exchange primarily between hydroxyl (–OH) protons attached to the proteoglycan GAG molecules and bulk water. Finally, motion and diffusion effects are also seen. The movement of water protons between different local environments, due to processes including flow, translation, and diffusion, primarily results in T_2 relaxation proportional to the square of the gradient field and the length of time the spin moves within these gradients.

10.3.1 T_2 Mapping

T_2-weighted imaging is a primary workhorse for clinical musculoskeletal imaging, and T_2 relaxivity in cartilage is believed to be primarily driven by a combination of dipolar coupling and chemical exchange mechanisms. However, it is largely dominated by dipolar coupling effects as indicated by T_2 being strongly influenced by cartilage collagen fiber orientation. That is, residual dipolar coupling results in strong magic angle effects, as shown in Figure 10.18. Accordingly, cartilage has a laminar appearance on T_2-weighted images, with signal intensity depending both on depth and orientation. T_2 images can thus provide important clinical information regarding both collagen structure and orientation.

The major advantage of T_2 contrast is that it is widely available with multiple MRI techniques. Challenges include magic angle effects and the hypothesis that the earliest OA-driven changes in cartilage proteoglycan content may be hard to detect on T_2-weighted images or T_2 maps. Examples of both T_2 and $T_{1\rho}$ (to be discussed in later sections) maps from patients with OA are shown in Figure 10.19.

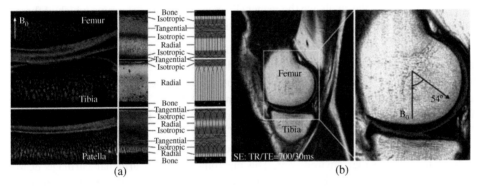

Figure 10.18 Laminar appearance of cartilage on T_2-weighted MRI. (a) Representative images, and (b) magic angle effects. Source: Adapted with permission from Gründer (2006) Figure 1 and 6.

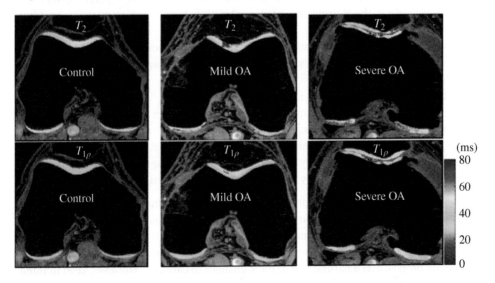

Figure 10.19 Representative T_2 and $T_{1\rho}$ maps of articular cartilage in a control subject and patients with mild and severe OA. Note, no spin-lock frequency was reported for the $T_{1\rho}$ maps. Source: Adapted with permission from Li and Majumdar (2013). Figure 8.

10.3.2 DWI

Diffusion is the primary transport mechanism for water in cartilage and is affected by the structure and composition of the collagen matrix. Hence, DWI, which can measure parameters including the ADC, MD, and FA, has an important role in the clinical assessment of cartilage. The diffusion parameters are believed to correlate with both the collagen network as well as proteoglycan concentrations. The primary advantage of DWI is the wide availability of this endogenous contrast mechanism. Challenges include low SNR due to the required long echo times for diffusion encoding, in combination with the relatively short T_2 of cartilage, particularly in the deeper layers.

10.3.3 $T_{1\rho}$ Mapping and Dispersion

There are multiple reports in the MRI literature using $T_{1\rho}$ to study cartilage GAG content (Choi and Gold 2011). However, there is considerable confusion in the use and interpretation of $T_{1\rho}$ maps or $T_{1\rho}$-weighted images. Furthermore, the spin-lock frequency is often not specified. For example, the following are ten different statements from published literature (mostly from review papers) regarding $T_{1\rho}$ and cartilage. We challenge the reader (Exercise E.10.1) to determine which of these statements are true, which are false, and which are partially true or ambiguous.

1. "In $T_{1\rho}$...the magnetization undergoes relaxation in the presence of the applied B_1..."
2. "The measurements of $T_{1\rho}$ probe molecular fluctuations in the kHz range because of the dependence on the Rf-generated magnetic field (B_1), whereas T_2 probes fluctuations in the MHz range because of the dependence on the static magnetic field (B_0)."
3. "The B_1 field attenuates the effect of dipolar relaxation, static dipolar coupling, chemical exchange, and background gradients on the signal."
4. "... $T_{1\rho}$ probes the slow motion interactions between motion-restricted water molecules and their local macromolecular environment. The macromolecules in articular cartilage ECM restrict the motion of water molecules. Changes to the ECM, such as PG loss, therefore, can be reflected in measurements of $T_{1\rho}$."
5. "...In $T_{1\rho}$ quantification experiments, the spin-lock techniques reduce dipolar interactions and, therefore, reduce the dependence of the relaxation time constant on collagen fiber orientation..."
6. "...It is the difference between low and high locking fields that reflects the contributions of exchanging protons, not the absolute values of T_2 ($\approx$ the low field value of $T_{1\rho}$) or the value at high field."
7. "$T_{1\rho}$ reflects interaction of collagen with water."
8. "$T_{1\rho}$ is sensitive to low-frequency exchange interactions between water molecules and the large, slow tumbling macromolecules."
9. "$T_{1\rho}$ may be more sensitive to the initial changes in the cartilage ECM associated with PG depletion, whereas T_2 is sensitive only to later changes in the collagen network."
10. "$T_{1\rho}$ values appear to be unaffected by the laminar structure of cartilage."

As discussed in Chapter 8, one of the primary targets for $T_{1\rho}$ imaging is the study of chemical exchange processes. A simple model of cartilage MR relaxation consists of both (1) a dipolar coupling component from restricted water tumbling via interactions with collagen fibers, and (2) a chemical exchange process between water and the hydroxyl groups on GAG molecules. Hence, $T_{1\rho}$ imaging has been proposed as a natural tool to track changes in cartilage. However, keep in mind that in order to most effectively use $T_{1\rho}$ for the study of GAG chemical exchange, the spin-lock frequency, ω_1, should be $\geq$ the associated exchange rate (thought to be around 1000 Hz for GAG –OH

(Ling et al. 2008)). Unfortunately, most human studies, for reasons of safety, are limited to more moderate values of ω_1 such as 500 Hz.

Spatial maps of $T_{1\rho}$ at lower spin-lock amplitudes (as well as T_2 maps) can indeed be sensitive to changes in GAG molecules in the ECM, such as hypothesized to occur in the early stages of OA. However, these changes cannot be isolated from dipolar-coupling-driven relaxation based on water mobility changes associated with collagen only using data acquired at a single value of ω_1 (see Exercise 8.6). Nonetheless, there are multiple literature reports studying cartilage $T_{1\rho}$ and changes with pathology. Changes in $T_{1\rho}$ have been shown to correlate with loss of GAG (Regatte et al. 2003; Wang et al. 2015), but $T_{1\rho}$ findings are often very similar to those reported using conventional T_2 methods. $T_{1\rho}$ images of cartilage, however, reduce magic angle effects, but do not fully eliminate them at typical values of spin-lock frequencies (Gründer 2006).

An advantage of $T_{1\rho}$ is that it is an endogenous GAG contrast mechanism. However, the major challenges are (1) spin-lock frequencies are typically limited by SAR restrictions and may not adequately decouple GAG changes from the collagen network dipolar-coupling interactions, (2) long scan times, and (3) $T_{1\rho}$ is not as specific to GAG content as the other MR methods to be discussed in the next section.

10.3.4 gagCEST

As discussed in Chapter 7, CEST allows the detection of metabolites normally invisible to MR, and hence has a potentially valuable role in cartilage imaging via the chemical exchange of protons between GAG and bulk water protons. The best developed technique, known as gagCEST, targets those GAG hydroxyl protons (–OH) that resonate ~1 ppm downfield from water. In general, the required slow-to-intermediate exchange condition may not be fully fulfilled at 3 T, hence 7 T scanners provide improved gagCEST images but may be limited by SAR (Brinkhof et al. 2018).

Although gagCEST has a high specificity to GAG, the method is also sensitive to pH, hydration, and collagen. Nonetheless, differences have been observed in both *ex vivo* PG-depleted cartilage and *in vivo* from focal lesions (Kogan et al. 2017). Additional advantages of gagCEST are that it does not need intravenous contrast agents or special MRI hardware. Challenges include the relatively small resonance frequency shifts between GAG hydroxyl protons and bulk water protons, sensitivity to B_0 inhomogeneities, and SAR limitations.

10.3.5 dGEMRIC

Delayed gadolinium-enhanced MRI of cartilage (dGEMRIC) is a contrast-enhanced MRI method exploiting the fact that the negatively charged contrast agent Gd-DTPA^{2-} will distribute in cartilage in inverse relation to the negatively charged GAG concentration (Williams et al. 2004). Consequently, changes in T_1 relaxation time after contrast administration can be used to quantify GAG concentration using this highly specific technique. The basic procedure is to (1) inject the Gd-DTPA, (2) have the subject exercise for 10–20 minutes, (3) wait ~60–80 minutes, and then (4) image using a T_1 mapping technique approximately 90–180 minutes after the initial contrast agent injection.

The very high sensitivity of dGEMRIC to GAG concentrations is a major strength of this high-SNR, high-resolution imaging approach. The specificity to GAG has also been well-validated (Choi and Gold 2011). However, there remain multiple challenges, including the diffusion of Gd-DTPA into the cartilage being dependent on collagen content, long study times, and the need for a Gd-based contrast agent that may pose health risks for individuals with renal insufficiency. Figure 10.20 shows a representative deGEMRIC study of early OA.

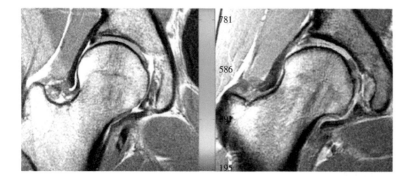

Figure 10.20 dGEMRIC images of weight-bearing hip cartilage in a healthy volunteer (left) and a patient with early OA (right) at 65 minutes after contrast injection. The color bar represents T_1 values in the presence of Gd (T_1-Gd). The mean T_1-Gd (the dGEMRIC index) for the ROI was 36% lower in the patient with early OA than in the healthy volunteer (381 and 591 ms, respectively). Source: Adapted with permission from Tiderius et al. (2007) Figure 1.

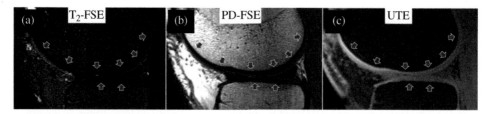

Figure 10.21 Comparison between the normal appearance of cartilage on conventional MR and UTE MR sequences. Sagittal (a) FSE T_2-weighted and (b) PD-weighted FSE MRI of the lateral compartment of the knee shows low signal intensity within the deep layers of cartilage and the osteochondral junction (OCJ) (arrows in (a) and (b)). (c) Sagittal UTE MRI of the same knee region shows increased signal intensity across all layers of cartilage, including the OCJ, represented by a bright line (arrows in (c)). Source: Adapted with permission from Lombardi et al. (2023) Figure 4.

10.3.6 Ultrashort TE (UTE) Imaging

Signals from cortical bone, tendons, ligaments, menisci, and deep radial and calcified layers of cartilage are typically difficult to detect using conventional MRI sequences but are detectable using ultrashort echo time (UTE) imaging methods (Afsahi et al. 2022). Hence, UTE may be the only method capable of evaluating the calcified layer of cartilage in OA (see Figure 10.21).

10.3.7 Sodium MRI

Sodium ion homeostasis is a vital cellular function, with intracellular and extracellular concentrations being about 10 and 150 mM, respectively. This high extracellular to intracellular ratio generally makes *in vivo* [23]Na MRI of intracellular processes quite challenging (Zaric et al. 2021). However, the interest in [23]Na MRI for the study of cartilage is driven by the fact that extracellular positively charged sodium ions exist in association with the negatively charged GAG side-chains, and can, thus, provide a direct measure of GAG content.

Nuclei with spin >1/2 possess an electric quadrupole moment due to the nuclear charge distribution no longer being spherically symmetric. In the case of [23]Na, which has spin-3/2, there are four equally spaced energy levels, resulting in three single quantum transition frequencies, with double and triple quantum transitions also possible. Time dependent perturbation theory yields

3 : 4 : 3 relative intensities for the corresponding spectral lines (Chapman et al. 2010). If the electric field (*E*-field) seen by the ^{23}Na nucleus is spatially homogeneous, all three transitions have the same frequency. However, in the presence of nonzero spatial electric field gradients, a quadrupolar coupling term appears in the corresponding spin Hamiltonian.

Static *E*-field gradients result in shifts of the resonance frequencies of the two outer transitions (to first order, the inner transitions are insensitive to *E*-field gradients), with the splitting depending on both the strength of *E*-field gradient and molecular orientation, and typically in the kHz to MHz range. *In vivo*, the range of splitting results in very broad outer lines, and, in practice, *in vivo* ^{23}Na has only a 40% NMR visibility. This effect is called "heterogeneous broadening."

Dynamic, that is time varying, *E*-field gradients are also present. These time depended quadrupolar splittings average out to zero, but fluctuations induce relaxation. The net result is that *in vivo*, the range of splittings causes broad outer lines. When the T_2 of the outer lines are much shorter than the inner line (typically the case *in vivo*), the effect is referred to as "homogeneous broadening." In practice, both static and dynamic quadrupolar effects are present *in vivo*. This results in a bi-exponential ^{23}Na T_2 with one component being very short.

In extracellular environments, such as an aqueous solution, the observed ^{23}Na T_2 is monoexponential with a time constant on the order of 30–40 ms, and in fast isotropic tumbling regime, there is only a single quantum coherence. Intracellularly, the rapid exchange between free ^{23}Na and ^{23}Na bound to slowly tumbling macromolecular sites contribute to the biexponential T_2 decay with a short component of 1–3 ms. Sodium ions in this motion-restricted tumbling regime also exhibit both double- and triple-quantum coherences.

With respect to ^{23}Na imaging of cartilage, the sodium of interest is all extracellular. Negatively charged proteoglycans attract cations (mainly sodium), which draw in water by osmosis to provide cushioning and increased load-bearing capacity. Furthermore, in contrast to ^{23}Na in muscle for example which has a concentration of 15–30 mM, articular cartilage ^{23}Na concentrations are in the 250–350 mM range and ~150 mM in synovial fluid. Even with these relatively high concentrations, the SNR for ^{23}Na MRI is only around 9% that for ^{1}H nuclei, and high-field scanners (*e.g.*, 7 T) are typically preferred (Zaric et al. 2021). Imaging is helped somewhat by relatively short ^{23}Na T_1s in the range of 15–55 ms, which allows short repetition times, and short or ultrashort echo times are also used (see Figure 10.22).

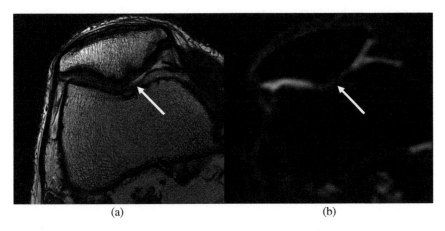

(a)   (b)

Figure 10.22 ^{23}Na MRI. (a) Proton density MRI of a patient with a lesion in the patellar cartilage showing early-stage degeneration of articular cartilage. (b) Corresponding area on the sodium image. Source: Adapted with permission from Zaric et al. (2021) Figure 2.

Table 10.1 Summary of the MRI techniques used to image cartilage.

Technique	Outcome measure	Biochemical correlates	Advantages	Challenges
T_2 mapping	T_2 relaxation time	Collagen content/ orientation	Widely available Endogenous contrast	Magic angle effects May not capture initial biochemical changes with OA
Diffusion MRI	ADC, MD, FA	Collagen network PG content	Widely available Multi-parametric Endogenous contrast	Low SNR Poor visibility of deep layers
$T_{1\rho}$	$T_{1\rho}$ relaxation time	GAG content	Endogenous contrast	High Rf power (SAR limits) Not specific to GAG
GagCEST	CEST asymmetry	GAG content	High specificity Endogenous contrast	Difficult at 3 T and requires advanced post processing
dGEMRIC	T_1 relaxation time	GAG content	High SNR High specificity Well validated	Invasive (contrast agent injection) Long scan times Expensive
UTE imaging	Ultrashort T_2 materials	Osteochondrial junction	Unique contrast Endogenous contrast	Long scan times Difficulty in slice selection
^{23}Na imaging	Sodium signal/ concentration	GAG content	High specificity Endogenous contrast	Low SNR Long scan times Specialized hardware

Because ^{23}Na ions interacting with macromolecules exhibit restricted tumbling, double- or triple-quantum imaging may be highly specific to pathological changes, as these methods are potentially more sensitive to subtle changes in the macromolecule arrangements in the ECM (Borthakur et al. 1999). The primary limitation of such multiple quantum imaging methods is low SNR (Zaric et al. 2021).

10.3.8 Summary

Table 10.1 lists the primary strengths and weaknesses of the multiple cartilage MRI methods discussed in this section.

10.4 Synopsis

Congratulations, you reached the end of this journey (see Figure 10.23)! We hope you have learned something about the source of the MR signals observable *in vivo* (from both classical and quantum mechanical perspectives), the associated image–contrast mechanisms,

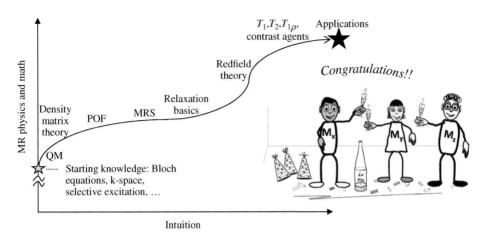

Figure 10.23 Time to celebrate reaching the end of this roadmap, and hopefully the beginning of many new journeys into the world of *in vivo* MR.

and how such information may be leveraged to provide important biological and clinical information.

This book started with an overview of historical developments in the evolution of MR. We then reviewed the analysis of MRI based on classical physics, whereby the use of linear gradient fields led to the powerful imaging concepts of Fourier encoding and selective excitation. More advanced imaging topics such as phased-array Rf coils and parallel imaging are left to other texts. Instead, the focus here was on developing methods for the analysis of interactions between spins and their local environment. This led to an analysis of MR relaxation processes which help answer the fundamental questions of why water molecules within different tissues generate such widely varying image contrasts and how intrinsic contrast can be enhanced via the use of specially designed MR contrast agents. Finally, we presented examples of these MR phenomenon from different in vivo tissues. Congratulations on making it to the end!!

Exercises

E.10.1 $T_{1\rho}$ **– The Most Misunderstood MRI Relaxation Time**

Which of the following 10 statements are true, which are false, and which are partially true or ambiguous? Why?

a) "In $T_{1\rho}$...the magnetization undergoes relaxation in the presence of the applied B_1..."

b) "The measurements of $T_{1\rho}$ probe molecular fluctuations in the kHz range because of the dependence on the Rf-generated magnetic field (B_1), whereas T_2 probes fluctuations in the MHz range because of the dependence on the static magnetic field (B_0)."

c) "The B_1 field attenuates the effect of dipolar relaxation, static dipolar coupling, chemical exchange, and background gradients on the signal."

d) "...$T_{1\rho}$ probes the slow motion interactions between motion-restricted water molecules and their local macromolecular environment. The macromolecules in articular cartilage ECM restrict the motion of water molecules. Changes to the ECM, such as PG loss, therefore, can be reflected in measurements of $T_{1\rho}$."

e) "...In $T_{1\rho}$ quantification experiments, the spin-lock techniques reduce dipolar interactions and, therefore, reduce the dependence of the relaxation time constant on collagen fiber orientation..."

f) "...It is the difference between low and high locking fields that reflects the contributions of exchanging protons, not the absolute values of T_2 ($\approx$ the low field value of $T_{1\rho}$) or the value of at high field."

g) "$T_{1\rho}$ reflects interaction of collagen with water."

h) "$T_{1\rho}$ is sensitive to low-frequency exchange interactions between water molecules and the large, slow tumbling macromolecules."

i) "$T_{1\rho}$ may be more sensitive to the initial changes in the cartilage ECM associated with PG depletion, whereas T_2 is sensitive only to later changes in the collagen network."

j) "$T_{1\rho}$ values appear to be unaffected by the laminar structure of cartilage."

E.10.2 It is party time!

As a final challenge, Figure 10.23 contains a subtle MR quantum mechanics joke that summarizes this textbook. Can you find it?

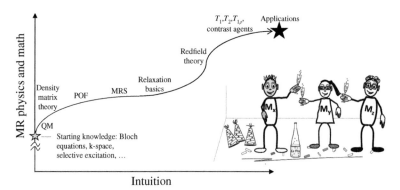

Further Readings

Quantum Mechanics

Miller, D. A. B. 2008. *Quantum Mechanics for Scientists and Engineers* (Cambridge University Press).

Spin Physics

Abragam, A. 1983. *Principles of Nuclear Magnetism* (International Series of Monographs on Physics, Volume 32) (Clarendon Press).

Ernst, R. R. 1987. *Principles of Nuclear Magnetic Resonance (NMR) in One and Two Dimensions* (Clarendon Press).

Golman, M. 1991. *Quantum Description of High-Resolution NMR in Liquids, Reprint Edition* (Clarendon Press).

Levitt, M. H. 2008. *Spin Dynamics: Basics of Nuclear Magnetic Resonance (NMR), 2nd Edition* (Wiley).

Slichter, C. P. 1996. *Principles of Magnetic Resonance* (Springer Series in Solid-State Sciences, Volume 1) (Springer).

van de Ven, F. J. M. 1996. *Multidimensional NMR in Liquids: Basic Principles and Experimental Methods, 1st Edition* (Wiley-VCH).

Magnetic Resonance Imaging (MRI)

Bernstein, M. A., K. F. King, and X. J. Zhou. 2004. *Handbook of MRI Pulse Sequences, 1st Edition* (Academic Press).

Nishimura, D. G. 2010. *Principles of Magnetic Resonance Imaging (MRI)* (LuLu).

In vivo Magnetic Resonance Spectroscopy

de Graaf, R. 2007. *In Vivo NMR Spectroscopy: Principles and Techniques, 2nd Edition* (Wiley).

Fundamentals of In Vivo Magnetic Resonance: Spin Physics, Relaxation Theory, and Contrast Mechanisms, First Edition.
Daniel M. Spielman and Keshav Datta.
© 2024 John Wiley & Sons, Inc. Published 2024 by John Wiley & Sons, Inc.
Companion website: www.wiley.com/go/Spielman

Relaxation Theory

Kowalewski, J., and L. Mäler. 2006. *Nuclear Spin Relaxation in Liquids: Theory, Experiments, and Applications*. Chemical Physics (Taylor & Francis).

Clinical MRI

Barker, P. B., A. Bizzi, N. De Stenfano, R. Gullapalli, and D. M. Lin. 2015. *Clinical MR Spectroscopy Techniques and Applications* (Cambridge University Press).

References

Abragam, A. 1983. *Principles of Nuclear Magnetism* (Clarendon Press).

Abragam, A., and M. Goldman. 1978. 'Principles of dynamic nuclear polarization', *Rep Prog Phys*, 41: 395.

Ackerman, J. J., G. E. Soto, W. M. Spees, Z. Zhu, and J. L. Evelhoch. 1996. 'The NMR chemical shift pH measurement revisited: analysis of error and modeling of a pH dependent reference', *Magn Reson Med*, 36: 674–83.

Adalsteinsson, E., and D. M. Spielman. 1999. 'Spatially resolved two-dimensional spectroscopy', *Magn Reson Med*, 41: 8–12.

Adalsteinsson, E., D. M. Spielman, G. A. Wright, J. M. Pauly, C. H. Meyer, and A. Macovski. 1993. 'Incorporating lactate/lipid discrimination into a spectroscopic imaging sequence', *Magn Reson Med*, 30: 124–30.

Adalsteinsson, E., P. Irarrazabal, S. Topp, C. Meyer, A. Macovski, and D. M. Spielman. 1998. 'Volumetric spectroscopic imaging with spiral-based *k*-space trajectories', *Magn Reson Med*, 39: 889–98.

Adalsteinsson, E., R. E. Hurd, D. Mayer, N. Sailasuta, E. V. Sullivan, and A. Pfefferbaum. 2004. 'In vivo 2D J-resolved magnetic resonance spectroscopy of rat brain with a 3-T clinical human scanner', *NeuroImage*, 22: 381–6.

Afsahi, A. M., Y. Ma, H. Jang, S. Jerban, C. B. Chung, E. Y. Chang, and J. Du. 2022. 'Ultrashort echo time magnetic resonance imaging techniques: met and unmet needs in musculoskeletal imaging', *J Magn Reson Imaging*, 55: 1597–612.

Aime, S., and P. Caravan. 2009. 'Biodistribution of gadolinium-based contrast agents, including gadolinium deposition', *J Magn Reson Imaging*, 30: 1259–67.

Akella, S. V., R. R. Regatte, A. J. Wheaton, A. Borthakur, and R. Reddy. 2004. 'Reduction of residual dipolar interaction in cartilage by spin-lock technique', *Magn Reson Med*, 52: 1103–9.

Anik, M., K. Hossain, I. Hossain, A. Mahfuz, T. Rahman, and I. Ahmed. 2021. 'Recent progress of magnetic nanoparticles in biomedical applications: a review', *Nano Select*, 2: 1146–86.

Ardenkjaer-Larsen, J. H., B. Fridlund, A. Gram, G. Hansson, L. Hansson, M. H. Lerche, R. Servin, M. Thaning, and K. Golman. 2003. 'Increase in signal-to-noise ratio of >10,000 times in liquid-state NMR', *Proc Natl Acad Sci U S A*, 100: 10158–63.

Arnold, D. L., P. M. Matthews, and G. K. Radda. 1984. 'Metabolic recovery after exercise and the assessment of mitochondrial function in vivo in human skeletal muscle by means of 31P NMR', *Magn Reson Med*, 1: 307–15.

Ayyala, R. S., L. A. Teot, and J. M. Perez Rossello. 2017. 'Gaucher disease in the liver on hepatocyte specific contrast agent enhanced MR imaging', *Pediatr Radiol*, 47: 484–87.

Fundamentals of In Vivo Magnetic Resonance: Spin Physics, Relaxation Theory, and Contrast Mechanisms, First Edition. Daniel M. Spielman and Keshav Datta.
© 2024 John Wiley & Sons, Inc. Published 2024 by John Wiley & Sons, Inc.
Companion website: www.wiley.com/go/Spielman

Bammer, R., M. Auer, S. L. Keeling, M. Augustin, L. A. Stables, R. W. Prokesch, R. Stollberger, M. E. Moseley, and F. Fazekas. 2002. 'Diffusion tensor imaging using single-shot SENSE-EPI', *Magn Reson Med*, 48: 128–36.

Bashir, M. R., L. Bhatti, D. Marin, and R. C. Nelson. 2015. 'Emerging applications for ferumoxytol as a contrast agent in MRI', *J Magn Reson Imaging*, 41: 884–98.

Bloch, F. 1946. 'Nuclear induction', *Phys Rev*, 70: 460–74.

Bloch, F. 2007. 'Bloch, Felix: The Principle of Nuclear Induction Nobel Lecture, December 11, 1952 © Le Prix Nobel, 1952.' in R. K. Harris and R. L. Wasylishen (eds.), *eMagRes* (Wiley).

Bloch, F., W. W. Hansen, and M. Packard. 1946. 'The nuclear induction experiment', *Phys Rev*, 70: 474–85.

Bloembergen, N., E. M. Purcell, and R. V. Pound. 1948. 'Relaxation effects in nuclear magnetic resonance absorption', *Phys Rev*, 73: 679–712.

Boesch, C., and R. Kreis. 2001. 'Dipolar coupling and ordering effects observed in magnetic resonance spectra of skeletal muscle', *NMR Biomed*, 14: 140–8.

Bonm, A. V., R. Ritterbusch, P. Throckmorton, and J. J. Graber. 2020. 'Clinical imaging for diagnostic challenges in the management of gliomas: a review', *J Neuroimaging*, 30: 139–45.

Borthakur, A., I. Hancu, F. E. Boada, G. X. Shen, E. M. Shapiro, and R. Reddy. 1999. 'In vivo triple quantum filtered twisted projection sodium MRI of human articular cartilage', *J Magn Reson*, 141: 286–90.

Borthakur, A., E. Mellon, S. Niyogi, W. Witschey, J. B. Kneeland, and R. Reddy. 2006. 'Sodium and $T_{1\rho}$ MRI for molecular and diagnostic imaging of articular cartilage', *NMR Biomed*, 19: 781–821.

Bottomley, P. A. 1987. 'Spatial localization in NMR spectroscopy in vivo', *Ann N Y Acad Sci*, 508: 333–48.

Brayda-Bruno, M., M. Tibiletti, K. Ito, J. Fairbank, F. Galbusera, A. Zerbi, S. Roberts, E. Wachtel, Y. Merkher, and S. S. Sivan. 2014. 'Advances in the diagnosis of degenerated lumbar discs and their possible clinical application', *Eur Spine J*, 23 Suppl 3: S315–23.

Brinkhof, S., R. Nizak, V. Khlebnikov, J. J. Prompers, D. W. J. Klomp, and D. B. F. Saris. 2018. 'Detection of early cartilage damage: feasibility and potential of gagCEST imaging at 7T', *Eur Radiol*, 28: 2874–81.

de Broglie, L. 1923. 'Waves and quanta', *C R Hebd Seances Acad Sci*, 177: 507–10.

Broocks, G., H. Leischner, U. Hanning, F. Flottmann, T. D. Faizy, G. Schön, P. Sporns, G. Thomalla, S. Kamalian, M. H. Lev, J. Fiehler, and A. Kemmling. 2020. 'Lesion age imaging in acute stroke: water uptake in CT versus DWI-FLAIR mismatch', *Ann Neurol*, 88: 1144–52.

Brown, T. R., R. Stoyanova, T. Greenberg, R. Srinivasan, and J. Murphy-Boesch. 1995. 'NOE enhancements and T_1 relaxation times of phosphorylated metabolites in human calf muscle at 1.5 Tesla', *Magn Reson Med*, 33: 417–21.

Buntkowsky, G., F. Theiss, J. Lins, Y. A. Miloslavina, L. Wienands, A. Kiryutin, and A. Yurkovskaya. 2022. 'Recent advances in the application of parahydrogen in catalysis and biochemistry', *RSC Adv*, 12: 12477–506.

Bydder, M., A. Rahal, G. D. Fullerton, and G. M. Bydder. 2007. 'The magic angle effect: a source of artifact, determinant of image contrast, and technique for imaging', *J Magn Reson Imaging*, 25: 290–300.

Canitano, R., and R. Palumbi. 2021. 'Excitation/inhibition modulators in autism spectrum disorder: current clinical research', *Front Neurosci*, 15: 753274.

Cember, A. T. J., N. E. Wilson, L. J. Rich, P. Bagga, R. P. R. Nanga, S. Swago, A. Swain, D. Thakuri, M. Elliot, M. D. Schnall, J. A. Detre, and R. Reddy. 2022. 'Integrating ^{1}H MRS and deuterium labeled glucose for mapping the dynamics of neural metabolism in humans', *NeuroImage*, 251: 118977.

Chapman, B. E., C. Naumann, D. J. Philp, U. Eliav, G. Navon, and P. W. Kuchel. 2010. 'z-Spectra of ^{23}Na$^+$ in stretched gels: quantitative multiple quantum analysis', *J Magn Reson*, 205: 260–8.

Choi, J. A., and G. E. Gold. 2011. 'MR imaging of articular cartilage physiology', *Magn Reson Imaging Clin N Am*, 19: 249–82.

Cobb, J. G., J. Xie, and J. C. Gore. 2011. 'Contributions of chemical exchange to $T_{1\rho}$ dispersion in a tissue model', *Magn Reson Med*, 66: 1563–71.

Compton, A. H. 1923. 'A quantum theory of the scattering of X-rays by light elements', *Phys Rev*, 21: 0483–502.

Corbett, R. J., A. R. Laptook, G. Tollefsbol, and B. Kim. 1995. 'Validation of a noninvasive method to measure brain temperature in vivo using ^{1}H NMR spectroscopy', *J Neurochem*, 64: 1224–30.

de Crespigny, A. J., H. E. D'Arceuil, T. Engelhorn, and M. E. Moseley. 2000. 'MRI of focal cerebral ischemia using ^{17}O-labeled water', *Magn Reson Med*, 43: 876–83.

Cunningham, C. H., J. Y. Lau, A. P. Chen, B. J. Geraghty, W. J. Perks, I. Roifman, G. A. Wright, and K. A. Connelly. 2016. 'Hyperpolarized ^{13}C metabolic MRI of the human heart: initial experience', *Circ Res*, 119: 1177–82.

Daksh, S., A. Kaul, S. Deep, and A. Datta. 2022. 'Current advancement in the development of manganese complexes as magnetic resonance imaging probes', *J Inorg Biochem*, 237: 112018.

Daldrup-Link, H. E. 2017. 'Ten things you might not know about iron oxide nanoparticles', *Radiology*, 284: 616–29.

Datta, K., and D. M. Spielman. 2017. 'Doublet asymmetry for estimating polarization in hyperpolarized ^{13}C–pyruvate studies', *NMR Biomed*, 30: e3670.

De Feyter, H. M., and R. A. de Graaf. 2021. 'Deuterium metabolic imaging – back to the future', *J Magn Reson*, 326: 106932.

De Feyter, H. M., K. L. Behar, Z. A. Corbin, R. K. Fulbright, P. B. Brown, S. McIntyre, T. W. Nixon, D. L. Rothman, and R. A. de Graaf. 2018. 'Deuterium metabolic imaging (DMI) for MRI-based 3D mapping of metabolism in vivo', *Sci Adv*, 4: eaat7314.

Deelchand, D. K., A. Berrington, R. Noeske, J. M. Joers, A. Arani, J. Gillen, M. Schär, J. F. Nielsen, S. Peltier, N. Seraji-Bozorgzad, K. Landheer, C. Juchem, B. J. Soher, D. C. Noll, K. Kantarci, E. M. Ratai, T. H. Mareci, P. B. Barker, and G. Öz. 2021. 'Across-vendor standardization of semi-LASER for single-voxel MRS at 3T', *NMR Biomed*, 34: e4218.

DeFeo, E. M., and L. L. Cheng. 2010. 'Characterizing human cancer metabolomics with ex vivo ^{1}H HRMAS MRS', *Technol Cancer Res Treat*, 9: 381–91.

Dirac, P. A. M. 1928. 'The quantum theory of the electron', *Proc R Soc Lond Series A-Contain Pap Math Phys Charact*, 117: 610–24.

Duvvuri, U., A. D. Goldberg, J. K. Kranz, L. Hoang, R. Reddy, F. W. Wehrli, A. J. Wand, S. W. Englander, and J. S. Leigh. 2001. 'Water magnetic relaxation dispersion in biological systems: the contribution of proton exchange and implications for the noninvasive detection of cartilage degradation', *Proc Natl Acad Sci U S A*, 98: 12479–84.

Einstein, A. 1905a. 'Generation and conversion of light with regard to a heuristic point of view', *Ann Phys*, 17: 132–48.

Einstein, A. 1905b. 'On the movement of small particles suspended in stationary liquids required by the molecular-kinetic theory of heat', *Ann Phys*, 322: 549–60.

Eisberg, R., and R. Resnik. 1974. *Quantum Physics of Atoms, Molecules, Solids, Nuclei, and Particles* (Wiley: New York).

Elsayed, H., J. Karjalainen, M. J. Nissi, J. Ketola, A. W. Kajabi, V. Casula, Š. Zbýň, M. T. Nieminen, and M. Hanni. 2023. 'Assessing post-traumatic changes in cartilage using $T_{1\rho}$ dispersion parameters', *Magn Reson Imaging*, 97: 91–101.

Ernst, R. R. 1987. *Principles of Nuclear Magnetic Resonance in One and Two Dimensions* (Clarendon Press).

Ernst, R. R., and W. A. Anderson. 1966. 'Application of Fourier transform spectroscopy to magnetic resonance', *Rev Sci Instrum*, 37: 93–102.

Essig, M., M. S. Shiroishi, T. B. Nguyen, M. Saake, J. M. Provenzale, D. Enterline, N. Anzalone, A. Dörfler, A. Rovira, M. Wintermark, and M. Law. 2013. 'Perfusion MRI: the five most frequently asked technical questions', *AJR Am J Roentgenol*, 200: 24–34.

Farrell, B. T., B. E. Hamilton, E. Dósa, E. Rimely, M. Nasseri, S. Gahramanov, C. A. Lacy, E. P. Frenkel, N. D. Doolittle, P. M. Jacobs, and E. A. Neuwelt. 2013. 'Using iron oxide nanoparticles to diagnose CNS inflammatory diseases and PCNSL', *Neurology*, 81: 256–63.

Foss, B. J., and J. Krane. 2004. 'Structural elucidation by 1D and 2D NMR of three isomers of a carotenoid lysophosphocholine and its synthetic precursors', *Magn Reson Chem*, 42: 373–80.

Frahm, J., H. Bruhn, M. L. Gyngell, K. D. Merboldt, W. Hänicke, and R. Sauter. 1989. 'Localized proton NMR spectroscopy in different regions of the human brain in vivo. Relaxation times and concentrations of cerebral metabolites', *Magn Reson Med*, 11: 47–63.

Fraum, T. J., D. R. Ludwig, M. R. Bashir, and K. J. Fowler. 2017. 'Gadolinium-based contrast agents: a comprehensive risk assessment', *J Magn Reson Imaging*, 46: 338–53.

Freeman, D. M., and R. Hurd. 1997. 'Decoupling: theory and practice. II. State of the art: in vivo applications of decoupling', *NMR Biomed*, 10: 381–93.

Freeman, R., and E. Kupce. 1997. 'Decoupling: theory and practice. I. Current methods and recent concepts', *NMR Biomed*, 10: 372–80.

Fullerton, G. D., J. L. Potter, and N. C. Dornbluth. 1982. 'NMR relaxation of protons in tissues and other macromolecular water solutions', *Magn Reson Imaging*, 1: 209–26.

Gerlach, W., and O. Stern. 1922. 'The experimental evidence of direction quantistion in the magnetic field', *Z Phys*, 9: 349–52.

Gilani, I. A., and R. Sepponen. 2016. 'Quantitative rotating frame relaxometry methods in MRI', *NMR Biomed*, 29: 841–61.

de Graaf, R. 2019. 'Spectral Editing and 2D NMR.' in R. de Graaf (ed.), *In Vivo NMR Spectroscopy: Principles and Techniques* (Wiley).

de Graaf, R. A., G. F. Mason, A. B. Patel, K. L. Behar, and D. L. Rothman. 2003. 'In vivo ^{1}H-[^{13}C]-NMR spectroscopy of cerebral metabolism', *NMR Biomed*, 16: 339–57.

de Graaf, R. A., D. L. Rothman, and K. L. Behar. 2011. 'State of the art direct ^{13}C and indirect ^{1}H-[^{13}C] NMR spectroscopy in vivo. A practical guide', *NMR Biomed*, 24: 958–72.

Gribbestad, I. S., S. B. Petersen, H. E. Fjøsne, S. Kvinnsland, and J. Krane. 1994. '^{1}H NMR spectroscopic characterization of perchloric acid extracts from breast carcinomas and non-involved breast tissue', *NMR Biomed*, 7: 181–94.

Gründer, W. 2006. 'MRI assessment of cartilage ultrastructure', *NMR Biomed*, 19: 855–76.

Gupta, A. 2023. 'Cardiac ^{31}P MR spectroscopy: development of the past five decades and future vision—will it be of diagnostic use in clinics?', *Heart Fail Rev*, 28: 485–532.

Haacke, E. M., S. Mittal, Z. Wu, J. Neelavalli, and Y. C. Cheng. 2009. 'Susceptibility-weighted imaging: technical aspects and clinical applications, Part 1', *AJNR Am J Neuroradiol*, 30: 19–30.

Haase, A., J. Frahm, W. Hänicke, and D. Matthaei. 1985. '^{1}H NMR chemical shift selective (CHESS) imaging', *Phys Med Biol*, 30: 341–4.

Haller, S., G. Zaharchuk, D. L. Thomas, K. O. Lovblad, F. Barkhof, and X. Golay. 2016. 'Arterial spin labeling perfusion of the brain: emerging clinical applications', *Radiology*, 281: 337–56.

Haller, S., E. M. Haacke, M. M. Thurnher, and F. Barkhof. 2021. 'Susceptibility-weighted imaging: technical essentials and clinical neurologic applications', *Radiology*, 299: 3–26.

Hamilton, G., M. S. Middleton, M. Bydder, T. Yokoo, J. B. Schwimmer, Y. Kono, H. M. Patton, J. E. Lavine, and C. B. Sirlin. 2009. 'Effect of PRESS and STEAM sequences on magnetic resonance spectroscopic liver fat quantification', *J Magn Reson Imaging*, 30: 145–52.

Hancu, I., E. A. Zimmerman, N. Sailasuta, and R. E. Hurd. 2005. '^{1}H MR spectroscopy using TE averaged PRESS: a more sensitive technique to detect neurodegeneration associated with Alzheimer's disease', *Magn Reson Med*, 53: 777–82.

Haris, M., S. K. Yadav, A. Rizwan, A. Singh, K. Cai, D. Kaura, E. Wang, C. Davatzikos, J. Q. Trojanowski, E. R. Melhem, F. M. Marincola, and A. Borthakur. 2015. 'T1rho MRI and CSF biomarkers in diagnosis of Alzheimer's disease', *Neuroimage Clin*, 7: 598–604.

Harris, A. D., M. G. Saleh, and R. A. Edden. 2017. 'Edited ^{1}H magnetic resonance spectroscopy in vivo: methods and metabolites', *Magn Reson Med*, 77: 1377–89.

Hectors, S. J., R. P. Moonen, G. J. Strijkers, and K. Nicolay. 2015. '$T_{1\rho}$ mapping for the evaluation of high intensity focused ultrasound tumor treatment', *Magn Reson Med*, 73: 1593–601.

Henkelman, R. M., G. J. Stanisz, and S. J. Graham. 2001. 'Magnetization transfer in MRI: a review', *NMR Biomed*, 14: 57–64.

Hennig, J., and H. Friedburg. 1988. 'Clinical applications and methodological developments of the RARE technique', *Magn Reson Imaging*, 6: 391–5.

Hesse, F., A. J. Wright, F. Bulat, V. Somai, F. Kreis, and K. M. Brindle. 2022. 'Deuterium MRSI of tumor cell death in vivo following oral delivery of ^{2}H-labeled fumarate', *Magn Reson Med*, 88: 2014–20.

Hindman, J. C. 1966. 'Proton resonance shift of water in gas and liquid states', *J Chem Phys*, 44: 4582-92.

Hirschler, L., N. Sollmann, B. Schmitz-Abecassis, J. Pinto, F. Arzanforoosh, F. Barkhof, T. Booth, M. Calvo-Imirizaldu, G. Cassia, M. Chmelik, P. Clement, E. Ercan, M. A. Fernández-Seara, J. Furtner, E. Fuster-Garcia, M. Grech-Sollars, N. T. Guven, G. H. Hatay, G. Karami, V. C. Keil, M. Kim, J. A. F. Koekkoek, S. Kukran, L. Mancini, R. E. Nechifor, A. Özcan, E. Ozturk-Isik, S. Piskin, K. Schmainda, S. F. Svensson, C.-H. Tseng, S. Unnikrishnan, F. Vos, E. Warnert, M. Y. Zhao, R. Jancalek, T. Nunes, K. E. Emblem, M. Smits, J. Petr, and G. Hangel. 2023. 'Advanced MR techniques for preoperative glioma characterization: Part 1', *J Magn Reson Imaging*, 57: 1655–75.

Hoult, D. I., and P. C. Lauterbur. 1979. 'The sensitivity of the zeugmatographic experiment involving human samples', *J Magn Reson*, 34: 425–33.

Hu, J., Y. Yu, C. Juhasz, Z. Kou, Y. Xuan, Z. Latif, K. Kudo, H. T. Chugani, and E. M. Haacke. 2008. 'MR susceptibility weighted imaging (SWI) complements conventional contrast enhanced T1 weighted MRI in characterizing brain abnormalities of Sturge-Weber Syndrome', *J Magn Reson Imaging*, 28: 300–7.

Hurd, R. E., and D. Freeman. 1991. 'Proton editing and imaging of lactate', *NMR Biomed*, 4: 73–80.

Hurd, R. E., D. Gurr, and N. Sailasuta. 1998. 'Proton spectroscopy without water suppression: the oversampled J-resolved experiment', *Magn Reson Med*, 40: 343–7.

Hurd, R., N. Sailasuta, R. Srinivasan, D. B. Vigneron, D. Pelletier, and S. J. Nelson. 2004. 'Measurement of brain glutamate using TE-averaged PRESS at 3T', *Magn Reson Med*, 51: 435–40.

Hurd, R. E., Y. F. Yen, A. Chen, and J. H. Ardenkjaer-Larsen. 2012. 'Hyperpolarized ^{13}C metabolic imaging using dissolution dynamic nuclear polarization', *J Magn Reson Imaging*, 36: 1314–28.

Hurd, R. E., D. Spielman, S. Josan, Y. F. Yen, A. Pfefferbaum, and D. Mayer. 2013. 'Exchange-linked dissolution agents in dissolution-DNP ^{13}C metabolic imaging', *Magn Reson Med*, 70: 936–42.

International Commission on Non-Ionizing Radiation Protection (ICNIRP). 2004. 'Medical magnetic resonance (MR) procedures: protection of patients', *Health Phys*, 87: 197–216.

Jolesz, F. A., and N. McDannold. 2008. 'Current status and future potential of MRI-guided focused ultrasound surgery', *J Magn Reson Imaging*, 27: 391–9.

Jones, J. A., P. Hodgkinson, A. L. Barker, and P. J. Hore. 1996. 'Optimal sampling strategies for the measurement of spin–spin relaxation times', *J Magn Reson B*, 113: 25–34.

Jones, K. M., A. C. Pollard, and M. D. Pagel. 2018. 'Clinical applications of chemical exchange saturation transfer (CEST) MRI', *J Magn Reson Imaging*, 47: 11–27.

Jung, J. A., F. V. Coakley, D. B. Vigneron, M. G. Swanson, A. Qayyum, V. Weinberg, K. D. Jones, P. R. Carroll, and J. Kurhanewicz. 2004. 'Prostate depiction at endorectal MR spectroscopic imaging: investigation of a standardized evaluation system', *Radiology*, 233: 701–8.

Kantarci, K. 2013. 'Proton MRS in mild cognitive impairment', *J Magn Reson Imaging*, 37: 770–7.

Keenan, K. E., T. F. Besier, J. M. Pauly, R. L. Smith, S. L. Delp, G. S. Beaupre, and G. E. Gold. 2015. '$T_{1\rho}$ dispersion in articular cartilage: relationship to material properties and macromolecular content', *Cartilage*, 6: 113–22.

Khuu, A., J. Ren, I. Dimitrov, D. Woessner, J. Murdoch, A. D. Sherry, and C. R. Malloy. 2009. 'Orientation of lipid strands in the extracellular compartment of muscle: effect on quantitation of intramyocellular lipids', *Magn Reson Med*, 61: 16–21.

Kirschstein, A., C. Herbst, K. Riedel, M. Carella, J. Leppert, O. Ohlenschläger, M. Görlach, and R. Ramachandran. 2008. 'Broadband homonuclear TOCSY with amplitude and phase-modulated RF mixing schemes', *J Biomol NMR*, 40: 227–37.

Kleinloog, J. P. D., S. Mandija, F. D'Agata, H. Liu, O. van der Heide, B. Koktas, J. W. Dankbaar, V. C. Keil, E. J. Vonken, S. M. Jacobs, C. A. T. van den Berg, J. Hendrikse, A. G. van der Kolk, and A. Sbrizzi. 2023. 'Synthetic MRI with magnetic resonance spin tomogrAphy in time-domain (MR-STAT): results from a prospective cross-sectional clinical trial', *J Magn Reson Imaging*, 57: 1451–61.

Knutsson, L., J. Xu, A. Ahlgren, and P. C. M. van Zijl. 2018. 'CEST, ASL, and magnetization transfer contrast: how similar pulse sequences detect different phenomena', *Magn Reson Med*, 80: 1320–40.

Kobus, T., A. J. Wright, E. Weiland, A. Heerschap, and T. W. Scheenen. 2015. 'Metabolite ratios in ^{1}H MR spectroscopic imaging of the prostate', *Magn Reson Med*, 73: 1–12.

Kogan, F., B. A. Hargreaves, and G. E. Gold. 2017. 'Volumetric multislice gagCEST imaging of articular cartilage: optimization and comparison with T1rho', *Magn Reson Med*, 77: 1134–41.

Kowalewski, J., and L. Mäler. 2006. *Nuclear Spin Relaxation in Liquids: Theory, Experiments, and Applications* (Taylor & Francis).

Krššák, M., L. Lindebom, V. Schrauwen-Hinderling, L. S. Szczepaniak, W. Derave, J. Lundbom, D. Befroy, F. Schick, J. Machann, R. Kreis, and C. Boesch. 2021. 'Proton magnetic resonance spectroscopy in skeletal muscle: experts' consensus recommendations', *NMR Biomed*, 34: e4266.

Krushelnitsky, A., and D. Reichert. 2005. 'Solid-state NMR and protein dynamics', *Prog Nucl Magn Reson Spectrosc*, 47: 1–25.

Kumar, V., K. Bishayee, S. Park, U. Lee, and J. Kim. 2023. 'Oxidative stress in cerebrovascular disease and associated diseases', *Front Endocrinol (Lausanne)*, 14: 1124419.

Kurhanewicz, J., M. G. Swanson, S. J. Nelson, and D. B. Vigneron. 2002. 'Combined magnetic resonance imaging and spectroscopic imaging approach to molecular imaging of prostate cancer', *J Magn Reson Imaging*, 16: 451–63.

Kurhanewicz, J., D. B. Vigneron, J. H. Ardenkjaer-Larsen, J. A. Bankson, K. Brindle, C. H. Cunningham, F. A. Gallagher, K. R. Keshari, A. Kjaer, C. Laustsen, D. A. Mankoff, M. E. Merritt, S. J. Nelson, J. M. Pauly, P. Lee, S. Ronen, D. J. Tyler, S. S. Rajan, D. M. Spielman, L. Wald, X. Zhang, C. R. Malloy, and R. Rizi. 2019. 'Hyperpolarized ^{13}C MRI: path to clinical translation in oncology', *Neoplasia*, 21: 1–16.

Larson, P. E. Z., and J. W. Gordon. 2021. 'Hyperpolarized metabolic MRI—acquisition, reconstruction, and analysis methods', *Metabolites*, 11, 386.

Lau, A. Z., A. P. Chen, N. R. Ghugre, V. Ramanan, W. W. Lam, K. A. Connelly, G. A. Wright, and C. H. Cunningham. 2010. 'Rapid multislice imaging of hyperpolarized ^{13}C pyruvate and bicarbonate in the heart', *Magn Reson Med*, 64: 1323–31.

Lau, J. Y., A. P. Chen, Y. P. Gu, and C. H. Cunningham. 2013. 'A calibration-based approach to real-time in vivo monitoring of pyruvate C_1 and C_2 polarization using the J_{CC} spectral asymmetry', *NMR Biomed*, 26: 1233–41.

Lauterbur, P. C. 1973. 'Image formation by induced local interactions - examples employing nuclear magnetic-resonance', *Nature*, 242: 190–91.

Le Fur, M., and P. Caravan. 2019. 'The biological fate of gadolinium-based MRI contrast agents: a call to action for bioinorganic chemists', *Metallomics*, 11: 240–54.

Lee, J., J. W. Hyun, J. Lee, E. J. Choi, H. G. Shin, K. Min, Y. Nam, H. J. Kim, and S. H. Oh. 2021. 'So you want to image myelin using MRI: an overview and practical guide for myelin water imaging', *J Magn Reson Imaging*, 53: 360–73.

Levitt, M. H. 2008. *Spin Dynamics: Basics of Nuclear Magnetic Resonance, 2nd Edition* (Wiley).

Lewis, M., S. Yanny, and P. N. Malcolm. 2012. 'Advantages of blood pool contrast agents in MR angiography: a pictorial review', *J Med Imaging Radiat Oncol*, 56: 187–91.

Li, X., and S. Majumdar. 2013. 'Quantitative MRI of articular cartilage and its clinical applications', *J Magn Reson Imaging*, 38: 991–1008.

Li, H., and T. J. Meade. 2019. 'Molecular magnetic resonance imaging with Gd(III)-based contrast agents: challenges and key advances', *J Am Chem Soc*, 141: 17025–41.

Li, K. C., L. R. Pelc, S. Puvvala, and G. A. Wright. 1998. 'Mesenteric ischemia due to hemorrhagic shock: MR imaging diagnosis and monitoring in a canine model', *Radiology*, 206: 219–25.

Liimatainen, T., D. J. Sorce, R. O'Connell, M. Garwood, and S. Michaeli. 2010. 'MRI contrast from relaxation along a fictitious field (RAFF)', *Magn Reson Med*, 64: 983–94.

Ling, W., R. R. Regatte, G. Navon, and A. Jerschow. 2008. 'Assessment of glycosaminoglycan concentration in vivo by chemical exchange-dependent saturation transfer (gagCEST)', *Proc Natl Acad Sci U S A*, 105: 2266–70.

Lombardi, A. F., M. Guma, C. B. Chung, E. Y. Chang, J. Du, and Y. J. Ma. 2023. 'Ultrashort echo time magnetic resonance imaging of the osteochondral junction', *NMR Biomed*, 36: e4843.

Lucas-Torres, C., H. Roumes, V. Bouchaud, A.-K. Bouzier-Sore, and A. Wong. 2021. 'Metabolic NMR mapping with microgram tissue biopsy', *NMR Biomed*, 34: e4477.

Lynch, L. 2019. 'Sad News: Alfred G. Redfield, Emeritus Professor of Physics and Biochemistry', Office of the Provost, Brandeis University. https://www.brandeis.edu/provost/letters/2018-2019/2019-07-26-alfred-redfield.html.

Madelin, G., and R. R. Regatte. 2013. 'Biomedical applications of sodium MRI in vivo', *J Magn Reson Imaging*, 38: 511–29.

Mansfield, P., and P. K. Grannell. 1973. 'NMR 'diffraction?' In solids?', *J Phys C-Solid State Phys*, 6: L422–L26.

Martinez-Heras, E., F. Grussu, F. Prados, E. Solana, and S. Llufriu. 2021. 'Diffusion-weighted imaging: recent advances and applications', *Semin Ultrasound CT MR*, 42: 490–506.

Mathur-De Vré, R. 1979. 'The NMR studies of water in biological systems', *Prog Biophys Mol Biol*, 35: 103–34.

Maudsley, A. A., O. C. Andronesi, P. B. Barker, A. Bizzi, W. Bogner, A. Henning, S. J. Nelson, S. Posse, D. C. Shungu, and B. J. Soher. 2021. 'Advanced magnetic resonance spectroscopic neuroimaging: experts' consensus recommendations', *NMR Biomed*, 34: e4309.

Mayer, D., Y. F. Yen, Y. S. Levin, J. Tropp, A. Pfefferbaum, R. E. Hurd, and D. M. Spielman. 2010. 'In vivo application of sub-second spiral chemical shift imaging (CSI) to hyperpolarized ^{13}C metabolic imaging: comparison with phase-encoded CSI', *J Magn Reson*, 204: 340–5.

McConnell, H. M. 1958. 'Reaction rates by nuclear magnetic resonance', *J Chem Phys*, 28: 430–31.

Meyerspeer, M., C. Boesch, D. Cameron, M. Dezortová, S. C. Forbes, A. Heerschap, J. A. L. Jeneson, H. E. Kan, J. Kent, G. Layec, J. J. Prompers, H. Reyngoudt, A. Sleigh, L. Valkovič, and G. J. Kemp. 2020. '^{31}P magnetic resonance spectroscopy in skeletal muscle: experts' consensus recommendations', *NMR Biomed*, 34: e4246.

Michaeli, S., D. J. Sorce, C. S. Springer, Jr.,, K. Ugurbil, and M. Garwood. 2006. 'T$_{1\rho}$ MRI contrast in the human brain: modulation of the longitudinal rotating frame relaxation shutter-speed during an adiabatic RF pulse', *J Magn Reson*, 181: 135–47.

Moffett, J. R., B. Ross, P. Arun, C. N. Madhavarao, and A. M. Namboodiri. 2007. 'N-Acetylaspartate in the CNS: from neurodiagnostics to neurobiology', *Prog Neurobiol*, 81: 89–131.

von Morze, C., G. D. Reed, Z. J. Wang, M. A. Ohliger, and C. Laustsen. 2021. 'Hyperpolarized carbon (^{13}C) MRI of the kidneys: basic concept', *Methods Mol Biol*, 2216: 267–78.

Mullins, P. G., D. J. McGonigle, R. L. O'Gorman, N. A. Puts, R. Vidyasagar, C. J. Evans, and R. A. Edden. 2014. 'Current practice in the use of MEGA-PRESS spectroscopy for the detection of GABA', *NeuroImage*, 86: 43–52.

Naressi, A., C. Couturier, J. M. Devos, M. Janssen, C. Mangeat, R. de Beer, and D. Graveron-Demilly. 2001. 'Java-based graphical user interface for the MRUI quantitation package', *MAGMA*, 12: 141–52.

Nelson, S. J., E. Graves, A. Pirzkall, X. Li, C. A. Antiniw, D. B. Vigneron, and T. R. McKnight. 2002. 'In vivo molecular imaging for planning radiation therapy of gliomas: an application of ^{1}H MRSI', *J Magn Reson Imaging*, 16: 464–76.

Neuhaus, D., and M. Williamson. 2000. *The Nuclear Overhauser Efffect in Structural and Conformational Analysis, 2nd Edition* (Wiley-VCH).

Niess, F., L. Hingerl, B. Strasser, P. Bednarik, D. Goranovic, E. Niess, G. Hangel, M. Krššák, B. Spurny-Dworak, T. Scherer, R. Lanzenberger, and W. Bogner. 2023. 'Noninvasive 3-dimensional ^{1}H-magnetic resonance spectroscopic imaging of human brain glucose and neurotransmitter metabolism using deuterium labeling at 3T: feasibility and interscanner reproducibility', *Invest Radiol*, 58: 431–7.

Novikov, D. S., E. Fieremans, S. N. Jespersen, and V. G. Kiselev. 2019. 'Quantifying brain microstructure with diffusion MRI: theory and parameter estimation', *NMR Biomed*, 32: e3998.

Oatridge, A., A. H. Herlihy, R. W. Thomas, A. L. Wallace, W. L. Curati, J. V. Hajnal, and G. M. Bydder. 2001. 'Magnetic resonance: magic angle imaging of the Achilles tendon', *Lancet*, 358: 1610–1.

Ogawa, S., T. M. Lee, A. S. Nayak, and P. Glynn. 1990. 'Oxygenation-sensitive contrast in magnetic resonance image of rodent brain at high magnetic fields', *Magn Reson Med*, 14: 68–78.

van Oorschot, J. W., H. El Aidi, S. J. Jansen of Lorkeers, J. M. Gho, M. Froeling, F. Visser, S. A. Chamuleau, P. A. Doevendans, P. R. Luijten, T. Leiner, and J. J. Zwanenburg. 2014. 'Endogenous assessment of chronic myocardial infarction with T$_{1\rho}$-mapping in patients', *J Cardiovasc Magn Reson*, 16: 104.

Osorio, J. A., E. Ozturk-Isik, D. Xu, S. Cha, S. Chang, M. S. Berger, D. B. Vigneron, and S. J. Nelson. 2007. '3D ^{1}H MRSI of brain tumors at 3.0 Tesla using an eight-channel phased-array head coil', *J Magn Reson Imaging*, 26: 23–30.

Overhauser, A. W. 1953. 'Polarization of nuclei in metals', *Phys Rev*, 92: 411–15.

Öz, G., D. K. Deelchand, J. P. Wijnen, V. Mlynárik, L. Xin, R. Mekle, R. Noeske, T. W. J. Scheenen, and I. Tkáč. 2020. 'Advanced single voxel ^{1}H magnetic resonance spectroscopy techniques in humans: experts' consensus recommendations', *NMR Biomed* 34: e4236.

Pagel, M. D. 2011. 'Responsive paramagnetic chemical exchange saturation transfer MRI contrast agents', *Imaging Med*, 3: 377–80.

Pala, S., N. E. Hänninen, O. Nykänen, T. Liimatainen, and M. J. Nissi. 2023. 'New methods for robust continuous wave $T_{1\rho}$ relaxation preparation', *NMR Biomed*, 36: e4834.

Park, I., P. E. Z. Larson, J. W. Gordon, L. Carvajal, H. Y. Chen, R. Bok, M. Van Criekinge, M. Ferrone, J. B. Slater, D. Xu, J. Kurhanewicz, D. B. Vigneron, S. Chang, and S. J. Nelson. 2018. 'Development of methods and feasibility of using hyperpolarized carbon-13 imaging data for evaluating brain metabolism in patient studies', *Magn Reson Med*, 80: 864–73.

Perron, S., and A. Ouriadov. 2023. 'Hyperpolarized ^{129}Xe MRI at low field: current status and future directions', *J Magn Reson*, 348: 107387.

Pike, G. B., B. S. Hu, G. H. Glover, and D. R. Enzmann. 1992. 'Magnetization transfer time-of-flight magnetic resonance angiography', *Magn Reson Med*, 25: 372–9.

Pike, G. B., G. H. Glover, B. S. Hu, and D. R. Enzmann. 1993. 'Pulsed magnetization transfer spin-echo MR imaging', *J Magn Reson Imaging*, 3: 531–9.

Planck, M. 1901. 'Law of energy distribution in normal spectra', *Ann Phys*, 4: 553–63.

Poetter-Lang, S., A. Messner, N. Bastati, K. I. Ringe, M. Ronot, S. K. Venkatesh, R. Ambros, A. Kristic, A. Korajac, G. Dovjak, M. Zalaudek, J. C. Hodge, C. Schramm, E. Halilbasic, M. Trauner, and A. Ba-Ssalamah. 2023. 'Diagnosis of functional strictures in patients with primary sclerosing cholangitis using hepatobiliary contrast-enhanced MRI: a proof-of-concept study', *Eur Radiol*, 33: 9022–37.

Polvoy, I., H. Qin, R. R. Flavell, J. Gordon, P. Viswanath, R. Sriram, M. A. Ohliger, and D. M. Wilson. 2021. 'Deuterium metabolic imaging—rediscovery of a spectroscopic tool', *Metabolites*, 11: 570.

Popadic Gacesa, J., F. Schick, J. Machann, and N. Grujic. 2017. 'Intramyocellular lipids and their dynamics assessed by ^{1}H magnetic resonance spectroscopy', *Clin Physiol Funct Imaging*, 37: 558–66.

Posse, S., S. R. Dager, T. L. Richards, C. Yuan, R. Ogg, A. A. Artru, H. W. Müller-Gärtner, and C. Hayes. 1997. 'In vivo measurement of regional brain metabolic response to hyperventilation using magnetic resonance: proton echo planar spectroscopic imaging (PEPSI)', *Magn Reson Med*, 37: 858–65.

Prince, M. R., H. Zhang, M. Morris, J. L. MacGregor, M. E. Grossman, J. Silberzweig, R. L. DeLapaz, H. J. Lee, C. M. Magro, and A. M. Valeri. 2008. 'Incidence of nephrogenic systemic fibrosis at two large medical centers', *Radiology*, 248: 807–16.

Provencher, S. W. 2001. 'Automatic quantitation of localized in vivo ^{1}H spectra with LCModel', *NMR Biomed*, 14: 260–4.

Purcell, E. M., H. C. Torrey, and R. V. Pound. 1946. 'Resonance absorption by nuclear magnetic moments in a solid', *Phys Rev*, 69: 37–38.

Rabi, II, J. R. Zacharias, S. Millman, and P. Kusch. 1938. 'A new method of measuring nuclear magnetic moment', *Phys Rev*, 53: 318–18.

Rae, C. D. 2014. 'A guide to the metabolic pathways and function of metabolites observed in human brain ^{1}H magnetic resonance spectra', *Neurochem Res*, 39: 1–36.

Ramsey, N. 1950. 'Magnetic shielding of nuclei in molecules', *Phys Rev*, 78: 699.

Redfield, A. G. 1955. 'Nuclear magnetic resonance saturation and rotary saturation in solids', *Phys Rev*, 98: 1787–809.

Regatte, R. R., S. V. Akella, A. Borthakur, and R. Reddy. 2003. 'Proton spin-lock ratio imaging for quantitation of glycosaminoglycans in articular cartilage', *J Magn Reson Imaging*, 17: 114–21.

Rothman, D. L., A. M. Howseman, G. D. Graham, O. A. Petroff, G. Lantos, P. B. Fayad, L. M. Brass, G. I. Shulman, R. G. Shulman, and J. W. Prichard. 1991. 'Localized proton NMR observation of [3-^{13}C] lactate in stroke after [1-^{13}C] glucose infusion', *Magn Reson Med*, 21: 302–7.

Ryan, C. N., A. Sorushanova, A. J. Lomas, A. M. Mullen, A. Pandit, and D. I. Zeugolis. 2015. 'Glycosaminoglycans in tendon physiology, pathophysiology, and therapy', *Bioconjug Chem*, 26: 1237–51.

Schrodinger, E. 1926. 'An undulatory theory of the mechanics of atoms and molecules', *Phys Rev*, 28: 1049–70.

Schroeder, M. A., A. Z. Lau, A. P. Chen, Y. Gu, J. Nagendran, J. Barry, X. Hu, J. R. Dyck, D. J. Tyler, K. Clarke, K. A. Connelly, G. A. Wright, and C. H. Cunningham. 2013. 'Hyperpolarized ^{13}C magnetic resonance reveals early- and late-onset changes to in vivo pyruvate metabolism in the failing heart', *Eur J Heart Fail*, 15: 130–40.

Shang, H., T. Skloss, C. von Morze, L. Carvajal, M. Van Criekinge, E. Milshteyn, P. E. Larson, R. E. Hurd, and D. B. Vigneron. 2016. 'Handheld electromagnet carrier for transfer of hyperpolarized carbon-13 samples', *Magn Reson Med*, 75: 917–22.

Sharma, U., and N. R. Jagannathan. 2022. 'Magnetic resonance imaging (MRI) and MR spectroscopic methods in understanding breast cancer biology and metabolism', *Metabolites*, 12: 295.

Sherry, A. D., and M. Woods. 2008. 'Chemical exchange saturation transfer contrast agents for magnetic resonance imaging', *Annu Rev Biomed Eng*, 10: 391–411.

Sherry, A. D., and Y. Wu. 2013. 'The importance of water exchange rates in the design of responsive agents for MRI', *Curr Opin Chem Biol*, 17: 167–74.

Shi, Y., D. Liu, Z. Kong, Q. Liu, H. Xing, Y. Wang, Y. Wang, and W. Ma. 2022. 'Prognostic value of choline and other metabolites measured using ^{1}H-magnetic resonance spectroscopy in gliomas: a meta-analysis and systemic review', *Metabolites*, 12: 1219.

Solomon, I. 1955. 'Relaxation processes in a system of two spins', *Phys Rev*, 99: 559.

Sorensen, O. W., G. W. Eich, M. H. Levitt, G. Bodenhausen, and R. R. Ernst. 1983. 'Product operatoro-formalism for the description of NMR pulsed experiments', *Prog Nucl Magn Reson Spectrosc*, 16: 163–92.

Spear, J. T., and J. C. Gore. 2016. 'New insights into rotating frame relaxation at high field', *NMR Biomed*, 29: 1258–73.

Spielman, D. M., M. Gu, R. E. Hurd, R. K. Riemer, K. Okamura, and F. L. Hanley. 2022. 'Proton magnetic resonance spectroscopy assessment of neonatal brain metabolism during cardiopulmonary bypass surgery', *NMR Biomed*, 35: e4752.

Stables, L. A., R. P. Kennan, A. W. Anderson, R. T. Constable, and J. C. Gore. 1999. 'Analysis of J coupling-induced fat suppression in DIET imaging', *J Magn Reson*, 136: 143–51.

Swanson, M. G., D. B. Vigneron, Z. L. Tabatabai, R. G. Males, L. Schmitt, P. R. Carroll, J. K. James, R. E. Hurd, and J. Kurhanewicz. 2003. 'Proton HR-MAS spectroscopy and quantitative pathologic analysis of MRI/3D-MRSI-targeted postsurgical prostate tissues', *Magn Reson Med*, 50: 944–54.

Takayama, Y., A. Nishie, Y. Asayama, Y. Ushijima, D. Okamoto, N. Fujita, K. Morita, K. Shirabe, K. Kotoh, Y. Kubo, T. Okuaki, and H. Honda. 2015. 'T$_{1\rho}$ Relaxation of the liver: a potential biomarker of liver function', *J Magn Reson Imaging*, 42: 188–95.

Tiderius, C. J., R. Jessel, Y. J. Kim, and D. Burstein. 2007. 'Hip dGEMRIC in asymptomatic volunteers and patients with early osteoarthritis: the influence of timing after contrast injection', *Magn Reson Med*, 57: 803–5.

Vander Heiden, M. G., L. C. Cantley, and C. B. Thompson. 2009. 'Understanding the Warburg effect: the metabolic requirements of cell proliferation', *Science*, 324: 1029–33.

Veksler, R., I. Shelef, and A. Friedman. 2014. 'Blood-brain barrier imaging in human neuropathologies', *Arch Med Res*, 45: 646–52.

van de Ven, F. J. M. 1996. *Multidimensional NMR in Liquids: Basic Principles and Experimental Methods, 1st Edition* (Wiley-VCH).

Wagner, M., S. Wagner, J. Schnorr, E. Schellenberger, D. Kivelitz, L. Krug, M. Dewey, M. Laule, B. Hamm, and M. Taupitz. 2011. 'Coronary MR angiography using citrate-coated very small superparamagnetic iron oxide particles as blood-pool contrast agent: initial experience in humans', *J Magn Reson Imaging*, 34: 816–23.

Wahsner, J., E. M. Gale, A. Rodríguez-Rodríguez, and P. Caravan. 2019. 'Chemistry of MRI contrast agents: current challenges and new frontiers', *Chem Rev*, 119: 957–1057.

Wang, P., J. Block, and J. C. Gore. 2015. 'Chemical exchange in knee cartilage assessed by $R_{1\rho}$ ($1/T_{1\rho}$) dispersion at 3T', *Magn Reson Imaging*, 33: 38–42.

Wang, Z. J., M. A. Ohliger, P. E. Z. Larson, J. W. Gordon, R. A. Bok, J. Slater, J. E. Villanueva-Meyer, C. P. Hess, J. Kurhanewicz, and D. B. Vigneron. 2019. 'Hyperpolarized ^{13}C MRI: state of the art and future directions', *Radiology*, 291: 273–84.

Wangsness, R. K., and F. Bloch. 1953. 'The dynamical theory of nuclear induction', *Phys Rev*, 89: 728–39.

Wassef, S. N., J. Wemmie, C. P. Johnson, H. Johnson, J. S. Paulsen, J. D. Long, and V. A. Magnotta. 2015. '$T_{1\rho}$ imaging in premanifest Huntington disease reveals changes associated with disease progression', *Mov Disord*, 30: 1107–14.

van der Weijden, C. W. J., E. Biondetti, I. W. Gutmann, H. Dijkstra, R. McKerchar, D. de Paula Faria, E. F. J. de Vries, J. F. Meilof, R. A. J. O. Dierckx, V. H. Prevost, and A. Rauscher. 2022. 'Quantitative myelin imaging with MRI and PET: an overview of techniques and their validation status', *Brain*, 146: 1243–66.

Welch, J. W., K. Bhakoo, R. M. Dixon, P. Styles, N. R. Sibson, and A. M. Blamire. 2003. 'In vivo monitoring of rat brain metabolites during vigabatrin treatment using localized 2D-COSY', *NMR Biomed*, 16: 47–54.

Williams, A., A. Gillis, C. McKenzie, B. Po, L. Sharma, L. Micheli, B. McKeon, and D. Burstein. 2004. 'Glycosaminoglycan distribution in cartilage as determined by delayed gadolinium-enhanced MRI of cartilage (dGEMRIC): potential clinical applications', *AJR Am J Roentgenol*, 182: 167–72.

Wilson, M., O. Andronesi, P. B. Barker, R. Bartha, A. Bizzi, P. J. Bolan, K. M. Brindle, I. Y. Choi, C. Cudalbu, U. Dydak, U. E. Emir, R. G. Gonzalez, S. Gruber, R. Gruetter, R. K. Gupta, A. Heerschap, A. Henning, H. P. Hetherington, P. S. Huppi, R. E. Hurd, K. Kantarci, R. A. Kauppinen, D. W. J. Klomp, R. Kreis, M. J. Kruiskamp, M. O. Leach, A. P. Lin, P. R. Luijten, M. Marjańska, A. A. Maudsley, D. J. Meyerhoff, C. E. Mountford, P. G. Mullins, J. B. Murdoch, S. J. Nelson, R. Noeske, G. Öz, J. W. Pan, A. C. Peet, H. Poptani, S. Posse, E. M. Ratai, N. Salibi, T. W. J. Scheenen, I. C. P. Smith, B. J. Soher, I. Tkáč, D. B. Vigneron, and F. A. Howe. 2019. 'Methodological consensus on clinical proton MRS of the brain: review and recommendations', *Magn Reson Med*, 82: 527–50.

Witney, T. H., M. I. Kettunen, and K. M. Brindle. 2011. 'Kinetic modeling of hyperpolarized ^{13}C label exchange between pyruvate and lactate in tumor cells', *J Biol Chem*, 286: 24572–80.

Wright, G. A., B. S. Hu, and A. Macovski. 1991a. 'Estimating oxygen saturation of blood in vivo with MR imaging at 1.5 T', *J Magn Reson Imaging*, 1: 275–83.

Wright, G. A., D. G. Nishimura, and A. Macovski. 1991b. 'Flow-independent magnetic resonance projection angiography', *Magn Reson Med*, 17: 126–40.

Wu, Y., S. Zhang, T. C. Soesbe, J. Yu, E. Vinogradov, R. E. Lenkinski, and A. D. Sherry. 2016. 'pH imaging of mouse kidneys in vivo using a frequency-dependent paraCEST agent', *Magn Reson Med*, 75: 2432–41.

Wuthrich, K. 1995. *NMR in Structural Biology: A Collection of Papers by Kurt Wüthrich* (World Scientific Publishing Company).

Yen, Y. F., S. J. Kohler, A. P. Chen, J. Tropp, R. Bok, J. Wolber, M. J. Albers, K. A. Gram, M. L. Zierhut, I. Park, V. Zhang, S. Hu, S. J. Nelson, D. B. Vigneron, J. Kurhanewicz, H. A. Dirven, and R. E. Hurd.

2009. 'Imaging considerations for in vivo ^{13}C metabolic mapping using hyperpolarized ^{13}C-pyruvate', *Magn Reson Med*, 62: 1–10.

Yu, Y., J. Wang, Y. Li, Y. Chen, and W. Cui. 2023. 'Cartilaginous organoids: advances, applications, and perspectives', *Adv NanoBiomed Res*, 3: 2200114.

Yue, K., A. Marumoto, N. Binesh, and M. A. Thomas. 2002. '2D JPRESS of human prostates using an endorectal receiver coil', *Magn Reson Med*, 47: 1059–64.

Zaric, O., V. Juras, P. Szomolanyi, M. Schreiner, M. Raudner, C. Giraudo, and S. Trattnig. 2021. 'Frontiers of sodium MRI revisited: from cartilage to brain imaging', *J Magn Reson Imaging*, 54: 58–75.

Zavoisky, E. K. 1945. 'Paramagnetic relaxation of liquid solutions for perpendicular fields', *Zhur Eksperiment i Theoret Fiz*, 15: 344–50.

Zhang, Q., Z. Bai, Y. Gong, X. Liu, X. Dai, S. Wang, and F. Liu. 2015. 'Monitoring glutamate levels in the posterior cingulate cortex of thyroid dysfunction patients with TE-averaged PRESS at 3T', *Magn Reson Imaging*, 33: 774–8.

Zhou, J., H. Y. Heo, L. Knutsson, P. C. M. van Zijl, and S. Jiang. 2019. 'APT-weighted MRI: techniques, current neuro applications, and challenging issues', *J Magn Reson Imaging*, 50: 347–64.

Zhou, J., M. Zaiss, L. Knutsson, P. Z. Sun, S. S. Ahn, S. Aime, P. Bachert, J. O. Blakeley, K. Cai, M. A. Chappell, M. Chen, D. F. Gochberg, S. Goerke, H. Y. Heo, S. Jiang, T. Jin, S. G. Kim, J. Laterra, D. Paech, M. D. Pagel, J. E. Park, R. Reddy, A. Sakata, S. Sartoretti-Schefer, A. D. Sherry, S. A. Smith, G. J. Stanisz, P. C. Sundgren, O. Togao, M. Vandsburger, Z. Wen, Y. Wu, Y. Zhang, W. Zhu, Z. Zu, and P. C. M. van Zijl. 2022. 'Review and consensus recommendations on clinical APT-weighted imaging approaches at 3T: application to brain tumors', *Magn Reson Med*, 88: 546–74.

van Zijl, P. C., and N. N. Yadav. 2011. 'Chemical exchange saturation transfer (CEST): what is in a name and what isn't?', *Magn Reson Med*, 65: 927–48.

van Zijl, P. C. M., K. Brindle, H. Lu, P. B. Barker, R. Edden, N. Yadav, and L. Knutsson. 2021. 'Hyperpolarized MRI, functional MRI, MR spectroscopy and CEST to provide metabolic information in vivo', *Curr Opin Chem Biol*, 63: 209–18.

Index

Fundamentals of In Vivo Magnetic Resonance: Spin Physics, Relaxation Theory, and Contrast Mechanisms, First Edition.
Daniel M. Spielman and Keshav Datta.
© 2024 John Wiley & Sons, Inc. Published 2024 by John Wiley & Sons, Inc.
Companion website: www.wiley.com/go/Spielman

Printed and bound by CPI Group (UK) Ltd, Croydon, CR0 4YY

16/04/2025

14658428-0001